Preface

This is not a physics textbook; neither is it an encyclopaedia of 'how it works'. Rather this book demonstrates the relevance of basic scientific concepts in a real-world context. It thus encourages the study of physics as a *meaningful* part of general education for a wide range of students.

In the past many students have been 'turned off' physics by an academic treatment followed – time permitting – by a brief discussion of applications. In this book the traditional order of presentation has been reversed in the hope of increasing motivation. It first shows the applications of physics principles in familiar everyday things such as television, transport, telecommunications, domestic gadgets and health. The concepts are then introduced only when they are needed to make *sense* and to make *use* of the physical phenomena.

Each topic is discussed in an easily-read double page spread which includes, where necessary, the essential physics in a summary form.

The book encourages an activity-based methodology and suggestions are made throughout for experimental work and for a number of practical investigations. The appendix deals with units and symbols.

Thanks

So many people have contributed to the production of this book that it is impossible adequately to express our thanks to individuals for all the help we have received.

We are indebted to many firms and individuals for allowing their photographs to be reproduced.

We are especially grateful for the help received from schools, colleges, science advisers, the Examination Board and to the staff of Oxford University Press for their expertise, patience and good humour. In particular we express our gratitude to the science editor, Chris Ray, for his vision and skill without which this book could not have been published.

Jim Jardine, Autumn 1988

TELECOMMUNICATIONS

USING ELECTRICITY

HEALTH PHYSICS

ELECTRONICS

PHYSICS THROUGH APPLICATIONS

Writers

Ken Stewart
Ian Bothwell
Derrick Hannan
Alastair Steven
Jim Wilson
Ross Dawson
Jim Jardine
Arthur Gibbons
John Sharkey

Editor

Jim Jardine

Oxford University Press 1989

Oxford University Press, Walton Street, Oxford OX2 6DP

Oxford New York Toronto
Delhi Bombay Calcutta Madras Karachi
Petaling Jaya Singapore Hong Kong Tokyo
Nairobi Dar es Salaam Cape Town
Melbourne Auckland

and associated companies in
Berlin Ibadan

Oxford is a trade mark of Oxford University Press

© Jim Jardine, Ken Stewart, Ian Bothwell,
Derrick Hannan, Alastair Steven, Jim Wilson,
Ross Dawson, Arthur Gibbons, John Sharkey.

ISBN 0 19 914280 7

Typeset in Plantin light & Helvetica condensed by
MS Filmsetting Limited, Frome, Somerset
Printed in Singapore

Acknowledgements

The Publishers wish to thank the following for permission to reproduce photographs:

Acuson: p 71 (top right); **Agema Infrared Systems Ltd:** p 85 (top right); **Allsport Photographic Library:** pp 10 (bottom), 124 (right), 215, 216, 214 (bottom), **/Bob Martin 214 (top); Amersham International plc:** pp 84 (top left), 89 (top); **Amstrad plc:** p 46 (left); **Argentum:** pp 74, 75, 76; **Associated Press:** p 108 (left); **Barnaby's Photo Library:** p 169 (top); **Boxmag Rapid:** p 60 (bottom left); **Bruce Coleman Ltd:** p 125; **Paul Brierley:** p 24 (left); **British Gas:** p 156 (centre); **Carters:** p 91 (top & bottom); **Castle Associates:** p 72; **Central Electricity Generating Board:** pp 131, 159, 170, 171; **Chevron Petroleum:** p 132 (top); **Clairol Appliances:** p 37; **Bruce Coleman Ltd:** pp 125, 229 (left); **Coloursport:** pp 125 (top), 126 (top right); **Condor Ltd:** p 135 (top right); **Daneglow International:** p 84 (top right); **De Vere:** p 195; **EMI Records:** p 206 (centre); **European Space Agency/Met Office:** pp 240 (top), 241 (top); **Ferguson:** p 94 (centre); **Fiat:** p 152 (bottom); **Ford:** p 135 (bottom); **Graystad:** p 73; **IBA:** p 23 (top right); **Impact Photos/Jeremy Nicholl** 212 (top), **/Homer Sykes** 229 (top left); **Institute of Urology, London:** p 85 (bottom); **Jim Jardine:** p 198; **King's College Hospital:** p 84 (left); **Kodak Ltd:** p 188 (bottom); **Andrew Lambert:** pp 120, 199 (all); **London Transport Museum:** p 66; **London College of Printing:** p 22 (top right); **Maranello Concessionaires/ Ferrari:** p 152 (centre); **Marconi:** p 114; **NASA:** pp 220 (all), 221, 222, 232 (bottom), 233 (top), 233 (bottom), 236, 237, 239, 242 (top), 244 (top), 244 (bottom), 245 (top right), 245 (centre); **NEI Peebles Ltd:** p 173; **New Scientist/Jerry Mason:** p 115 (centre & bottom); **Novosti Press Agency:** p 232 (centre); **Omega Electronics:** p 118; **Oxford Scientific Films/Sean Morris:** p 196 (left & right); **Philips Electronics:** pp 51, 95 (top right); **Picker/GEC:** p 228 (top); **Picturepoint:** p 11; **"Schwinn Air-Dyne",** ***courtesy*** **of Polaris International:** p 144; **Quadrant Picture Library:** pp 7 (top right), 132 (bottom right), 134 (centre); **Racal-Vodac:** p 20 (top right); **Rex Features Ltd:** p 150; **Rover Group Ltd:** p 152 (top); **Royal Astronomy Observatory, Cambridge/Dr Guy Pooley:** p 230 (bottom); **Royal Observatory, Edinburgh:** p 226 (top); **Royal National Institute for the Deaf:** p 6 (top right); **Schlumberger Industries Electricity Management:** p 48, **Science Photo Library:** pp 6 (right), 28; **SPL/CNRI** 71 (bottom right), **SPL/Martin Dohrn** 78 (right), **SPL/David Parker** 78 (top), **SPL/Alexander Tsiaras** 81 (top and bottom), **SPL/CNRI** 88 (bottom), **SPL/Elscint** 88 (top), **SPL/Lowell Georgia** 167 (right), **SPL/Dr Tony Brain** 206 (bottom); **SPL/Jean Lorre** 223 (top), **SPL/CRNI** 228 (right), 225 (bottom right), **SPL/John Bova** 225 (top right); **Science Museum:** pp 8 (right), 197 (top right & centre), 202; **Shearman Mark:** p 7 (right); **Shell:** pp 132 (centre), 143; **Space Frontiers:** pp 240 (bottom), 241 (top); **Spectrum:** p 206 (top); **Sportsphoto/Stewart Kendall:** p 26 (bottom right); **Tony Stone Photo Library:** pp 7 (left), **/John McDermott** 80; **Swiss National Tourist Office:** p 145; **Survival Anglia/Mike Price:** pp 159 (right), **/Annie Price** 167 (left); **Jeff Tabberner:** p 126 (bottom left); **Tandy:** p 229 (top right & right); **Tass:** pp 159 (bottom right), 232 (top), 234; **Tefal:** p 176; **Telefocus:** pp 18 (top right & right), 27, 28, 31 (top right), 77 (bottom right); **The Mansell Collection:** p 227 (top); **The National Motor Museum, Beaulieu:** p 124 (left); **Tildawn Electronics Ltd:** p 119; **Transport & Road Research Laboratory:** p 137 (top right); **United Kingdom Atomic Energy Authority:** pp 83 (right), 86, 87, 90, 162, 163; **Volvo Concessionaires:** p 136 (top right); **John Watney:** pp 69, 84 (right); **Wind Energy Group:** p 166 (centre right); **Zefa Photographic Library:** pp 30 (top right), **Zefa/Teasy** 83 (left), 94 (top), **Zefa/Benser** 145 (right), **Zefa/Clive Sawyer** 153 (centre right), 156 (centre), 159 (right), 160, **Zefa/J Pfaff** 164, **Zefa/H Grathwohl** 165, 166, 168, 198, 224, 230 (top).

The illustrations are by:

Bill Donohoe, Chris Duggan, Jones Sewell, ML Design, Martin Newton, Kridon Panteli, Mark Rogerson, Simon Roulstone, Mike Saunders, Tony Townsend, Borin Van Loon, Malcolm Walker and **Galina Zolfaghari.**

TRANSPORT

ENERGY MATTERS

LEISURE

SPACE PHYSICS

About this book

Physics rules

What is physics? Why study physics? What are we up to when we are doing physics?

Watching a game of American football is a bit like studying physics. At first glance there seems to be chaos and confusion on the field. But once we have a knowledge of the rules of the game and can understand these rules we are better able to appreciate the finer points of the play. Knowing the rules helps to make sense of the game.

Physics is about the game that Mother Nature plays. Nature's game has rules – or so we think. But what are they? One thing is sure, we will have a better chance of understanding these rules if we join in the play.

Physics is about trying to find rules and relationships which describe the way nature behaves.

Aspects of physics

Looking for the rules which describe the way the physical world works is just one aspect of physics. Another is concerned with using and applying the rules in a practical way. For example it is hard to find an appliance in the home that is not based on the principles of physics. Light bulbs, electric irons, TV sets, telephones, microwave ovens, record players, radios, recorders and burglar alarms are all applications of physics.

But physics has many applications in other areas beyond the home – in telecommunication, sport, leisure, health, transport, and space. This book is all about these applications.

Physics through applications

You can see physics in action all around you and you can therefore learn about physics principles through a study of its applications.

For example, by examining a telephone you can find out about electricity and how it is used to transmit information.

Studying an electronic alarm system could help you come to terms with transistors and silicon chips.

Investigating how a camera works will involve you in learning about the properties of light.

Problem solving

Through a study of these applications you will acquire more than just physics facts. Your study will help you to use your knowledge to solve problems.

Some of these problems you will be able to solve without even having to move from your armchair!

For example, using your knowledge of physics and data supplied by the Department of Energy you could compare the cost of heating water by gas and by electricity. In attempting to solve problems like this you will be doing calculations and processing information.

Sometimes, however, you will have to solve a practical problem. For example, how could you check the power output of a microwave oven? You will never solve this problem simply by doing calculations from the comfort of your armchair. No amount of theorising will give you the answer to this problem! You have to carry out a practical investigation.

In trying to find a solution you will have to plan how you will tackle the problem. You will need to think about the apparatus you will use and how you will set it up. You will be involved in taking and recording measurements and making a report. Your report will include the conclusions you draw from your results. Solving practical and theoretical problems is what physics is all about.

Progress

You should use this book to improve your knowledge and your ability to solve problems. And you can check your progress by trying the Questions and Activities as you work your way through it.

TELECOMMUNICATIONS

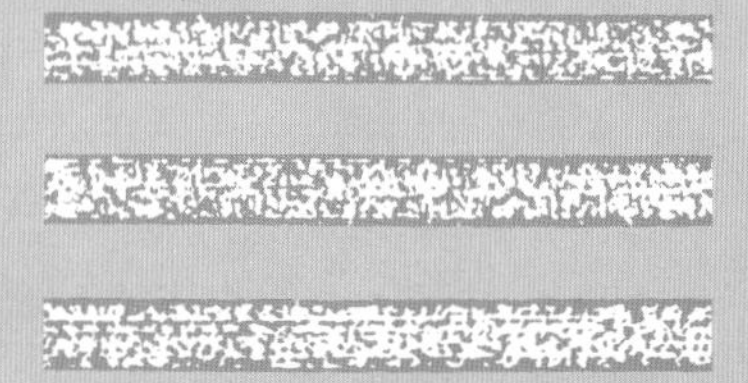

Getting the message across

Communicating with wires

Wave models

Telephone links

Radio waves

ELF to UHF

Microwave links

Cellphones

Black and white T.V.

Colour T.V.

Reflection

Optical fibres

Communication satellites

Getting the message across

Sight

Whatever form of communication is used we will eventually receive the message through one or other of our five senses.

For most people sight is one of the most important detectors of information. During all our waking hours we are continually receiving visual signals from the light reflected from different parts of the environment.

Light travels at 3×10^8 m/s (300 000 000 metres in 1 second).

Touch

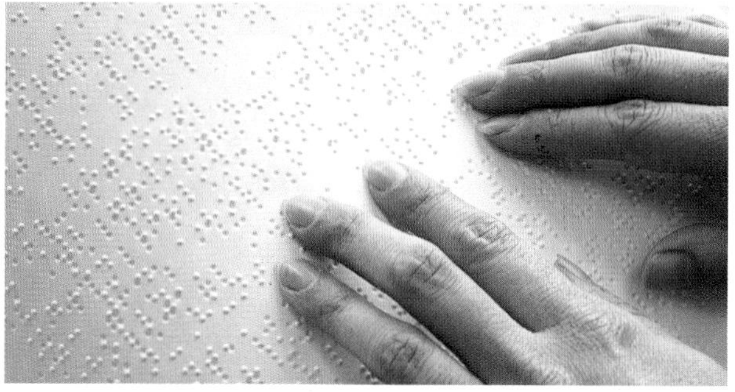

Many messages – from kicks to kisses – can be conveyed by physical contact. The invention by Louis Braille, in 1834, of a system of raised dots to represent different letters in the alphabet has enabled blind people, ever since, to read by touch.

Taste

The food we like may not always be the food that is best for us but at least our taste buds warn us if it is rancid or rotten.

Smell

A leak of gas in the kitchen or of petrol in a car can alert us to the danger. The skunk can signal very effectively when it is attacked or disturbed.

Hearing

This is another major detector. Most of our day to day communication with each other depends on the ability to hear and to interpret the never ending stream of sound entering our ears. But sound travels slowly compared to light which travels about a million times faster.

Mach 1

Concorde can fly at Mach 2 – that is at twice the speed of sound in air.

Thunder and lightning are produced together during a thunderstorm. Look at the cartoon and its caption and then calculate the speed of sound in air (Mach 1). You can ignore the time taken for light to travel one kilometre. It is less than a millionth of a second!

A puff of smoke and a loud bang are produced together when the starter's gun is fired at an athletics meeting. At such meetings two steps must be taken to allow for the slow speed of sound in air. First, the starter is positioned so that all the althetes in a race hear the gun at (nearly) the same time. Secondly, timekeepers are careful to start their watches at the moment the gun is fired and not when the sound reaches them.

$$\text{average speed} = \frac{\text{distance moved}}{\text{time taken}} \qquad \bar{v} = \frac{d}{t}$$

When the thunderstorm was 1 km away the time interval between the lightning and the thunder was 3 seconds.

›› ***A*** *Devise a way of finding the speed of sound in air using a starting pistol and a stop watch. State any sources of inaccuracy.*

B *Draw a line graph of distance travelled against time, for sound in air. Use a time scale up to five seconds.*

1 Calculate Concorde's top flying speed in metres per second.

2 Describe what measurements you would take to decide whether a thunderstorm is coming closer or not.

3 A race timekeeper starts his watch when he hears the sound of the gun rather than when the gun fires. Explain whether his measured race time will be too long or too short.

4 Why are loudspeakers attached to the starting blocks?

Communicating with wires

Enter electricity

In the 18th and 19th centuries communication was totally changed by a number of discoveries. These include electric currents, electric batteries and the link between current and magnetism. It became possible to have a current flowing in a circuit, and to detect this current by the magnetic effect it produced. This, in turn, led to the electric telegraph. Signals could be sent over long distances quickly, cheaply and with almost total privacy.

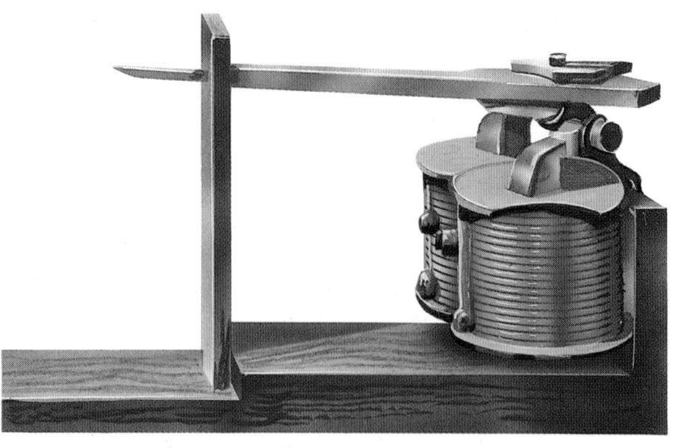

An early electric telegraph.

One telegraph developed in Britain by Cooke and Wheatstone (1837) used a system of five magnetized needles. These were turned by electromagnets to point to different letters and numbers. Another system developed in America in the 1830s by Samuel Morse marked a paper tape with dots and dashes in a coded signal.

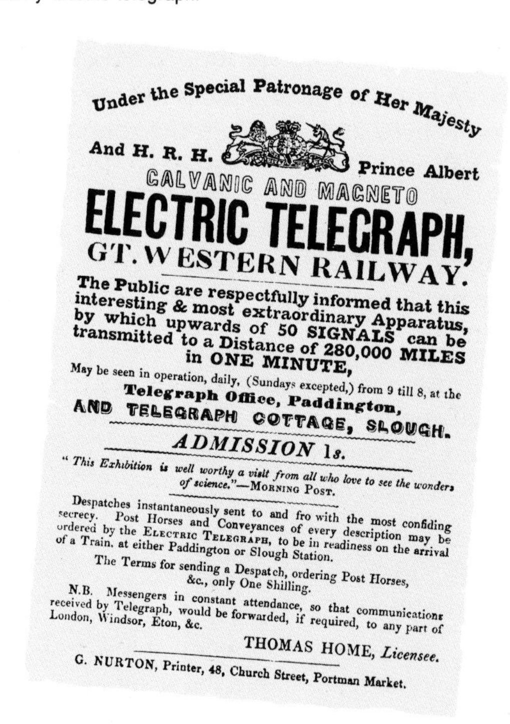

Under the Special Patronage of Her Majesty

And H. R. H. Prince Albert

GALVANIC AND MAGNETO

ELECTRIC TELEGRAPH,

GT. WESTERN RAILWAY.

The Public are respectfully informed that this interesting & most extraordinary Apparatus, by which upwards of 50 SIGNALS can be transmitted to a Distance of 280,000 MILES in ONE MINUTE,

May be seen in operation, daily, (Sundays excepted,) from 9 till 8, at the

Telegraph Office, Paddington,

AND TELEGRAPH COTTAGE, SLOUGH.

ADMISSION 1s.

" *This Exhibition is well worthy a visit from all who love to see the wonders of science.*"—MORNING POST.

Despatches instantaneously sent to and fro with the most confiding secrecy. Post Horses and Conveyances of every description may be ordered by the ELECTRIC TELEGRAPH, to be in readiness on the arrival of a Train, at either Paddington or Slough Station.

The Terms for sending a Despatch, ordering Post Horses, &c., only One Shilling.

N.B. Messengers in constant attendance, so that communications received by Telegraph, would be forwarded, if required, to any part of London, Windsor, Eton, &c.

THOMAS HOME, *Licensee.*

G. NURTON, Printer, 48, Church Street, Portman Market.

Before this time, accidents on the railway were frequent because there was no signal which could travel faster than the train and give warning of its approach! Much of the early development of the telegraph in Britain was linked with the railways and telegraph wires were laid by the side of the track. Railway safety and timekeeping improved dramatically!

The use of the telegraph by the public followed, particularly in America, where Morse had used a 65 km telegraph line to carry the signals.

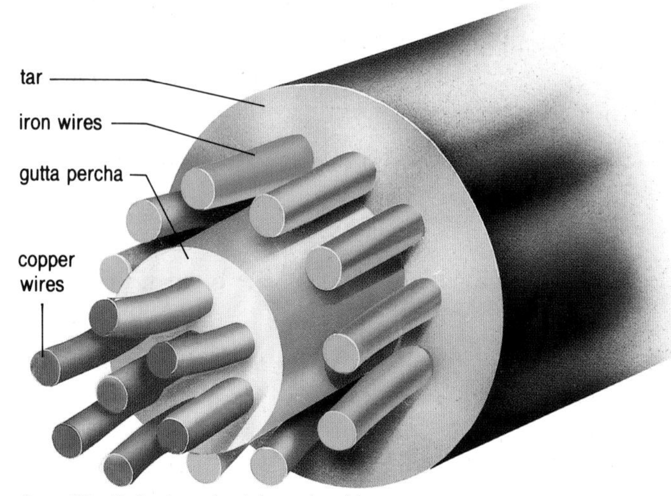

One of the first submarine telegraph cables.

Length	4000km approx
Mass	1kg per m approx
Central cable	7 wires purest copper
Gutta percha	withstands sea water
Outer case	10 iron wires
Tar	excludes sea water
Copper	less resistance to current than iron
Iron	much stronger than copper

Telegraph circuits

The simplest method of sending information by electricity is to use a direct current (d.c.) circuit. The earliest electrical communication system used a circuit similar to the one shown.

When the key is closed, bursts of current pass along the circuit to the receiving end. There they meet an energy changer (transducer), such as a bulb or a buzzer. Its purpose is to change the current pulses into a form of energy which can be detected by the person at the receiver. In this way, information can be transmitted by opening and closing the key according to an agreed code, such as Morse code.

Problems arise when very long telegraph wires are used. The circuit resistance gets large and the current becomes small. So, the received signal gets very weak and the coded pulses become difficult to detect. The rate of sending pulses must therefore be reduced. Transmission rates at first were about 12 words per minute but later they improved to 200 words per minute.

The telegraph was the first of our modern electrical communication systems. The railways used a signalling system, similar to the needle telegraph, for over a hundred years.

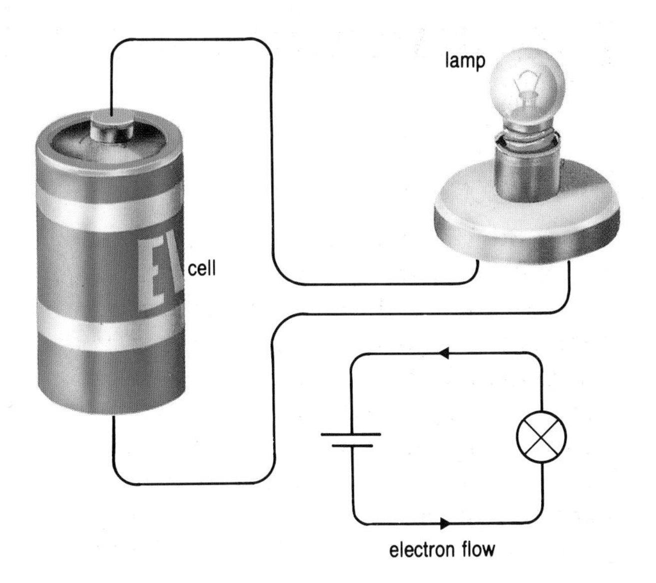

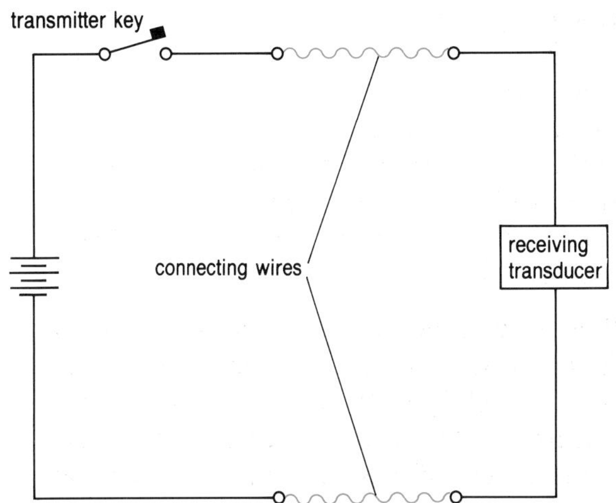

The circuit of an electric telegraph: the receiving transducer changes energy from one form to another.

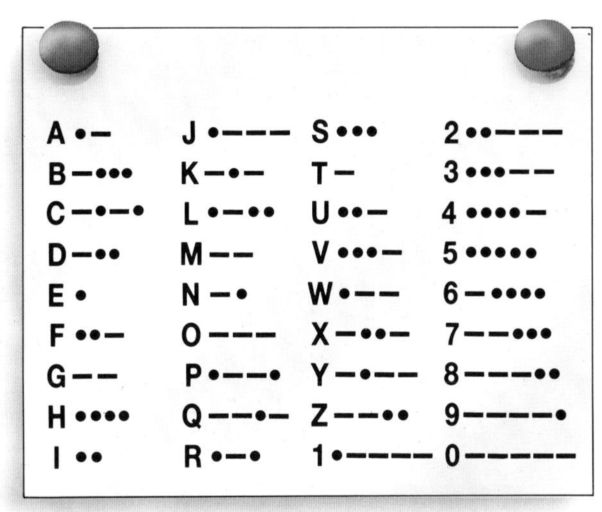

A •–	J •–––	S •••	2 ••–––
B –•••	K –•–	T –	3 •••––
C –•–•	L •–••	U ••–	4 ••••–
D –••	M ––	V •••–	5 •••••
E •	N –•	W •––	6 –••••
F ••–	O –––	X –••–	7 ––•••
G ––	P •––•	Y –•––	8 –––••
H ••••	Q ––•–	Z ––••	9 ––––•
I ••	R •–•	1 •––––	0 –––––

Morse code.

» *Set up a Morse telegraph system and transmit a simple coded message. Measure your transmission rate, in words per minute.*

1 a) What energy changes take place in (i) a bulb, (ii) a buzzer?
b) Why are such energy changers needed at the receiver of the electric telegraph?

2 Calculate the average signal speed advertised for the telegraph. Express the answer in km per second. (1 mile = 1.61 km)

3 What was the main purpose of each part of the submarine cable, shown in the diagram?

4 a) Draw the circuit for a telegraph system using a bulb as the 'receiver'.
b) Explain how the system works.
c) What difficulties might arise if the telegraph wires were made very long?

Wave models

Models

What is your idea of an atom, a gas or a galaxy? Can you picture each of them in your mind? Is an atom, for example, a big ball with lots of tiny balls flying around it? Scientists call such 'mind pictures' **models**. They are like the 'real thing' in a number of ways but they are different in others. Models may help you to understand how things work but you must treat them with care. Use them if they help – but remember there are differences between a model and the 'real thing'.

Waves on a shallow pond. Waves: amplitude 1 cm; speed 30 cm/s; wavelength 4 cm.

Waves

Information is transmitted in many ways: by sound, by light, by radio, by television. In each case we speak about waves: sound waves – light waves – radio waves – television waves. It is difficult to imagine these waves but not so difficult to imagine a water wave. We can therefore use a water wave as a 'model' to help us think about other waves too.

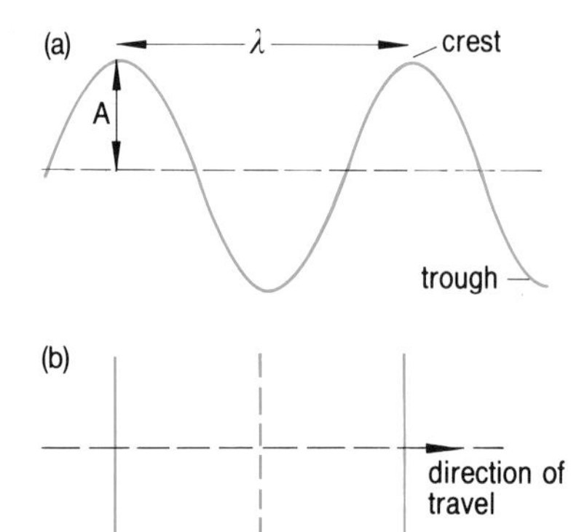

Two different ways of showing waves on a diagram: a) shows a side view; and b) shows a view from above the waves.

How useful is the water wave model? Water waves carry energy along in the direction in which they move. Other waves do the same. Water waves are reflected when they hit something like a sea wall. Other waves are also reflected from hard surfaces. The energy carried by water waves is absorbed by a gently sloping beach. The energy carried by other waves is also absorbed when they strike various substances. And there are many other things which all waves have in common – see the panel.

Surf-riding off Australia. Waves: size 5 m; speed 8 m/s; wavelength 80 m; interval between waves 10 s.

Communicating information

Water waves can 'tell' us a number of things: a fishing float making bigger waves 'tells' us that a fish is nibbling the bait – sea waves gettting bigger 'tell' us that a storm is approaching – a tidal wave 'tells' us that an earthquake has occurred in the sea bed. In short, waves can communicate information. They do this by changes in their height or amplitude (A), or their frequency (f).

We communicate with each other in many different ways, but in almost every case we use waves. In fact, it is difficult to think of any method of communication, ancient or modern, which does not use waves of some kind. The information to be transmitted is 'printed' on the waves by changing their amplitude or their frequency. Examples of the use of waves in communication are shown in the table.

Method of Communication	Waves transmitted	Receiver used
Speech, music	sound	our ears, microphone
Light signals	light	our eyes, film
Radio signals	radio	radio receiver

As waves reach the shallower water near a beach, they slow down and get closer together.

Properties of waves

In order to think about waves let us take a simple example – the water wave.

The main terms used to describe such a wave are described below.

- The amplitude (A) is the maximum movement of a particle of the water from its undisturbed position. Units: metres.
- The wavelength (λ) is the distance between neighbouring crests (or other corresponding points). Units: metres.
- The wave frequency (f) is the number of complete waves in one second. Units: hertz. 1 hertz (Hz) is one wave per second.
- The wave speed or velocity (v) is the rate at which the wave crests and troughs move through the water. Units: metres per second (m/s).

A runner taking 3 steps per second (f) each of 0.9 metres (λ) will be running at a speed (v) of: 3×0.9 ($f \times \lambda$) = 2.7 metres per second

The same relationship holds for waves:

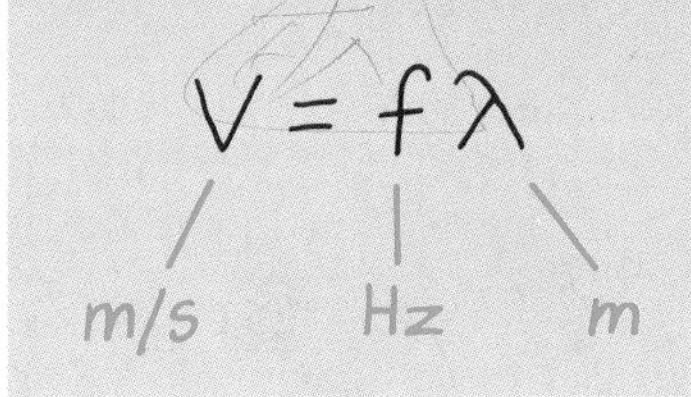

›› Make a ball bob up and down regularly on a water surface. Draw the wave pattern, as viewed from above, that is set up on the water.

1 Waves are shown on a pond and being used by a surfer. For these waves compare the size of the following: amplitude, frequency, speed, and wavelength.

2 Two wave sources produced the waves shown in the same time. Compare the amplitude, wavelength, frequency and speed of these waves from the red and blue sources.

3 Draw a diagram to show the wave pattern of waves approaching a beach, as viewed from above. For simplicity, show each wave as a straight line.

Telephone links

Speaking along wires!

The telephone can be used to pass a message quickly and clearly to almost any part of the world. But sound energy cannot travel far along the thin wires which are connected to most telephones. How then does this telephone link work?

Although sound energy does not travel far, electrical energy can travel long distances through wires. You can speak to someone thousands of miles away because your telephone turns sound signals into electrical signals.

All telephones require four main parts:

1 a microphone to change sound signals into electrical signals;

2 a supply of electrical energy to boost (amplify) the electrical signals;

3 a way of transmitting the signals to a distant subscriber;

4 a receiver (earphone) to change the electrical signals back into sound.

Answer. Please!

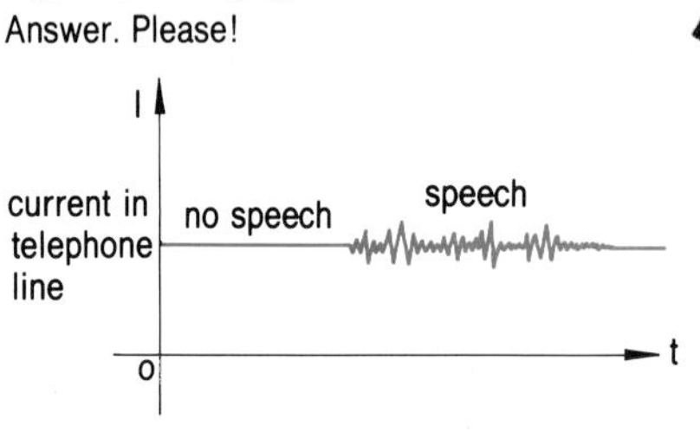

Current variations caused by speech.

A look at the parts

Sound waves make the diaphragm vibrate and change the electrical resistance of the carbon microphone. As a result, the current in the circuit varies to match the sound (speech) waves.

Small changes in sound energy can cause much larger changes in electrical energy in the circuit (about 1000 times larger). These changes in electrical energy can then be transmitted through the circuit to the distant subscriber.

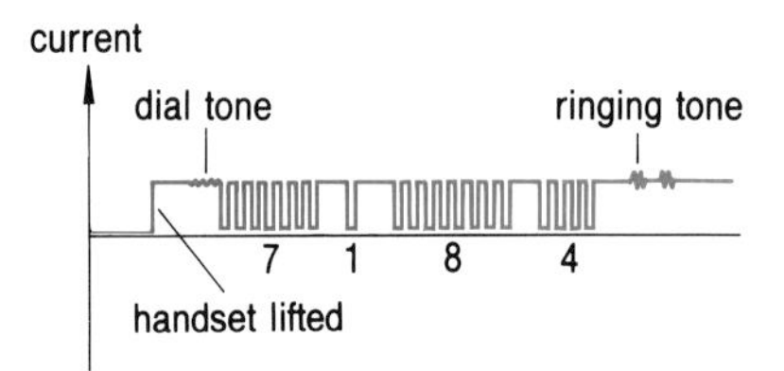

Dialled number as pulses at the exchange.

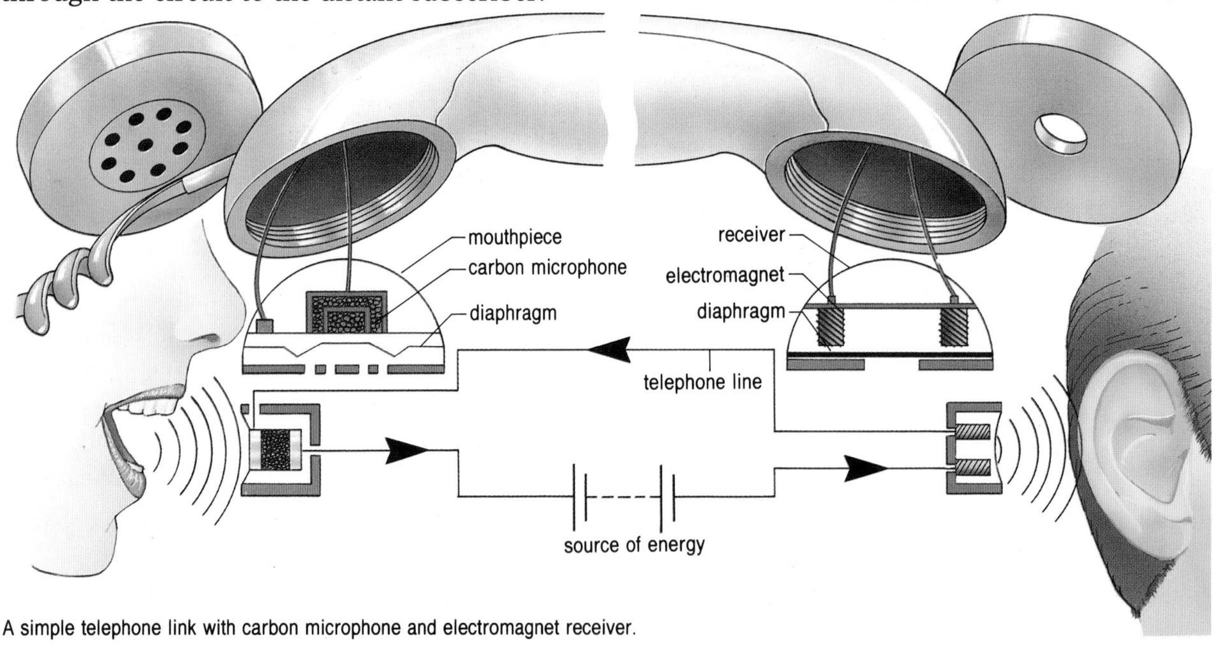

A simple telephone link with carbon microphone and electromagnet receiver.

The link

When you call a number a coded signal is sent to the exchange. The code for each number is recognised by selector switches which direct the call to the correct subscriber.

In the past, selection and switching have been done by electromagnetic switches (relays). Now more and more exchanges are being changed to electronic switches. Calls can be connected faster by these new electronic systems which are based on transistors and integrated circuit 'chips'.

Energy losses

Energy is needed for all forms of communication. While passing from the transmitter (microphone) to the receiver (earphone or loudspeaker), some energy is always lost. Communications will be poor if this loss is too large.

For example, imagine that you are phoning a friend who lives next door. Your call might travel through an exchange 8 km away and so the electrical signals would then travel 16 km. The copper wires joining telephones to the local exchange have some electrical resistance. Because of this resistance, a lot of the electrical energy travelling from your telephone is lost. In fact, only about $\frac{1}{300}$ of the original energy reaches your friend.

In long distance communication particularly, the problems caused by energy loss must be overcome. To do this amplifiers are fitted at regular intervals in the telephone link. Some examples are given in the table.

» *Connect up, and use, a telephone link which allows you to speak to someone in the next room. Draw the circuit used.*

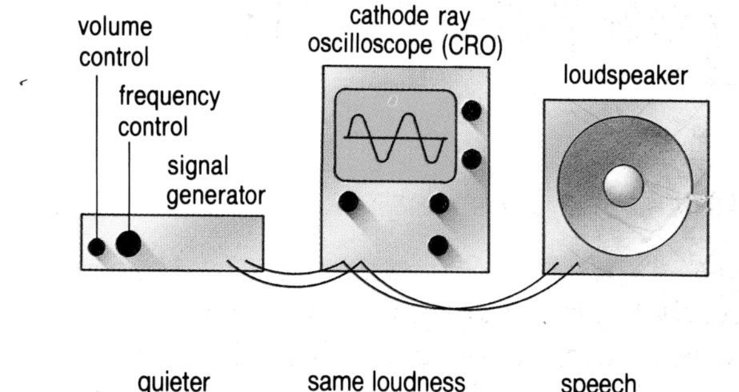

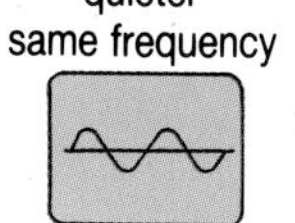

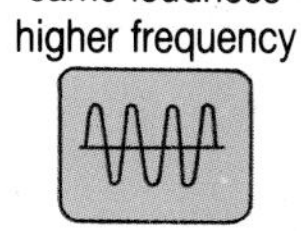

Looking at sound traces on a cathode ray oscilloscope.

Type of telephone link	Approximate amplifier gaps (km)
Coaxial cable (wire)	4
Microwave	45
Optical fibre	100
Satellite	36 000

Typical distances between amplifiers in different telephone links.

Smith's garage. Can we help?

1 If the controls of the oscilloscope shown above are not altered, draw the traces which would be produced by:

a) a higher frequency and a softer sound;

b) a lower frequency and a very loud sound.

2 In wires, a telephone signal travels at about 200 000 km/s. How long does it take an electrical signal to reach a receiver 16 km away?

Radio waves

On the air

In 1887 Heinrich Hertz was experimenting with a spark coil. He noticed that whenever it produced a spark, another spark appeared across the ends of an open metal ring several metres away. Radio waves, produced by the first spark, had been transmitted across the room. Later Guglielmo Marconi found a way of using these radio waves to carry sound. In 1907 he was able to transmit the first sound signal across the Atlantic. Marconi's discovery was the key to the widespread use of radio waves for communication.

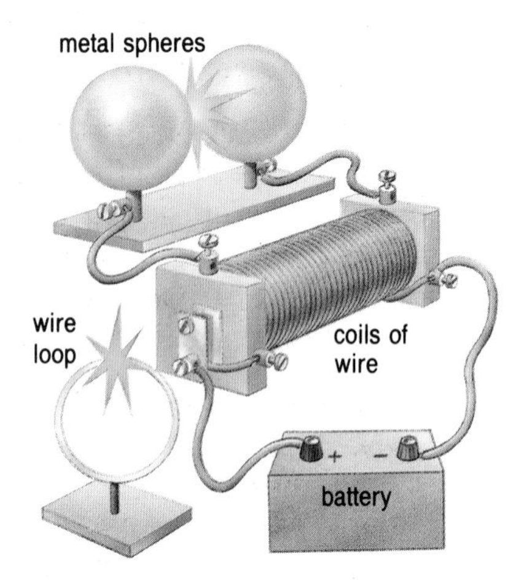

The battery and coil of wire combine to produce a spark between the spheres. From this spark, radio waves transmit energy across the room and make a spark appear at the ring.

Radio-wave communications

The air around us is filled with many different radio waves, all day, every day. Although many carry sound signals, we cannot hear them without special receivers. So how is it possible that radio waves carry sounds, silently and quickly, to all parts of the world and beyond?

The main stages of radio-wave communication are shown in the diagram.

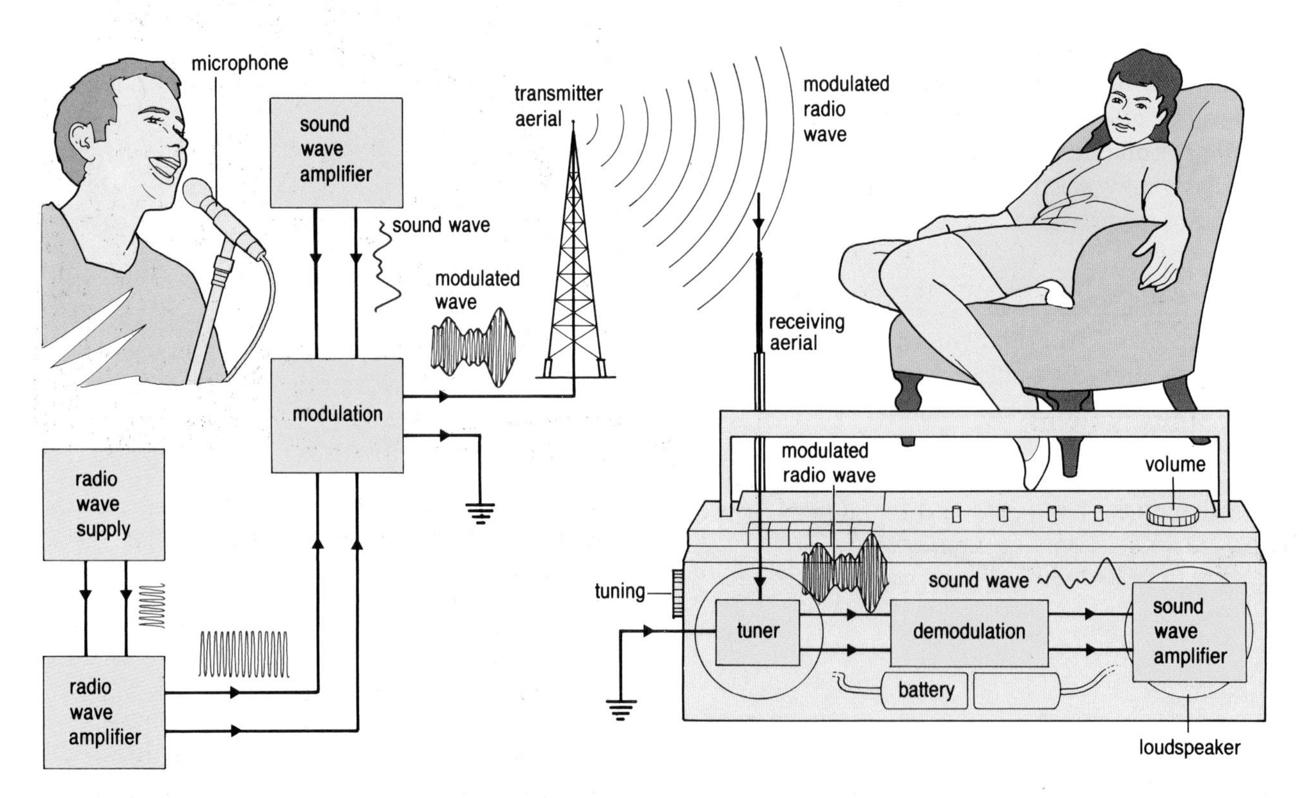

The radio wave supply produces a carrier wave of constant amplitude and frequency. This wave is then altered (modulated) in a way that matches the sound signal being transmitted. The modulated carrier wave finally leaves the transmitter aerial. Any receiver, tuned to the same frequency, picks up this modulated wave. The sound signal and the carrier wave are separated (demodulated) and an amplified sound signal goes to the loudspeaker.

Modulation

One way of adding sound to radio waves is to change the height (amplitude) at the frequency of the sound signal. This is called amplitude modulation (AM). The radio wave (carrier) frequency is unchanged.

Another way of transmitting sound by radio is to keep the amplitude of the radio waves constant and make the sound signals change frequency. This form of modulation – called frequency modulation (FM) – is used with very high frequency (VHF) radio waves.

There are two main advantages of VHF/FM over AM broadcasts. Firstly they do not suffer so much interference and secondly the sound quality is much better. For these reasons, FM is used where sound quality is important, such as with hi-fi.

Of course, there is always a snag! VHF transmissions can be received clearly over short distances only – perhaps 40 km or so – due to the curvature of the Earth's surface. They have a 'visual range' because, like light, they travel only in straight lines.

Many radio stations use both AM and FM transmissions. They broadcast AM signals on long and medium waves and the same programme on FM signals on very high frequency waves. This allows people with AM receivers to hear broadcasts in parts of the country which VHF/FM signals do not reach. It also gives high quality reception to others with VHF receivers.

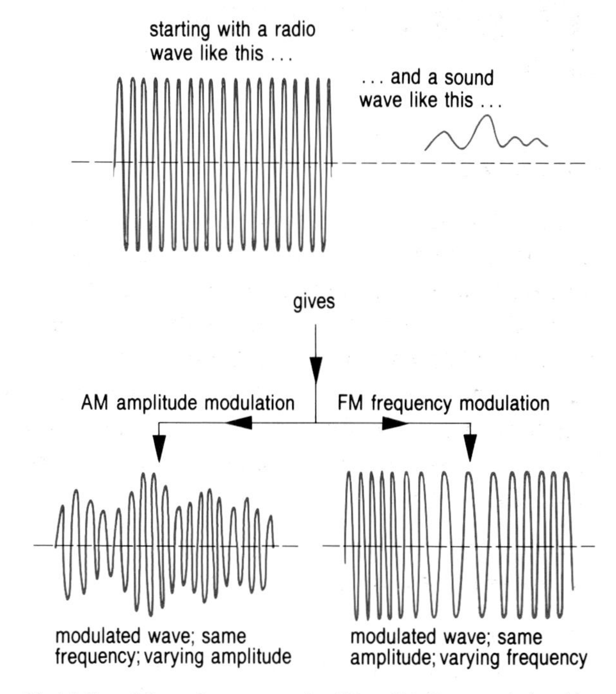

Modulation of the radio wave can be AM or FM. The sound signal is restored by removing the carrier wave from the modulated signal at the receiver.

RADIO FREQUENCIES

RADIO 1 1053kHz/285m, 1089kHz/275m. RADIO 2: 693kHz/433, 909kHz/330m/88.91vhf stereo. RADIO 3: 1215kHz/247m/90-92vhf. RADIO 4: 200kHz/1500m. RADIO SCOTLAND: 810kHz/370m/92-95vhf stereo. RADIO FORTH: 1548kHz/194m/97.3vhf stereo. RADIO TAY: Dundee, 1161kHv/258m/95.8vhf stereo; Perth, 1584kHz/189m/96.4vhf stereo. NORTHSOUND: 1035kHz/290m/96.9vhf. WEST SOUND: 1035kHz/290m/96.2vhf. MORAY FIRTH 271m, 95.9vhf

Silence is golden

Radio waves, whether AM or FM, are undetected by humans and so silence reigns, (unless you have a radio receiver).

1 Radio 4 uses a carrier wave frequency of 200 kHz with a wavelength of 1500 m. Use these values to calculate the speed of these (and all other) radio waves.

2 Copy and complete the following table for 5 radio station settings, using values listed in the newspaper cutting.

Radio station	Frequency	Wavelength
Forth	1548 kHz 97.3 MHz	194 m 3.12 m

›› *Slowly move the tuning control of a radio receiver over as wide a range as possible. Identify the stations which give strong signals. Find out where each of these transmitters is situated.*

3 a) Explain why different radio stations must use different radio frequencies.
b) Which receiver control is altered to receive these different stations?

ELF to UHF

Radio waves

Radio waves form part of the electromagnetic wave spectrum. Within the radio wave group itself there is a very large range – a spectrum – of radio-wave frequencies and properties. The diagram and table show some simple divisions of this spectrum.

Radio waves of different frequencies have different properties and so the waves in one division have different properties from those in the other divisions. Each part of the radio-wave spectrum has some advantages over the other parts. These advantages decide the main use for that particular part of the spectrum.

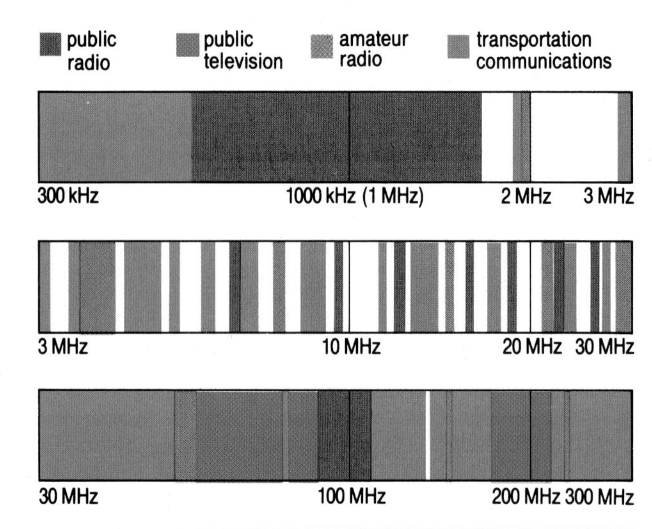

Frequency range	Frequency name	Wavelength range (m)	Wavelength name	Range approx. (km)	Main uses
30 Hz–3 kHz	extra low ELF	>100 000			links to submarines
3–30 kHz	very low VLF	100 000–10 000		>1500	long range navy, army use
30–300 kHz	low, LF	10 000–1000	long wave	>1500	long range navy, army use
300 kHz–3 MHz	medium, MF	1000–100	medium wave	<1500	sound broadcasts
3–30 MHz	high, HF	100–10	short wave	world wide	sound broadcasts
30–300 MHz	very high VHF	10–1	ultra short wave	just beyond horizon	high quality sound
300–3000 MHz	ultra high UHF	1–0.1		horizon	television or mobile link
>3000 MHz		<0.1	microwaves	36 000	microwave links, e.g. for satellites

One method of dividing the radio wave spectrum.

Which wave?

If all radio waves can be modulated to carry a sound signal, why use radio waves of such widely differing frequencies? The reason is they have different properties.

All electromagnetic waves travel through space at 3×10^8 m/s.

The **ionosphere** is a region of charged particles (ions) in the air. The energy to produce these ions comes from radiation reaching the Earth from the Sun.

Waves in the ELF band are the only ones capable of passing deep into the oceans. These waves are used to communicate with nuclear submarines moving deep in the water. Using waves in other groups, a submarine would have to be almost stationary and near the surface to receive a strong signal.

All BBC radio stations (and many others) are packed in the low, medium and high frequency bands. Some of the shorter waves in these bands travel large distances by reflecting from the ionosphere and are used for long distance communications. Longer waves travel close to the ground and are better suited for local radio.

VHF and UHF waves travel almost in straight lines and therefore they are not received at points which are hidden from 'sight' by the curvature of the Earth's surface. Even with this disadvantage, such waves are extremely valuable. VHF is used for high quality sound transmission, using FM signals. Such signals use at least an 18 kHz frequency range (bandwidth) to transmit hi-fi sound.

UHF waves, in bands ranging from about 470 to 850 MHz, are used for transmitting the signals needed for TV pictures and sound. The BBC and IBA television signals are sent out from hundreds of transmitting stations. So that the signals from one transmitter do not interfere with those from another, different transmission 'channels' are used for each station. Forty-four channels, each with a bandwidth of 8 MHz, are used for these TV transmissions. For example the Blackhill transmitter, near Glasgow, uses Channel 40 (623.25 to 631.25 MHz) for its BBC1 transmissions, while the transmitter at Craigkelly in Fife uses Channel 31 (551.25 to 559.25 MHz).

Microwaves are very useful as they pass through the ionosphere to link with satellites. Because of their very short wavelength, microwaves use much smaller aerials than other radio waves.

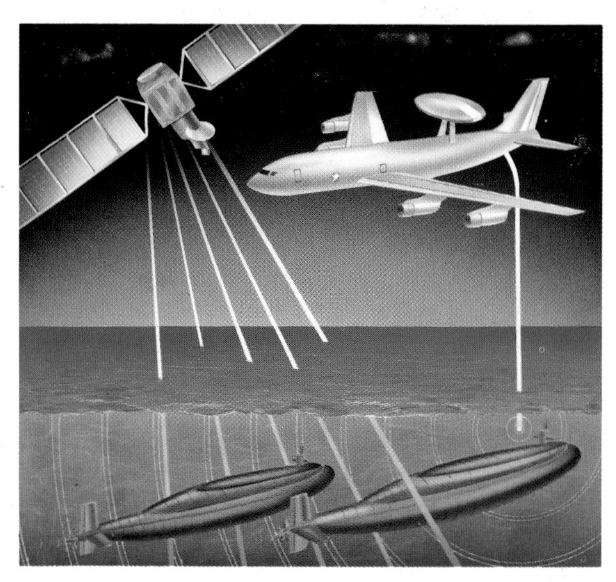

Communicating with submerged submarines using radio waves.

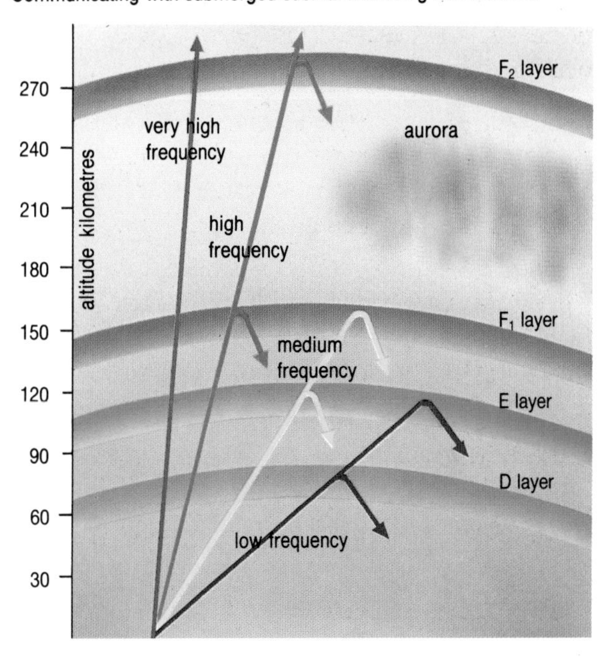

Details of the layers of the ionosphere showing where different radio waves are reflected.

1 Explain the following choices of radio waves:
a) microwaves for satellite links;
b) VHF for coastal radio stations where the range of communication is about 40 km;
c) most broadcasting stations use waves between 10 000 m and 100 m long.

2 The frequency range or bandwidth required to transmit speech is about 4 kHz. Explain why the transmission of high quality music needs more (18 kHz) and television very much more (8 MHz).

Microwave links

A tall story

Towering over the rest of London the Telecom Tower reaches a height of 189 metres. It was completed in 1965 and is now the centre of Britain's microwave communication system. A group of dish aerials near the top of the tower transmit and receive narrow beams of microwaves. From there, the microwaves are able to carry many signals, above the buildings and across the country. These narrow beams of microwaves provide paths of communication which replace expensive underground telephone cables.

Telecom tower.

Microwave pathways

Each radio transmission from the London Telecom Tower is directed accurately towards a repeater tower built on top of a high hill. The repeater receives the signal and passes it on to the next repeater and so on until it reaches its destination. This could be another Telecom tower or a ground station linked to a submarine cable or to a satellite.

A chain of repeater stations is necessary because the microwaves only travel in straight lines from the transmitter. This fact, and the curvature of the Earth, limit the range of transmission to about 40 km. It can be less if there are any hills in the path of the waves.

Repeater station.

Microwave paths along which signals are carried at 300 000 km/s.

Why microwaves?

The use of very high radio frequencies, such as 11 GHz, for a microwave link offers a number of advantages. These are:

1 small dish aerials, of about 2 m or 3 m diameter, can be used;

2 the narrow microwave beam between the radio towers ensures that less energy is lost by failure to hit the receiver dish;

3 each aerial can handle a very large number of signals at one time, perhaps as many as 16 000 telephone calls or 16 TV channels!

Signals galore

Music, television and speech all start as continuously changing signals. These are called **analogue signals** and they can be transmitted by modulating the microwave beam. There is, however, another way of sending the same information by microwaves. It is called **digital transmission**. In it the original analogue signals are changed into a series of numbers or digits.

The system has many advantages, particularly if the numbers are in binary form. In this form:

- transmission of the signal requires only a voltage to be switched on and off;
- many more signals can be transmitted at one time;
- the signals are of better quality.

To change an analogue system into a digital system, measurements have to be taken at regular time intervals. These measurements are then changed into numbers (digits). Provided enough measurements are taken, the digits can describe the original analogue signal reasonably well.

Decimal to binary

Numbers such as 5, 13, 2 are numbers expressed to the base 10. For example, 26 is 2 tens + 6 units.
Or put in a table:

Decimal number	Thousands, 10^3	Hundreds, 10^2	Tens, 10^1	Units, 10^0
5	0	0	0	5
13	0	0	1	3
26	0	0	2	6

The same decimal numbers could be expressed to the base 2. The table shows the conversion of decimal numbers into binary form (base two):

Decimal form	2^4 (= 16)	2^3 (= 8)	2^2 (= 4)	2^1 (= 2)	2^0 (= 1)
5 = 4 + 1	0	0	1	0	1
13 = 8 + 4+ 1	0	1	1	0	1
26 = 16 + 8 + 2	1	1	0	1	0

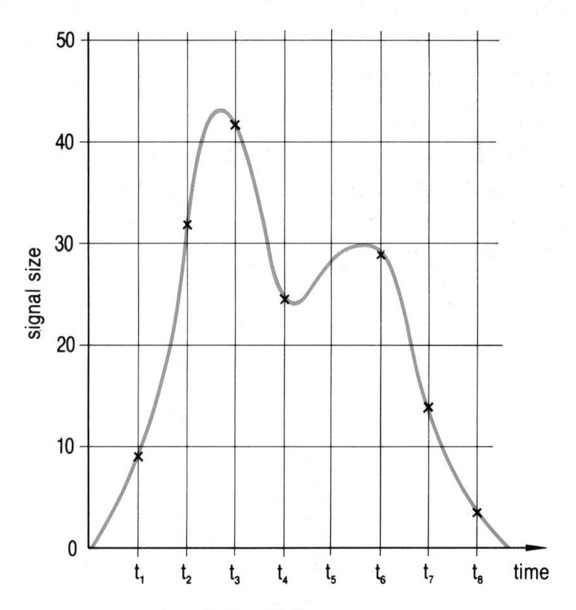

Analogue signal. Sample times t_1, t_2 ...

1 What is meant by a dish aerial?

2 Copy and complete the table showing the signal values at each of the sample times shown (i.e. t_1 ...... t_8).

Sample time	Decimal value	Binary value 8 bits used
t_1	9	00001001
t_2	32	00100000

Cellphones

Now we're speaking

Many people want to be 'tied to the telephone' even when they are travelling around the country. For them, a system called cellular radio has been devised. It divides the country into areas or cells, each of which is served by a low powered radio station. This can transmit and receive calls up to the cell boundary by using a two-way radio channel. Cells vary in width from 2 km in the city centre to 30 km in the countryside.

A cellphone in use – no visible link.

When the cellphone system is fully developed, the whole country will be covered by a network of low-powered radio transmitters, operating at around 900 MHz. Each will provide a link between any mobile cellphone operating in that area and the normal telephone network. Frequencies around 900 MHz were chosen for this link because no other radio services were using this frequency range.

Why cells?

A common arrangement of cells is shown on the right. Each cell is drawn as a hexagon, although in practice they may be roughly circular. A cell arrangement has been chosen for a number of reasons. First, it can handle many callers. Second, the same transmission frequencies can be used again and again in different cells.

Transmitters in cells that have the same number use identical radio frequencies.

Such cells are not close to each other and as their transmitters have only a small range there is no risk of interference between cells.

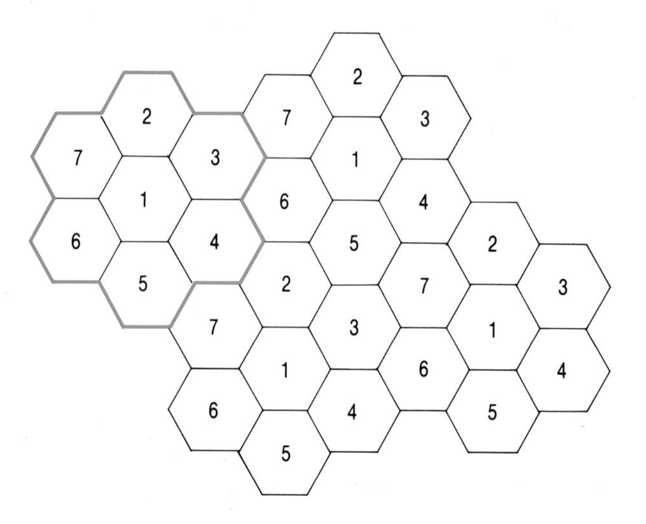

A moving telephone call

A doctor picks up the cellular phone in his car, dials the number of a hospital, 40 km away, and presses the 'send' button on the phone. A cellular phone is really a small radio transmitter/receiver and it 'broadcasts' the request for service. From the car, then, the signal travels to the nearest (cell) radio station where it is detected by a special control channel. A process then starts which takes,no longer than setting up a 'normal' telephone call. How will the telephone link with the hospital be made?

Electronic Mobile Exchanges

All cell transmitters are connected by underground cable to an Electronic Mobile Exchange (EMX) The name comes from their use with mobile phones, not because they move! The cell transmitter has available a number of two-way radio frequencies (channels), so one which is not in use is automatically selected and the car phone 'instructed' to tune to this frequency.

The outgoing telephone conversation is carried on this frequency from the car phone to the cell's receiver. There the carrier wave is removed and speech signal enters the normal telephone network through the EMX. From this point on, the link is the same as it would be for a 'normal' telephone. Incoming speech uses the same carrier frequency. But now it is transmitted from the cell transmitter and received by the cellular phone.

The electronic circuits at the exchange are the heart of the cellphone system. These circuits select a radio channel, route the call to its correct destination and even transfer the call to the next cell as the doctor drives along.

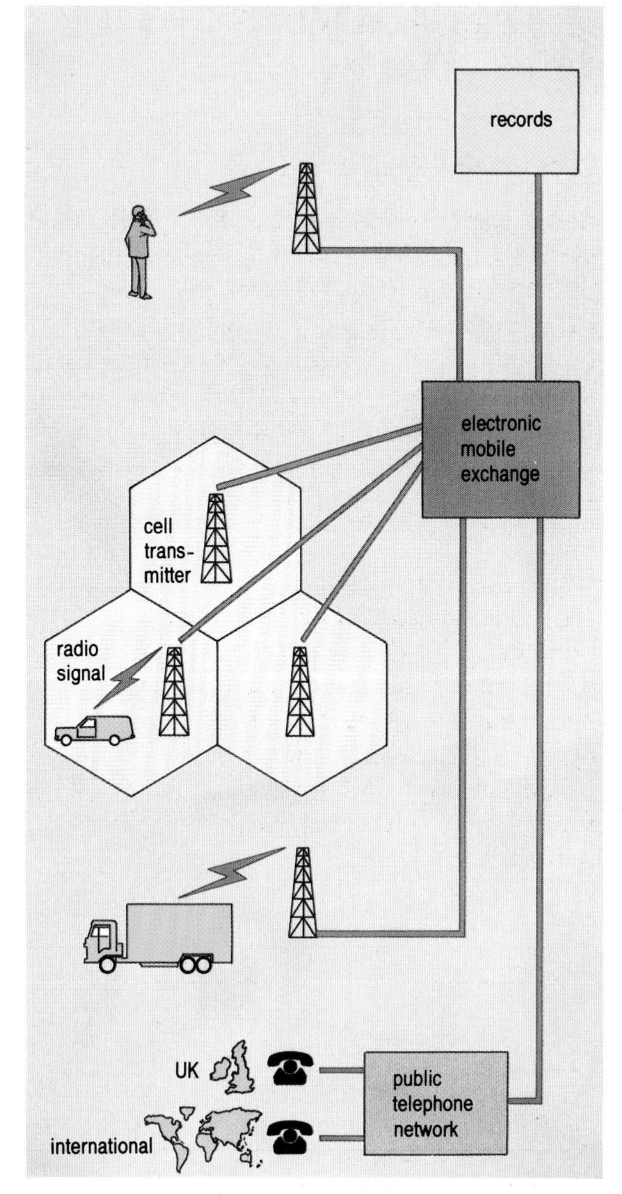

The central part played by the EMX in the cell system.

›› *Use a table of values of BBC radio frequencies to explain why cellphone transmitters do not use a transmission frequency of 200 kHz.*

1 Explain why the cell pattern shown will reduce interference between cellphone users.

2 Explain what is meant by 'a two-way radio channel'.

3 Explain what is meant by 'tuning' a radio receiver.

4 Why are cellphones not made more powerful?

Black and White TV

Moving pictures

When you watch television, the pictures you see on the screen are images of what is actually taking place at a distance – in a television studio in London for instance.

Look at the enlarged newspaper picture. It is made up of a large number of black dots of different sizes. Where the dots are small the picture is white and where they are large it is black. At normal viewing distances the individual dots cannot be seen, but they merge to form a single picture.

A black and white television picture is formed in a similar way. The TV screen is covered with many dots or elements. In this case, each dot is the same size but they can vary in brightness from black to white through many shades of grey. At normal TV viewing distances these individual dots cannot be seen as they also merge to form a single picture.

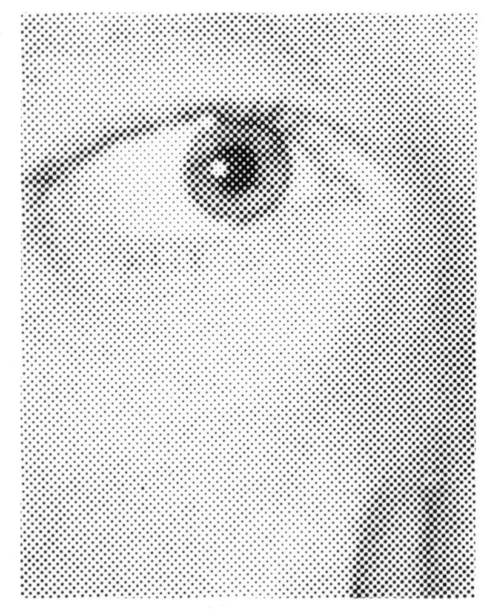

Part of a black and white newspaper photograph enlarged to show its dotty make-up.

The television camera

The operation of one type of television camera depends on the change in resistance of a semiconductor when light falls on it. The camera uses a cathode ray tube – called a picture tube – in which the fluorescent screen is replaced by a semiconductor plate. On the front surface of the plate there is a semi-transparent conducting layer which is wired to an external circuit as shown.

In total darkness the resistance of the semi-conductor is very high, so that very few electrons pass through it to the conducting layer. The current in the external circuit is therefore small and the voltage across R negligible. When light falls on the conducting layer it passes through to the semiconductor, reduces its resistance and allows electrons to pass through to the conducting layer. Current now flows in the external circuit and there is a voltage across R. The size of this voltage depends on the amount of light falling on the plate.

If a lens system now forms an image on the plate and if, at the same time, the electron beam is made to scan the semiconductor, as in the TV receiver, the output voltage will vary with the brightness of the picture area being scanned.

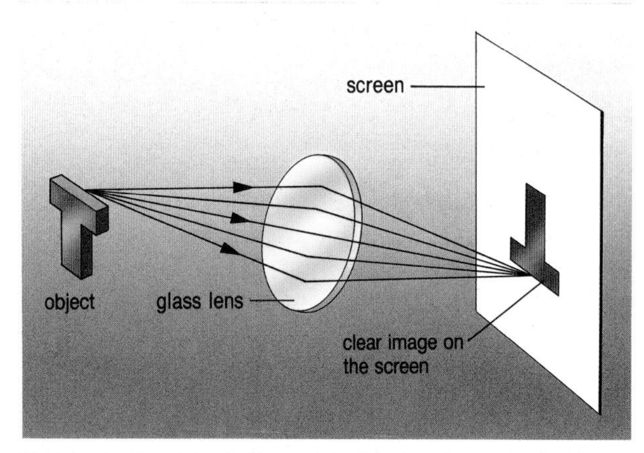

Light leaving the object is 'focused' by the glass lens to form a clear image on the screen.

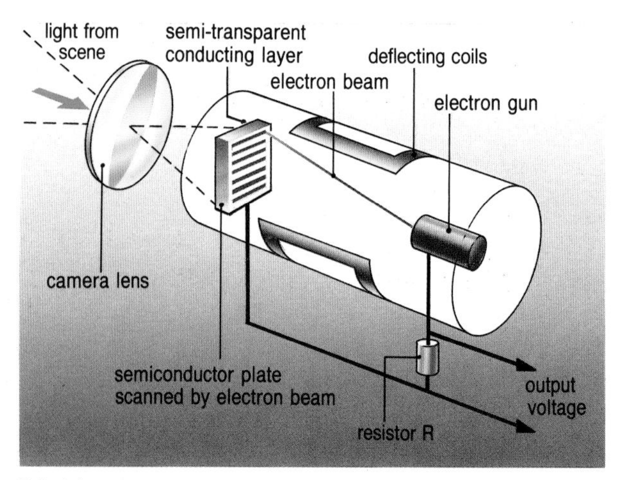

Television camera.

Transmission

The signal leaving the 'picture tube' is amplified and printed on the UHF carrier wave, ready for broadcasting from the aerial of the TV transmitting station. From there, the modulated electromagnetic waves radiate in all directions carrying information about each part of the picture.

Television transmitter aerial mast.

The TV receiver

At the receiver the TV aerial picks up the UHF wave. The picture signal is then separated from the carrier wave and converted back into a changing current. This is sent to the picture tube where it controls the current in an electron beam. While this happens, the beam is made to scan the screen exactly in time with the electron beam in the camera.

The inside of a TV screen is coated with a large number of dots arranged in 625 lines. Each dot is made from a phosphor, a material which gives out light when hit by electrons.

The picture on the TV screen is made to match the image in the TV camera because:

1 each line on the screen matches a line of squares in the camera picture tube;

2 the brightness of each dot on the screen varies in the same way as the light hitting the corresponding spot in the camera tube;

3 the build-up of the screen picture, line by line, is exactly the same as the scan used in the camera tube.

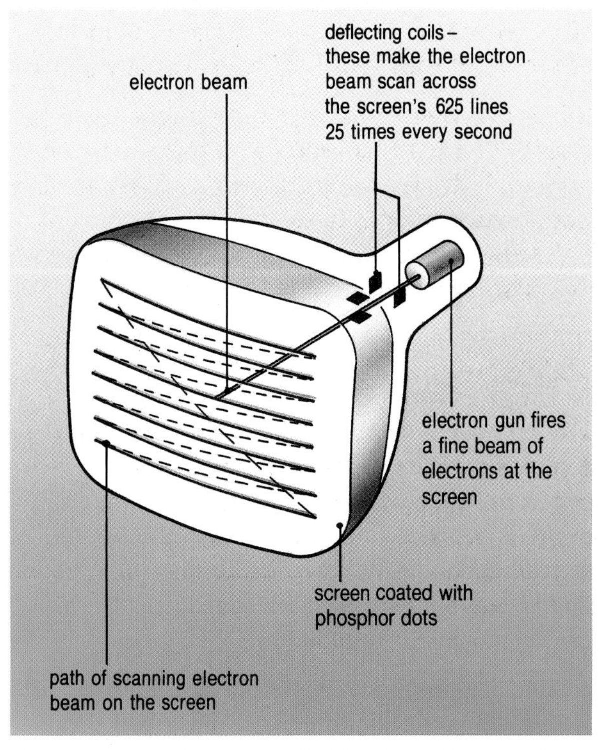

The picture signal alters the strength of the electron beam and thus the brightness of the phosphor dots.

›› *Take a piece of graph paper and shade in a number of its squares to form a picture. Use this to show the dots and line build-up for a black and white TV picture. Explain why some parts are made 'black' and others 'white'.*

1 a) In an ordinary camera, what is used to detect the image?
b) In what ways does this image differ from the image detected in the TV camera?

2 Each TV picture has 625 evenly spaced lines and each picture is scanned in about 0.04 seconds.
a) Estimate the time to scan 1 line.
b) If the length of 1 line on the TV screen is about 0.50 m, calculate the speed at which this line must be scanned.

Colour TV

Television camera

A colour television camera is really three cameras in one. Light from the object being televised enters the camera where an arrangement of mirrors and filters takes the light and splits it into the three primary colours – red, green and blue. The red part of the light is passed to one of the picture tubes which converts the red light pattern into electrical signals. The other two tubes do the same for the green and blue parts. The picture tubes are similar to the tube in the black and white camera. Before transmission, the electrical signals are coded to show dot colour and brightness for each dot and line in sequence.

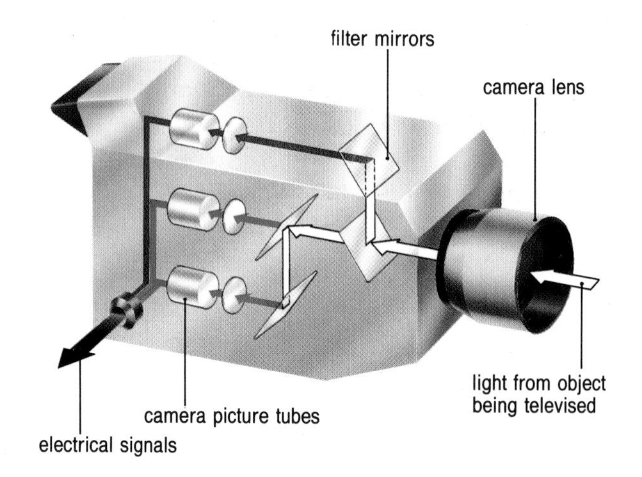

A colour-T.V. camera.

The colour-TV tube

At the receiver the TV aerial picks up the carrier wave with its coded information of colour, brightness and sound. The colour signal produced by the red light is fed to the electron beam responsible for producing the red colour on the screen. The signals for green and blue get a similar treatment.

The inside of a colour TV screen is coated with a large number of dots. The dots give out either red, green or blue light and from these three colours alone, it is possible to produce the whole range of colours seen on a colour TV screen.

A shadow mask, with many holes in it, is positioned close to the screen so that each electron beam hits only the correct dots.

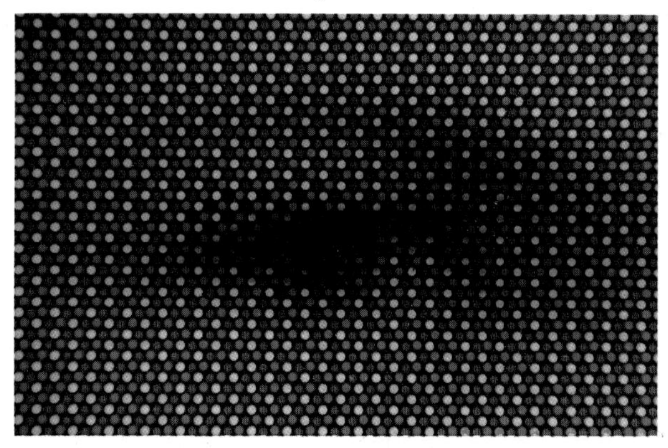

An enlarged view of a colour-TV screen showing its dotty make up.

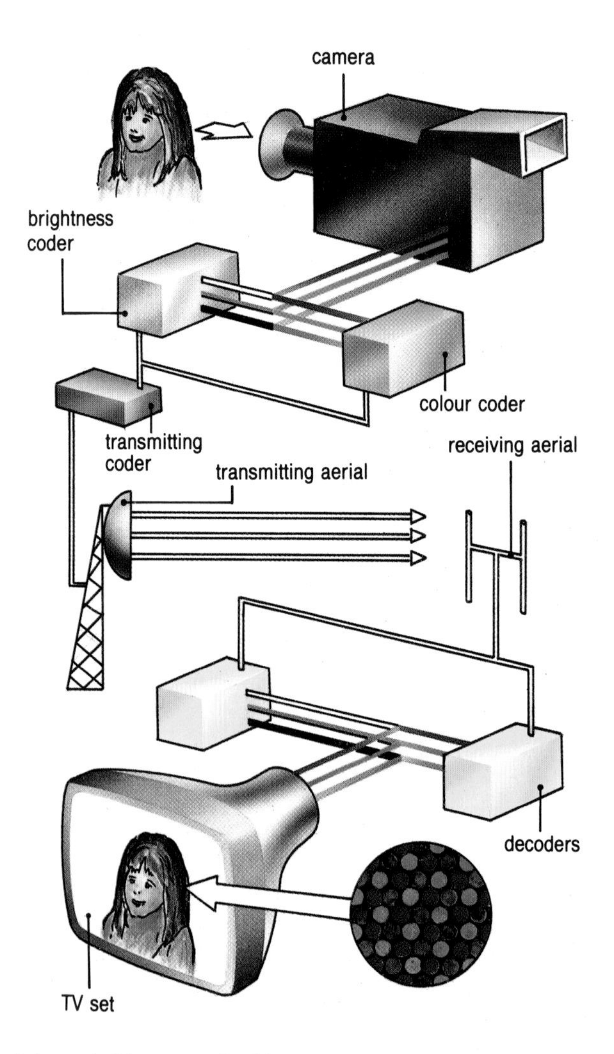

Colour television, transmitted from the camera to the receiver in your home.

Changing colours

Each electron beam hits one type of phosphor dot and each dot hit gives out a coloured light. The overall colour of a tiny part of the screen is decided by the brightness of several separate dots. To fill the screen with a coloured picture the electron beams must scan along one line of the screen making all the dots on this line glow. This process is then repeated for each line on the screen until the whole screen has been covered.

The scanning of the complete screen is repeated 25 times per second. The brightness of each dot on each line is controlled by the received video (light) signals.

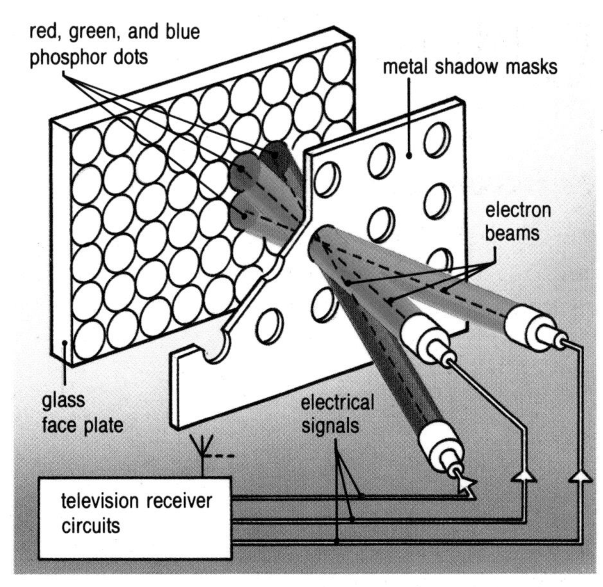

A view inside the tube of a colour-T.V. receiver.

Moving pictures

We do not see the screen as a series of little flashes of light, but rather as a complete, moving picture. Each still picture on the screen appears to run smoothly on to the next. The reasons for this are, first, that our eyes cannot detect the rapid changes between pictures (persistence of vision) and, second, that the phosphor dots continue to glow for a short time after the electron beam has passed.

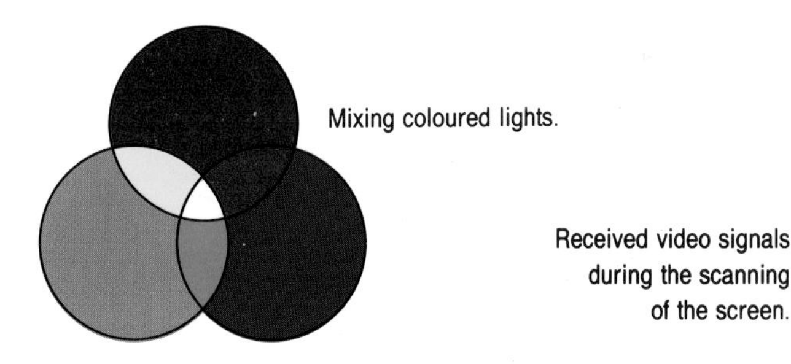

Mixing coloured lights.

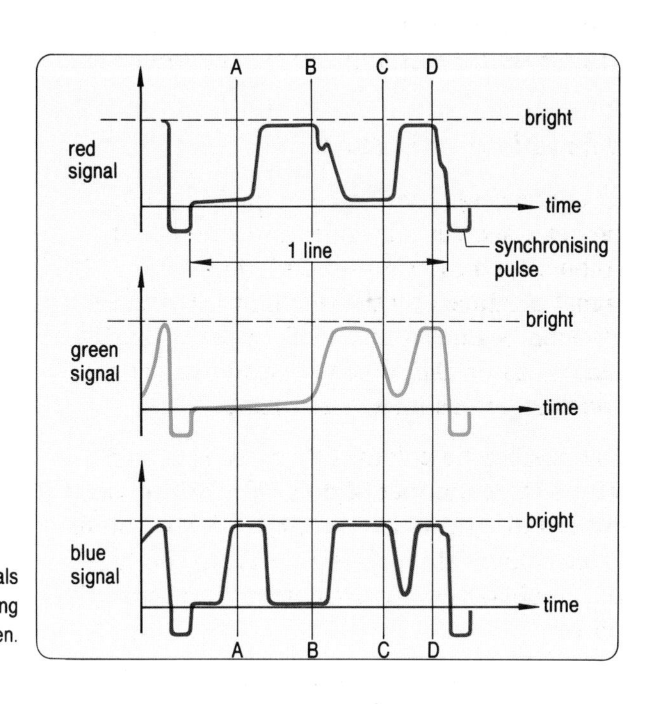

Received video signals during the scanning of the screen.

1 What overall screen colour will be produced at positions A, B, C and D with the video signals shown?

2 Most television transmission is now in digital form. Explain briefly what is meant by 'digital form'.

3 a) Draw an 8 by 6 grid as shown. Colour in each square with one of the colours listed below the grid.
b) Starting at the top left, and then as you normally read, write down the numbers which describe the colours.
c) Convert each number into digital form. (Each number only needs three bits.)
d) Draw, in digital form, the signal which would describe one line of your picture.

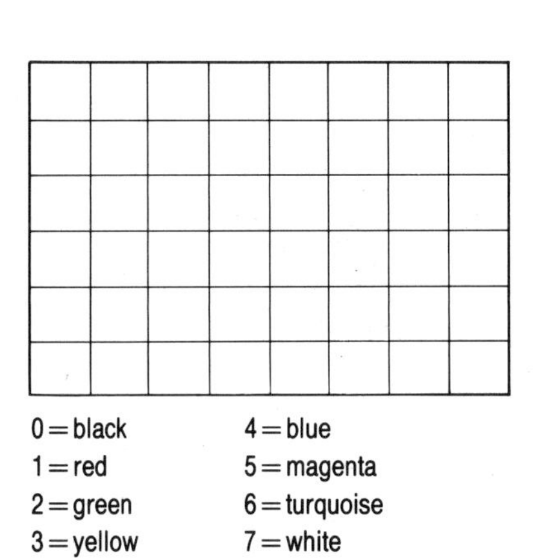

0 = black
1 = red
2 = green
3 = yellow
4 = blue
5 = magenta
6 = turquoise
7 = white

Reflection

Diffuse reflection

We see objects because they give out light or because they reflect some of the light falling on them.

Most surfaces, if viewed through a microscope, look slightly rough. Light shining on such surfaces is reflected in all directions. It is this diffuse reflection of light which allows us to see these objects or to photograph them. All kinds of waves are reflected in this irregular way if the surface they hit is sufficiently rough.

Regular reflection

The surfaces of mirrors or polished metals are very smooth and allow regular reflection to take place. The behaviour of this reflected wave follows certain rules, known as Laws of Reflection.

Look at the example of the periscope and see how the plane mirrors reflect the light so that it reaches the eye. A similar diagram can be produced to show how two plane mirrors could be used in a rear viewer.

A dish (parabolic) reflector has an interesting property because of its shape. Look at the diagram and you will see that if a plane wave XY is approaching the reflector the distances, from every point on the wave front to the focus by reflection, are equal. This means that all waves heading directly towards the reflector will be collected at the focus. A much bigger wave arrives at the focus than at any other point and a detector placed there will receive a large wave (signal).

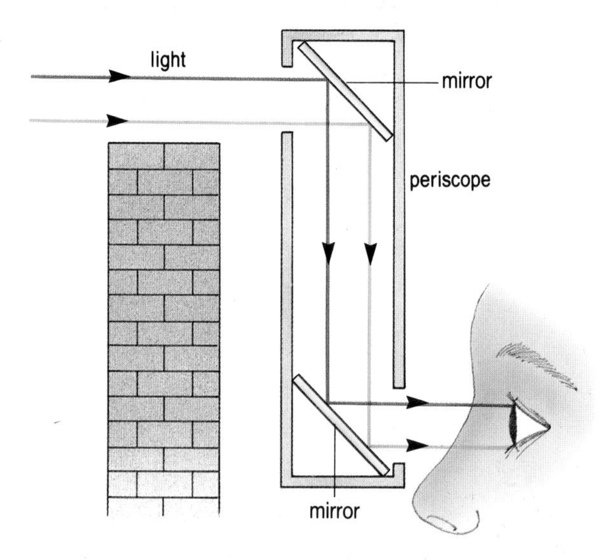

Light enters the periscope and after two reflections reaches the eye. The view over the wall is seen.

›› *Use a plane mirror to verify the laws of reflection.*

Build your own periscope using two plane mirrors.

Improving communications

A transmitter sends out a signal carried on a wave – it could be a radio wave or a microwave or any other wave. The wave and the signal get very much smaller the further they move away from the transmitter. Although small signals can cause problems for reception there are a number of ways of overcoming them.

The following methods are used in practice:

- the transmitter is made more powerful;
- the signal is directed more accurately at the receiver;
- the signal is amplified at points along its journey;
- the weak signal is greatly amplified at the end of its journey.

Look at the photograph of a typical transmitter and receiver of microwaves. Both parts have a dish aerial. These aerials play a large part in the attempts to improve communication. They are used:

- to collect the signal, and amplify the signal received by a detector placed at the focus;
- to shape the microwaves into a narrow beam by placing the transmitter at the focus.

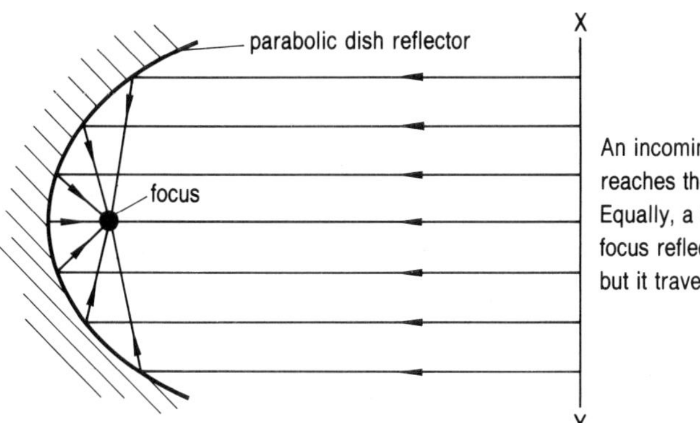

An incoming wave XY all reaches the focus by reflection. Equally, a wave starting at the focus reflects to become XY, but it travels outwards.

1 The gain of a dish receiver is a measure of how much it increases the signal received. The gain can be calculated from; gain $= 6(D/L)^2$, where D = the diameter of the dish and L = the wavelength of the waves used.
a) Calculate the gain of a 32 m diameter dish aerial using 0.075 m waves.
b) In what way is the gain changed by using waves of higher frequency with a dish receiver? Explain your answer.
c) Only a narrow beam of waves leaves a transmitter using a dish aerial. A measure of the beam width, in degrees, is given by: beamwidth $= 70\ L/D$.
Calculate the beamwidth for this dish and explain why such dishes must be accurately aimed.

Dishes used as transmitters and receivers of microwaves.

Reflection

When a wave, such as light, hits any object, some of the wave is reflected. Objects like mirrors are designed to reflect most of the light incident on them. The diagrams show regular wave reflection from a mirror surface. These diagrams and the laws of reflection apply to all mirrors and all waves. The normal is the line, at right angles to the surface, at the point where the wave strikes.

Laws of reflection

1 Angle of incidence, i = Angle of reflection, r

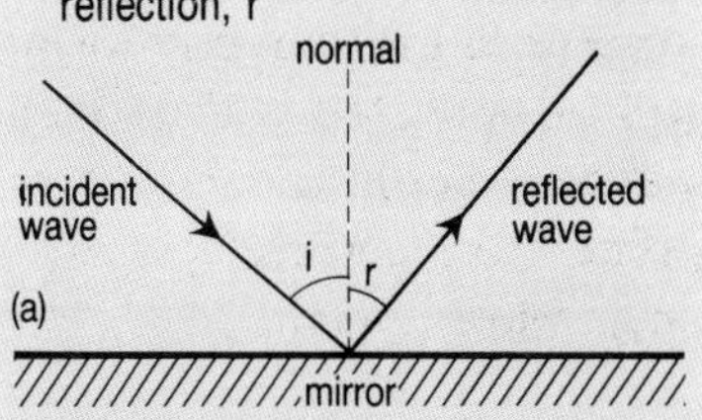

2 Wave speed, wave length and frequency remain unchanged by reflection.

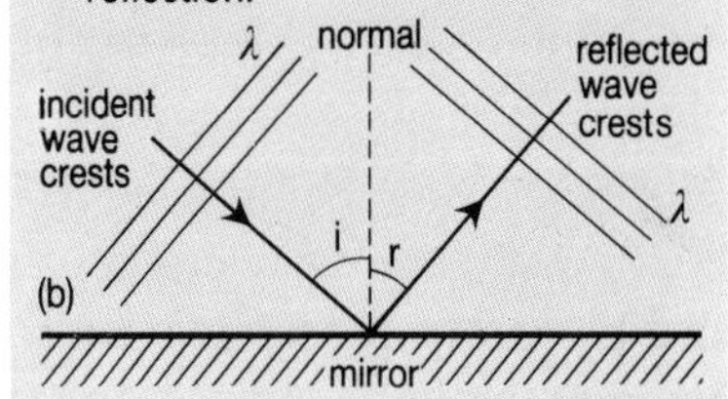

Optical fibres

Making light of communication

Light has been used for years for communication – beacons of fire, Morse code lamps and heliographs are a few examples. Each system uses flashes of light, and a suitable code, to pass on signals.

A number of recent developments has led to great advances in the use of light in communication. These include the production of:

1 very thin, very clear optical fibres of glass to transmit the light;

2 small and powerful light sources made from lasers;

3 very small microelectronic circuits to control communication systems.

An outline of a telephone communication system, using an optical fibre, is shown.

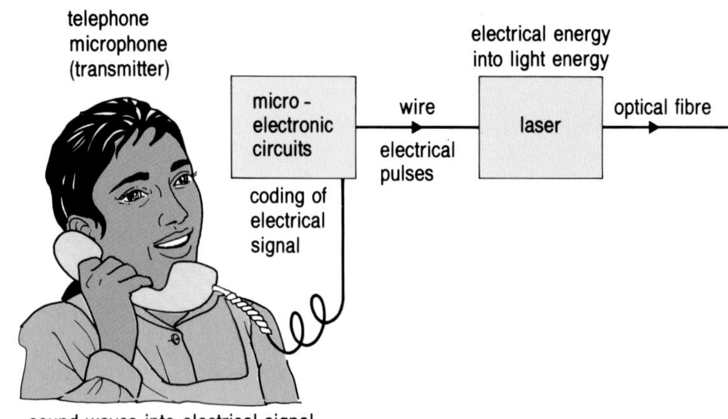

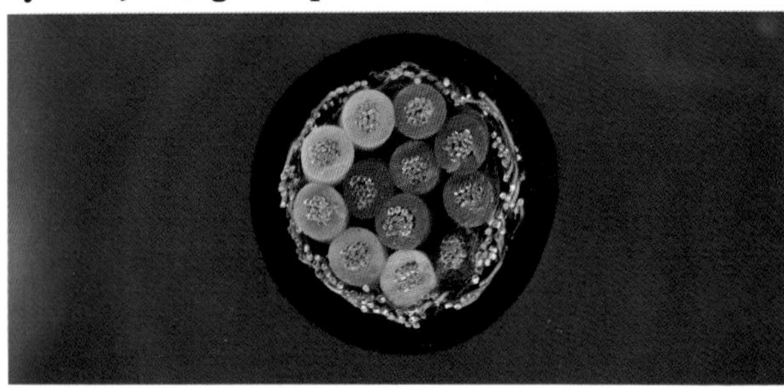

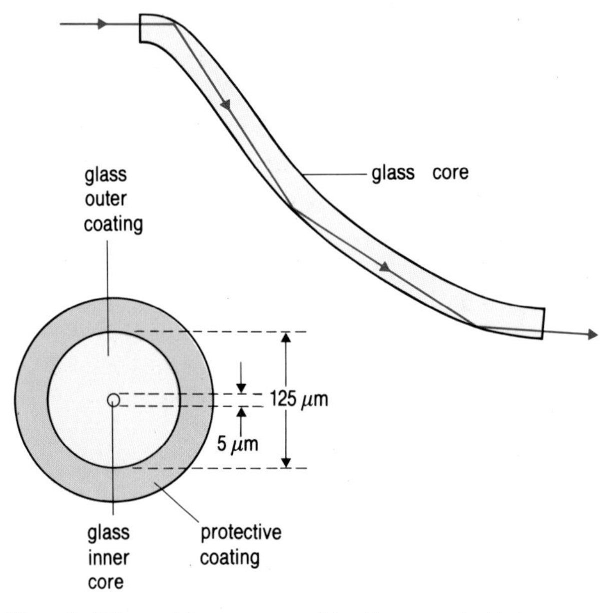

The optical fibre and the copper coaxial cable are each able to carry the same number of telephone calls.

Optical fibres

The clear hair-like threads of glass are called optical fibres. They are made from the purest glass and are about one tenth of a millimetre in diameter. One design of optical fibre is shown in the diagram.

A ray of light can be passed along one of these fibres, from one end to the other, even if the fibre is bent. A series of internal reflections of the light from the fibre walls make this possible. Flashes of light, and a suitable code, can be used to pass signals through these optical fibres.

Fibres are now being used to replace many heavy and expensive copper cables. The fibre and the cable shown can both carry about 10 000 telephone calls at one time.

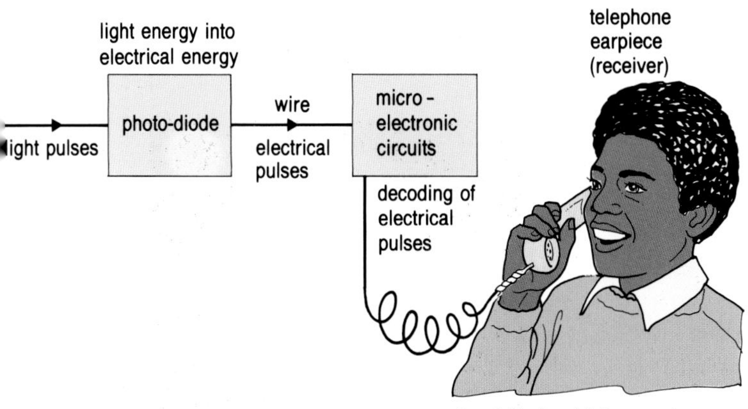

Refraction of light

Light travels in straight lines so long as it remains in the same material. When it passes from one material to another it is bent because the speed of light is different in the two materials. When light leaves glass and moves into air it bends away from the normal. This bending occurs for all angles of incidence, i, (except 0) up to a certain value. Beyond this critical value no light leaves the glass, instead, all light is reflected back into the glass. This effect is known as total internal reflection.

Designing an optical fibre system

The coding of signals in binary form uses only the digits 0 and 1. It is much simpler to use such a system of communication with light. 0 = 'light off' and 1 = 'light on'. Signalling then requires the laser source only to be switched off and on. The light flashes match the 'bit pattern' of the signal, once it has been converted into binary form. Using this method it is possible to transmit telephone conversations and television programmes along these glass threads.

Many optical fibre communication systems use a laser source which emits radiation with a wavelength of 1300 nm. This source is chosen because:

- a laser is small and bright, and can be switched off and on quickly;
- the glass fibre absorbs less radiation of this wavelength than other wavelengths.

Information in binary digits (bits) can be sent along these fibres at a rate of several million bits per second.

Modern optical fibres transmit signals with very little loss and can be used up to 100 km without use of repeater amplifiers. Copper cables on the other hand have a high signal loss and need amplifiers about every 4 km.

Optical fibres	Copper cables
Very high signal rate	High signal rate
Small cable size, 0.012 kg of glass each km	Large cable size, 30 kg of copper each km
Low cost of glass	High cost of copper
Repeaters up to 100 km apart	Repeaters about 4 km apart
No electrical interference	Signals affected by any electric cables
Signals 'safe' in the fibres	Signals may be 'stolen' from cables
Difficult to join	
More easily broken	

›› *Shine a ray of light into a 45° glass prism. Produce 2 examples of total internal reflection in the prism. Draw a diagram of each example. Suggest a use for each example.*

1 One optical fibre link in operation has the following details.
Cable length = 50 km; repeaters used = 1; wavelength = 1300 nm; information rate = 565 million bits per second (Mbits/s); speed of light in glass = 200 000 km/s.
a) Calculate the time taken by a signal to travel along the fibre.
b) Why is a 'repeater' used in the link?

2 State three reasons why optical fibres are used instead of copper cables. Explain each reason.

3 Describe what is meant by a digital signal and why such a signal is simple to use in optical fibres.

Communications satellites

Eyes and ears above us

This is the place where we all live – the Earth. The picture could have been taken by an astronaut but was in fact taken by a satellite. Apart from taking such photographs, satellites can be used for many other purposes – weather forecasting, examining rocks, examining crops, and communicating. Today, we send many kinds of communication by satellite from one side of the world to the other – telephone conversations, television pictures or drawings.

Telecommunication satellites

In 1945 before he became a famous science fiction writer Arthur C. Clarke wrote about a way of providing world-wide communications. He said that a satellite could be made to 'hover' above a point on the Earth's equator. To do this, the satellite would have to be put into the correct orbit, at the correct height and travelling at a speed which just matched the rate at which the Earth spins on its axis. You can use this graph to estimate the height of such an orbit.

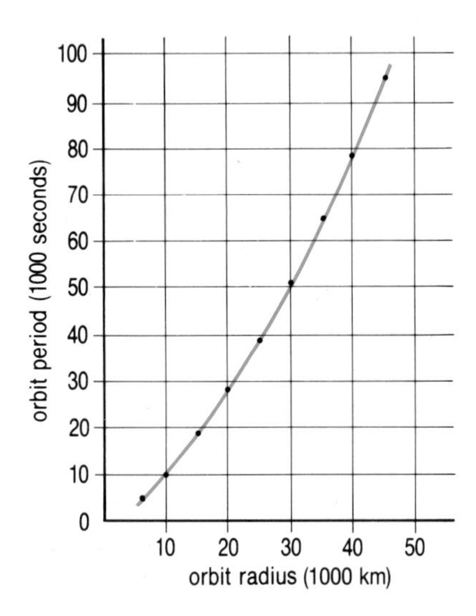

The period of a satellite orbit at different radii from the centre of the Earth.

Clarke's dream became a reality in 1965 when the world's first geostationary communications satellite was put in orbit above the Atlantic ocean. Within 4 years others were positioned – some over the Pacific and Indian oceans. World-wide communication via satellite then became possible. Today many modern satellites circle the Earth in geostationary orbits, each satellite supplying communication links with ground stations on different continents.

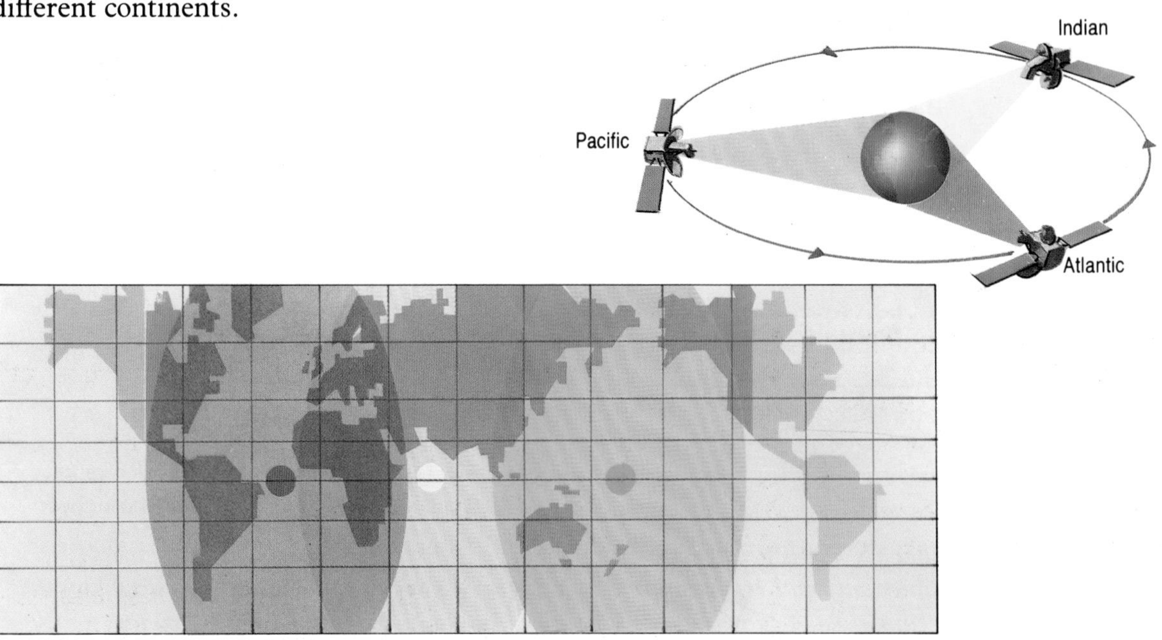

World-wide coverage by three geostationary satellites, equally spaced in orbit.

Dish aerials

The ground station dish aerial is a vital link in every satellite communication system. The aerial is carefully adjusted so that it points accurately at the satellite. This is important as a signal can be completely lost if the aerial is misdirected by as little as 0.16 degrees.

These aerials are vital for two reasons:

1 the transmitted wave is formed into a narrow beam by reflection from this aerial shape;

2 the received signal is amplified about 1 000 000 times by the aerial's ability to reflect incoming signals to a central detector.

A 32 m diameter dish aerial using 4 GHz and 6 GHz microwaves for communications.

Microwave carriers

A signal, such as a telephone call, may reach the main satellite ground station at Goonhilly in Cornwall. From there, it is transmitted on a microwave carrier to the geostationary satellite 'hovering' about 36 000 km above the Earth. The carrier wave frequency is either 6 GHz or 14 GHz.

The signal received at the satellite is much weaker than when it was transmitted – in fact, it is reduced to approximately $\frac{1}{10^{20}}$ of its original strength! Electrical energy generated on the satellite is used to amplify the signal before it is beamed back to Earth. A carrier wave frequency of 4 GHz or 11 GHz is used for the return signal to another ground station. In this way the satellite provides a link which allows communications to 'jump' natural barriers such as oceans and mountain ranges.

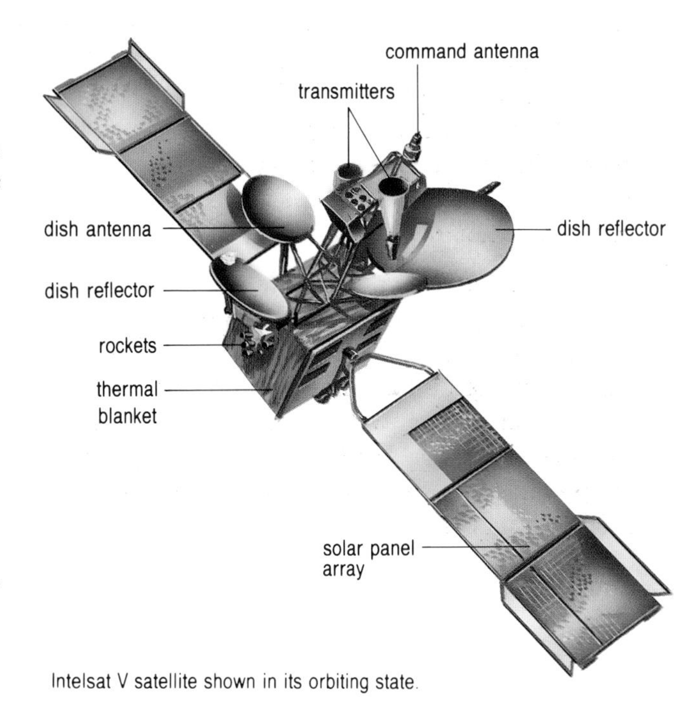

Intelsat V satellite shown in its orbiting state.

›› *Draw a scale diagram of the Earth and the orbit of geostationary satellites. Radius of the Earth = 6400 km and radius of orbit = 42 000 km approx. Show that three such satellites can give world-wide coverage.*

1 Many communications satellites now orbit the Earth. These satellites have many things in common. Explain the purpose of:
a) the solar panels;
b) the curved reflectors on their receiving aerials (antennae);
c) the curved reflectors on their transmitting aerials;
d) the small rocket motors;
e) the thermal blanket.

2 The microwave path from Britain to the USA via satellite is about 72 000 km. Calculate the time taken for these waves to travel this distance.

USING ELECTRICITY

Electricity in the home

Purgatory to paradise?

We depend on electricity for almost everything we do at work, home and leisure. Electricity provides the energy to help us vacuum clean floors, wash clothes, telephone friends, listen to the radio, watch television, keep food cool and so on. Homes and streets have electric lighting; machines big and small use electricity at work. Life without electricity seems almost unimaginable.

Yet the electrical age has been in the home for less than 100 years. Its success lies in the clean and instantaneous supply of energy at the flick of a switch! Instant cooking (microwave ovens), instant leisure (television etc) and many labour saving devices are here already.

What will the next 100 years bring? Robots to do all your household chores? If so what will it mean to the way you live?

Hard work without electricity.

Rating plates

Most appliances have fitted to them a plate which shows the technical information that you need to know and understand!

Several examples are given here.

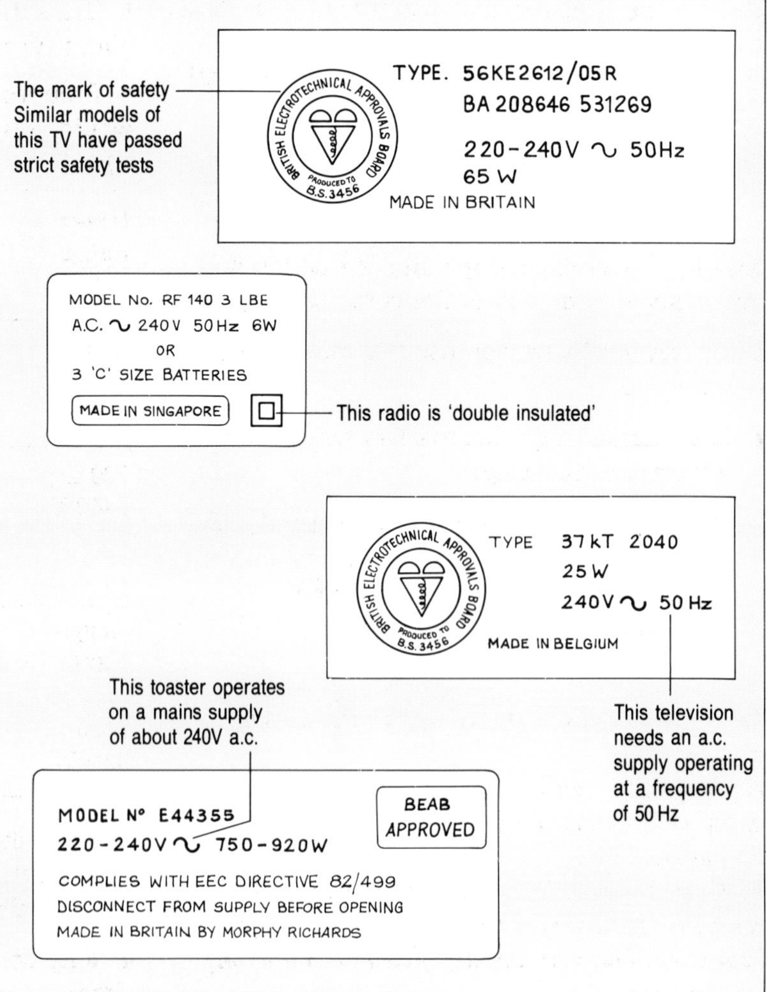

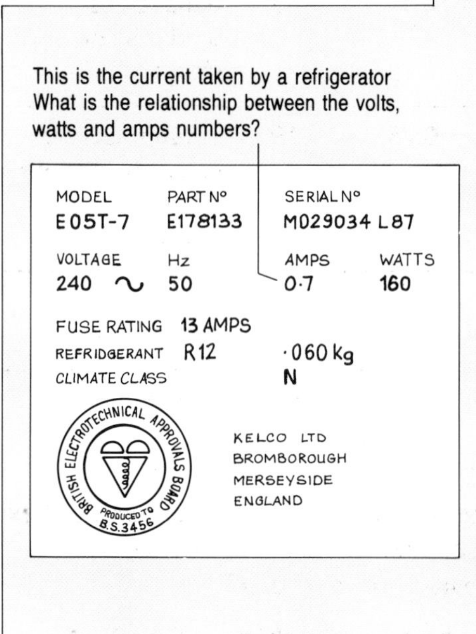

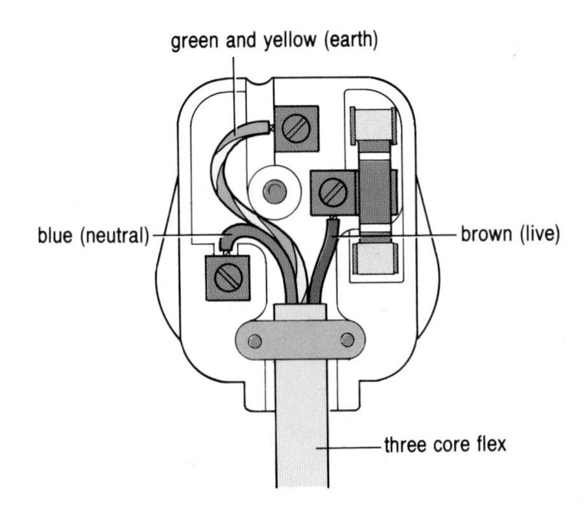

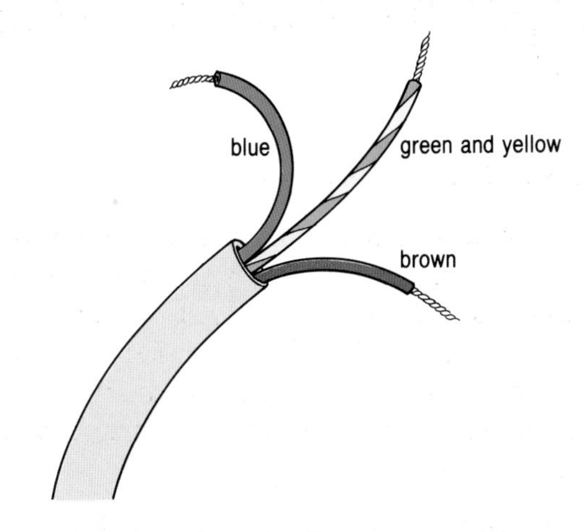

Three core flex connected to a plug. The cores are made from many strands of thin copper wire to permit movement and bending, hence the name of flexible cord or 'flex'. The insulation may be PVC, rubber or braided textile.

Flexes and fuses

Many sizes and types of flexible cord are used to connect household appliances to the mains. Appliances with bigger power ratings need thicker flexes and bigger fuses in the plugs. You can see why if you take hold of a flex which is supplying electrical energy to a 3 kW electric fire. The flex may feel warm. In fact all wires become slightly warm when there is a current in them, but the heating is greater with large currents.

Plug fuses provide some safety. If a fault develops which causes the current to become too large the fuse will melt and break the circuit. The current stops, the flex no longer overheats and the risk of fire is reduced.

For every appliance you use it is important therefore to:

- fit a thick enough flex to carry the current without overheating;
- choose the lowest value of fuse (usually 3A or 13A) which will carry the normal current but will melt if the current becomes accidently large.

›› *Examine various electrical appliances at school or at home.*
A *List those which are intended to provide mainly a) heat, b) movement, c) light, and d) sound.*
B *Copy the headings of the table shown and complete the table for the list of appliances you examined. Did you find any fitted with a wrong fuse or flex?*

1 Draw the symbols seen on the appliance plates. What does each symbol mean?

2 From the table shown, what plug fuse would you fit to appliances of power rating **a)** 400 W **b)** 2 kW **c)** 100 W?

Power rating (W)	Flex thickness (mm^2)	Plug fuse (A)	Typical appliance
up to 720W	0.5	3	table lamps food mixers clocks
720 to 1440W	0.75	13	vacuum cleaners colour televisions irons toasters
1440 to 240W	1.0	13	kettles fan heaters
2400 to 3240W	1.25	13	3kW fires

3 From the table shown, what thickness of flex would you connect to **a)** an iron, **b)** a vacuum cleaner, **c)** a toaster?

4 Why don't we use the thickest flex all of the time?

Safety first

The lamp does not light until there is a complete path for the current from one end of the battery to the other.

Materials which permit current are called conductors.
Materials which prevent current are called insulators.

Metals make good conductors.

Most non-metals make good insulators

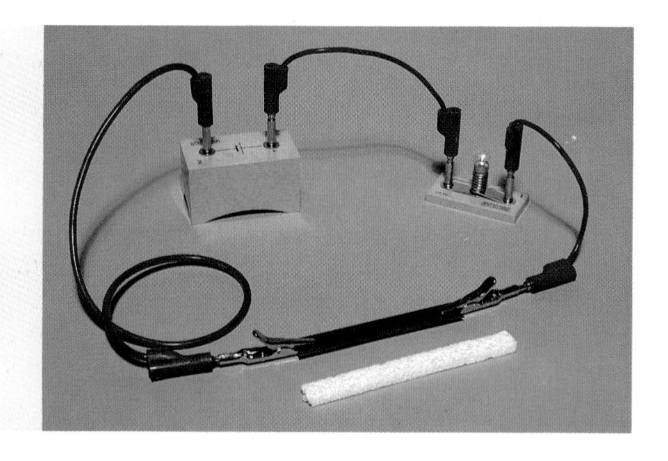

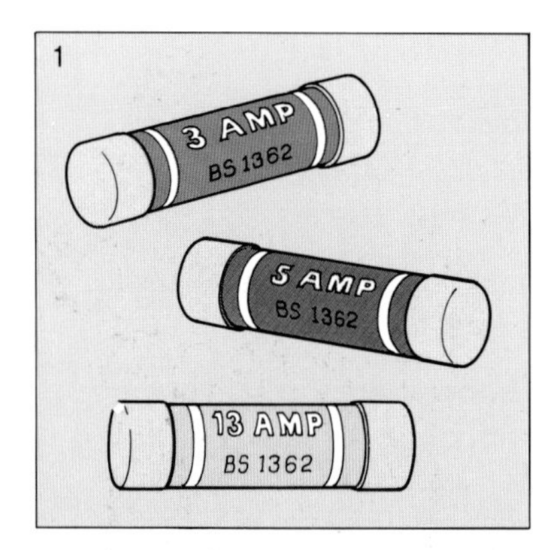

Rules for survival

We conduct. Even quite small currents can affect our hearts and electrocute us. In addition, the heating of flexes can cause fires. Unfortunately such accidents do happen. Nearly all of them can be avoided. Here are some hints for using electricity safely:

1 wire plugs properly and always fit the correct fuse;

2 check flexes and plugs regularly. Replace any which are worn or damaged;

3 never pull applicances by their flex or run flexes under carpets where they may be trodden on;

4 never use wet hands to switch appliances on or off;

5 never fill a kettle when it is plugged in;

6 use multiway adaptors as little as possible.

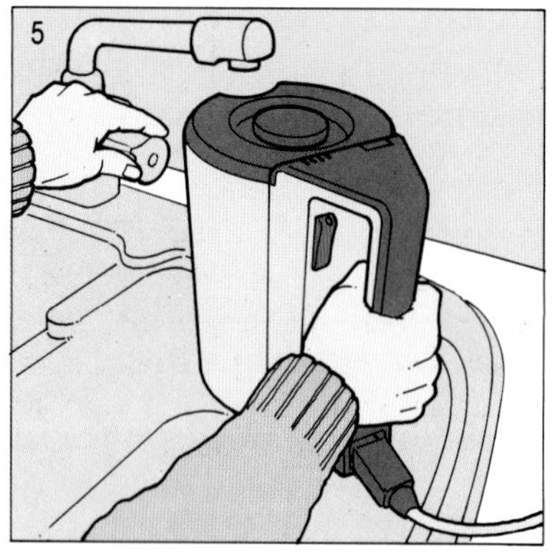

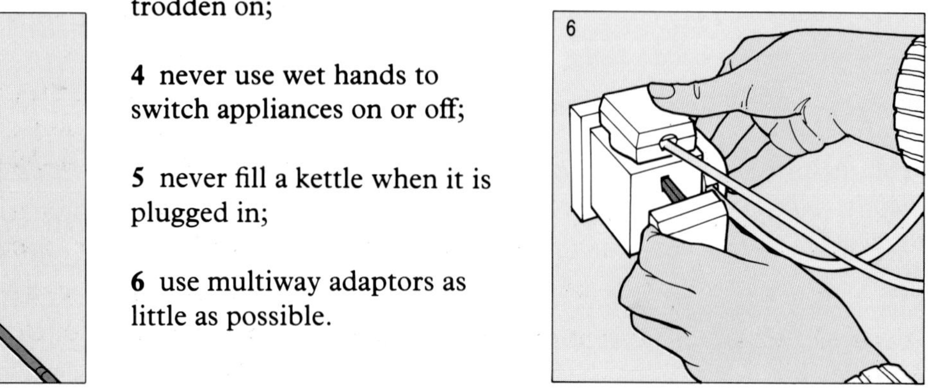

The need for earthing

Mains electricity can be used safely and without fear if the rules for survival are obeyed. Correct wiring may well involve an **earth wire**. This is included for our protection and safety. Electricians prefer to call it a *circuit protective conductor*.

The double insulated hair drier is safe to touch. Plastic is an insulator. There can be no conducting path through plastic and so an earth wire is not needed.

Double insulation

Many appliances such as lawnmowers and hair driers are now being made with no external metal parts. They are instead totally surrounded by plastic which is an insulator. The insulation is sufficient to protect you even if a fault occurs. Such appliances are **double insulated** and have the double insulation symbol marked on them. The flex fitted should be 2-core with live and neutral only. No earth wire is needed.

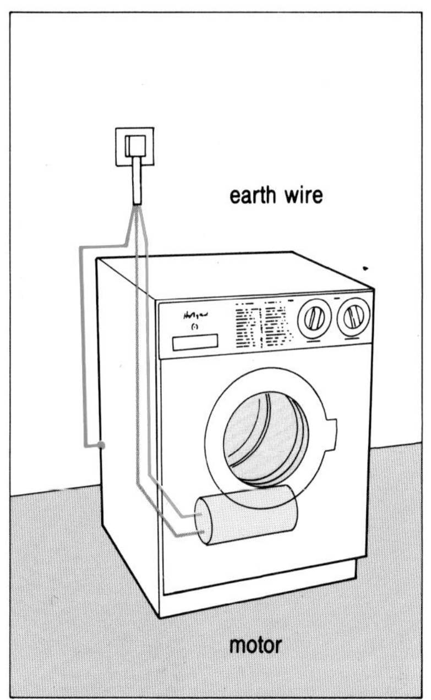

For example, when a washing machine is correctly wired and working properly, the complete electric circuit is made by the live and neutral wires. The earth wire carries no current and appears to be doing nothing.

With no earth wire. If the live wire was to vibrate loose and touch the appliance casing you, the user, could be electrocuted.

With earth wire and fault. A conducting path from the washing machine to the ground through the earth wire can be your life saver. A large current may cause a fuse to melt but at least you are safe.

1 Why should a plastic case be safe to touch at all times?

2 Look at each of the pictures of bad practice. Write down the danger in each case and explain why each is dangerous.

3 When might you not expect an earth wire to be fitted to an appliance?

4 Explain why the kitchen might be the most dangerous place for electrical accidents in the home.

›› *Use a 1.5 V cell and a sensitive microammeter to show that your body is a conductor. How are the results affected when your skin is moist?*

Never **do this experiment with any other source**

Working charges

Attractive electrons

Static electricity is familiar to us all. Examples include the crackle from a nylon sweater or the shock from a door handle after walking across a carpet. The rubbing between two materials, one of which must be an insulator, produces a steady transfer of charge which we often don't notice. This static electricity is not normally dangerous to us but it can be disastrous in industrial situations.

Dust and dirt may seem far removed from a study of electric charges. But they are not. Just as small pieces of paper can be attracted to a rubbed comb so dirt is attracted to electrically charged objects. Static on television screens attracts dirt, and so does static on gramophone records which must be removed if the record is to give good sound reproduction.

Dirt clings to carpets because of electrostatic attraction, even after you vacuum clean the floor. Yet the same attraction of dirt, dust and small particles can be put to good use.

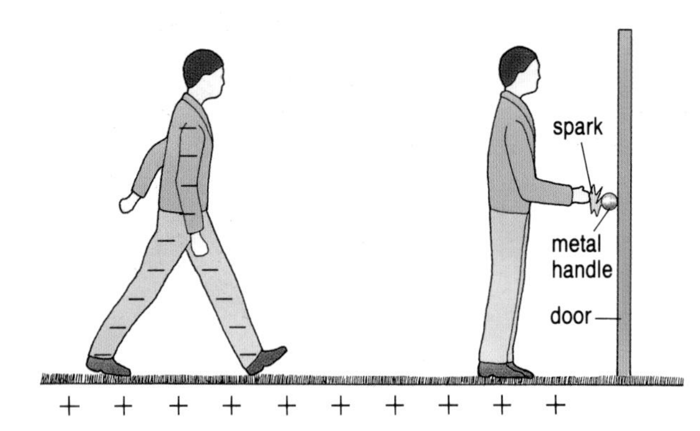

The build up of charge is hardly noticed until we approach a conductor. Discharge then occurs.

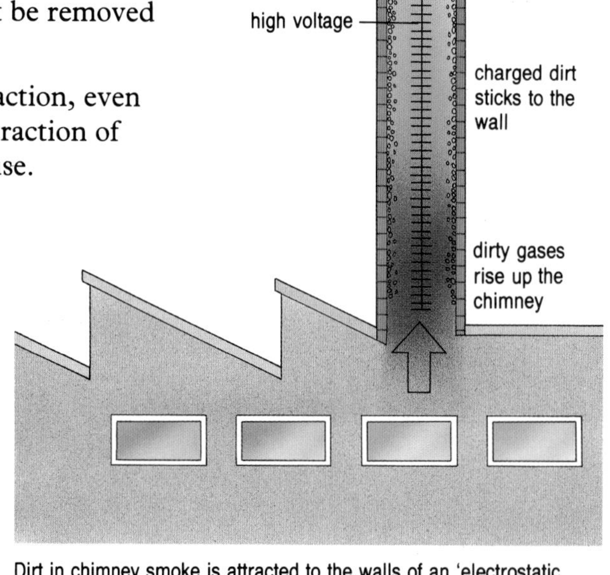

Dirt in chimney smoke is attracted to the walls of an 'electrostatic precipitator'. The removal of dirt from smoke produced by power stations and factories has done much to reduce atmospheric pollution and ensure cleaner air for us to breathe.

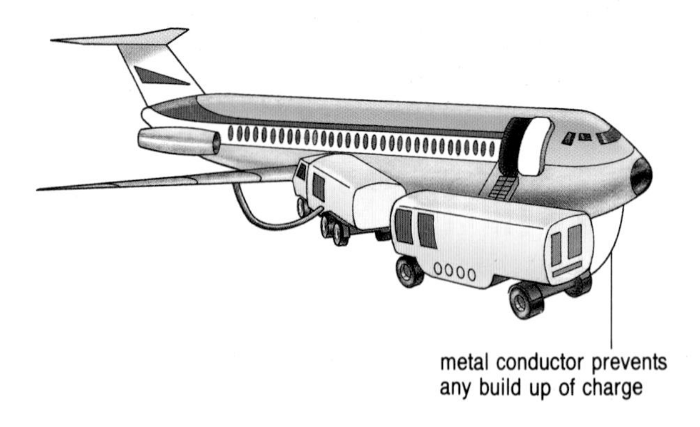

The transfer of fuel must be done very carefully. A build up of static electricity must be avoided as even a small spark could cause a major explosion.

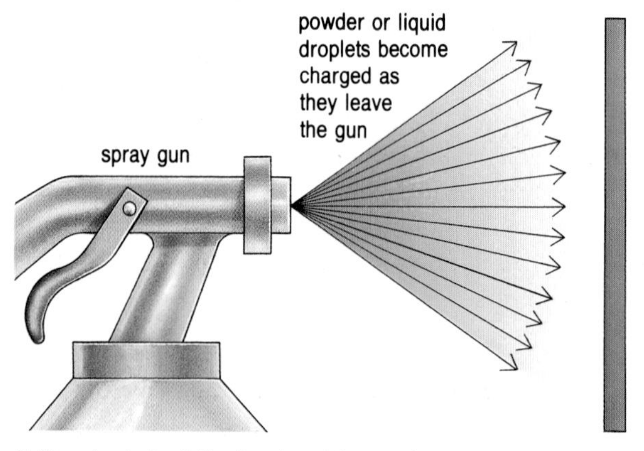

Deliberate electrostatic charging of the powder or liquid helps it to stick to the surface to be coated. This principle is used in paint spraying and crop spraying.

Current to the rescue

Charges don't remain static. They move. The movement of electric charges is called a current and is measured in amperes. An ampere is a rate of flow of 1 coulomb of charge every second.

Sparks and lightning are examples of moving charges in air. But charges also move along metal wires or through liquids.

One of the first practical uses of electricity was to send messages. Morse code was used long before lighting or heating were produced from electricity.

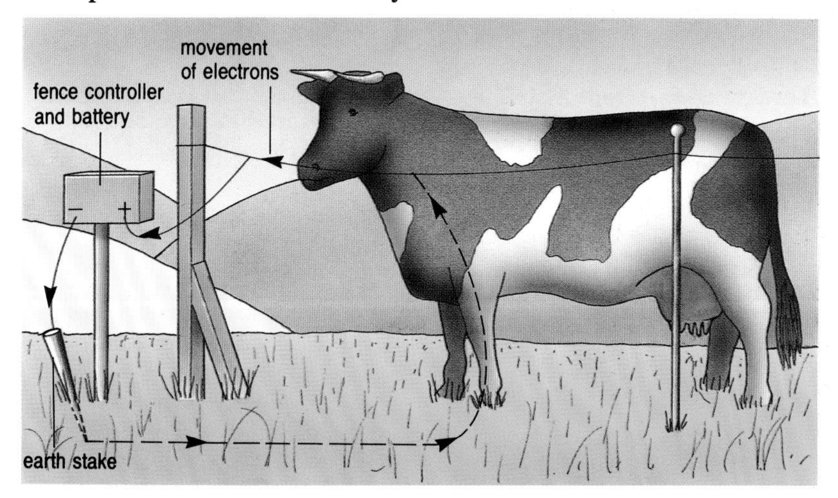

An electric fence in action. When an animal touches the wire the circuit is completed. Currents up to 20 mA are enough to give the animal a shock but not big enough to kill it

There are two types of charge called positive and negative. Atoms contain both types of charge.

Positive charges near the centre of the atom are called protons. The negative charges moving quickly around the outside are called electrons.

Friction can cause electrons to be knocked off the atoms of one material and onto the atoms of another.
A material with more electrons than protons is negatively charged.
A material with fewer electrons than protons is positively charged.

Two positive charges repel each other.
Two negative charges repel each other.
Positive and negative charges attract each other.

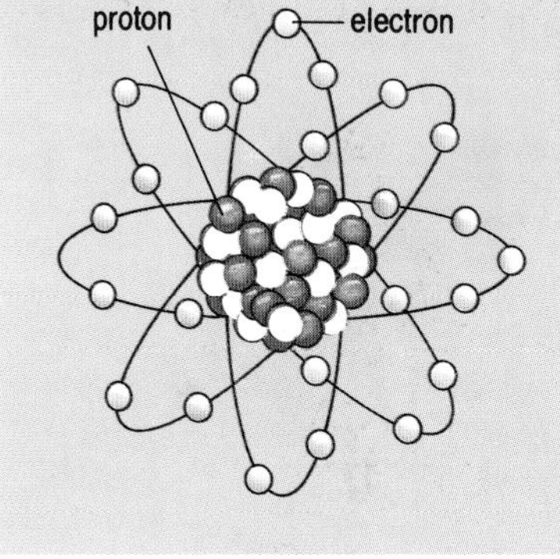

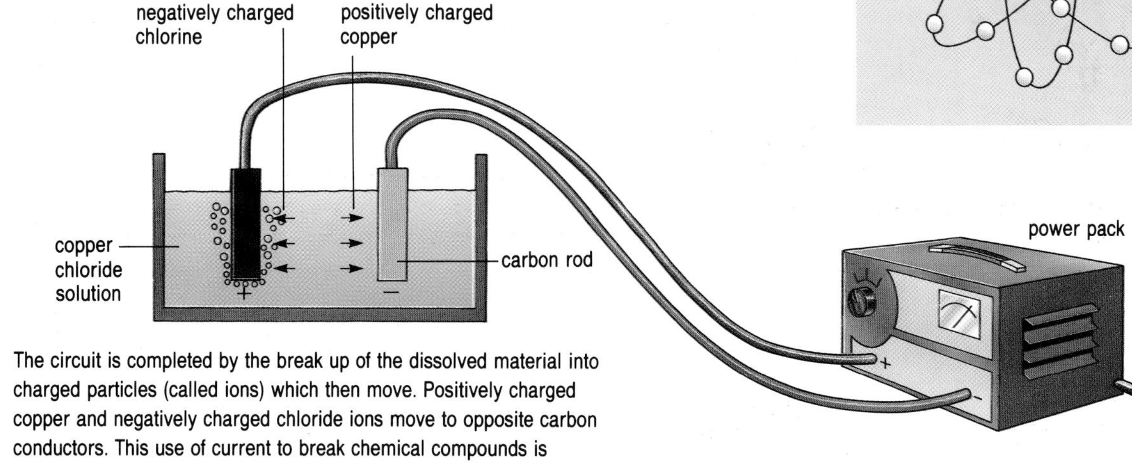

The circuit is completed by the break up of the dissolved material into charged particles (called ions) which then move. Positively charged copper and negatively charged chloride ions move to opposite carbon conductors. This use of current to break chemical compounds is called electrolysis

1 What is the current when 4 coulombs of charge pass a point in 0.2 seconds?

2 How much charge passes through a cow if it touches an electrified fence and receives a pulse of 20 mA for 0·1 seconds?

3 Explain why it is inadvisable to carry petrol in your car in plastic containers and why petrol pumps are always made from metal.

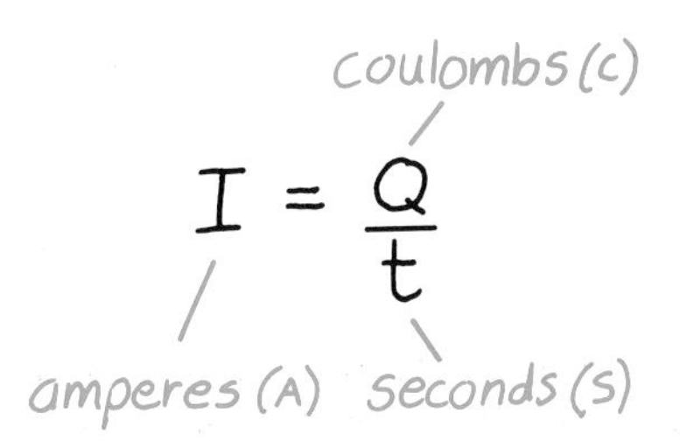

Voltage for the push

Naturally occurring

Electricity is found in all living animals, including ourselves. It makes our senses work. Nerve endings detect light, sound, smell and pain. They produce small electrical pulses which conduct through millions of other nerve cells to and from our brain. The nervous system is an information highway full of small voltage generators!

Millions of electrical pulses produce only a small amount of electrical energy in humans. Elsewhere, larger voltage generators can be found. Some fish for example can produce quite large voltages. The electric eel can produce up to 600 volts to kill fish.

The first man-made chemical cell was made by Alessandro Volta. He used copper and zinc plates separated by paper soaked in salt and water.

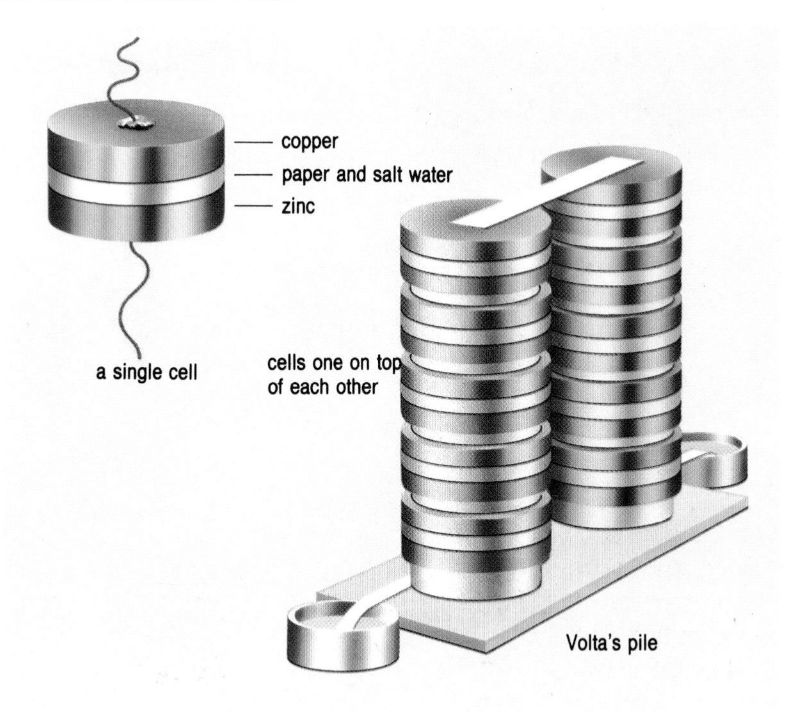

Volta's one volt cell gave a small electric current. When he connected a number of these cells together a bigger current was produced. This battery of cells became known as Volta's pile.

A group of cells is often needed to produce a large voltage. The extra 'push' can then provide bigger currents. Just as Volta connected cells one after the other, so do we today. A group of cells is called a battery, but the word 'battery' is often used loosely for all chemical supplies of electricity – even a single cell.

Charges don't suddenly start to move by themselves: a force is needed. The energy supply which produces that force might be a build up of static charges but is more likely to be a battery or dynamo.

When a battery is connected to the two ends of a conductor, an electric field is produced. In such a field, a force pushes and pulls electrically charged particles. If they are not held too tightly in the atoms of the conductor, the charges will move. In metals it has been found that only electrons move. They move towards the positive end of a battery and are jostled from atom to atom as they go. Energy is transferred to the atoms and the conductor becomes warm.

The source of the electric field is commonly called 'voltage'. It has the volt as its unit. A larger voltage produces a stronger field and makes each charge move faster. The current is greater and more energy is transferred to the atoms.

The 'voltage' is a measure of the amount of energy (in joules) transferred to the atoms by each coulomb of charge as electrons move through the conductor.

'volts = joules per coulomb'

is an expression which may become more familiar to you later.

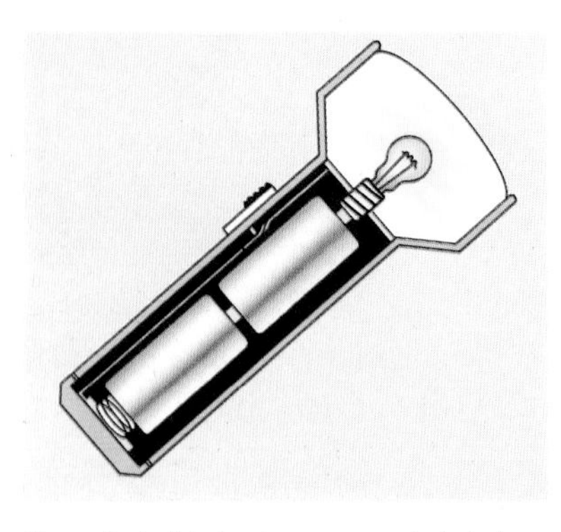

The cells in this torch are connected + to – to get more volts and current.

A pocketful of energy

Portable equipment gets energy from batteries. In any battery two conductors called the electrodes are separated by a chemical which reacts with the electrodes. The chemical reaction supplies energy by making one electrode short of electrons while the other has extra electrons. A few familiar cells or batteries are described.

A zinc-carbon cell. The positive electrode is in the centre of the cell and the negative electrode is the outer part of the cell beneath the plastic coating.

Zinc-carbon cells are used in the cheapest batteries. After a while, the zinc case dissolves and liquid leaks out.

Alkaline-manganese cells are used in the longer life batteries. They are more expensive but should last four times longer. The outer casing is steel and so no leakage occurs.

Nickel-cadmium cells are used in the very expensive rechargeable batteries. A separate mains unit is also needed to send electrons in the opposite direction to recharge these batteries.

Nickel-Cadmium (Rechargeable) batteries

Size	D	C	AA	PP9
Voltage (V)	1.25	1.25	1.25	8.4
Energy Capacity (Ah)	4	2	0·5	1·2
Cost	£5	£3	£1.50	£17

If cars could run off batteries the roads would be quieter and less polluted. Experimental vehicles have been built but the batteries they need are still too heavy to be satisfactory for most purposes. Batteries are suitable for milk floats and fork lift trucks. They travel slowly and the batteries can be recharged overnight.

1 Fork lift trucks, as used in warehouses, are almost always operated from batteries. Why are battery driven trucks preferred to petrol driven trucks?

2 Devise, carry out and describe an experiment to test which of two batteries is the longer lasting. Your answer should include:

a) details of your apparatus;
b) precautions needed to keep the test fair;
c) details of the observations you took; and
d) details of your conclusion.

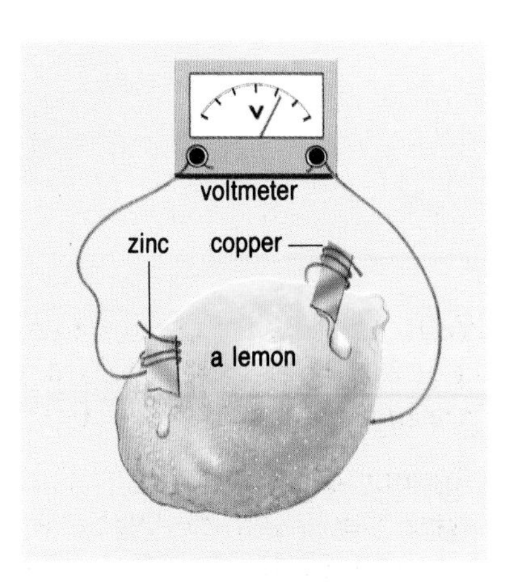

›› *Make your own cell as shown. What voltage does this cell generate?*

Battery or mains voltage?

Spot the difference

The electricity needed to operate household appliances comes either from batteries or from the mains supply. These different electrical sources can be compared by using a cathode ray oscilloscope (c.r.o.) to examine the voltages supplied. A battery can be connected directly to the oscilloscope but the mains voltage is too big to connect directly. However, the yellow terminals of a low voltage laboratory power pack can be connected to give a pattern similar to the mains voltage but at a lower, safer value.

The mains supply is a.c. to make it easy to send electricity to our homes. This is explained in a later unit. (page 174).

There is no particular reason for using a frequency of 50 Hz in Britain. Turbines at the power stations are simply made to rotate 50 times every second. In America the mains frequency is 60 Hz.

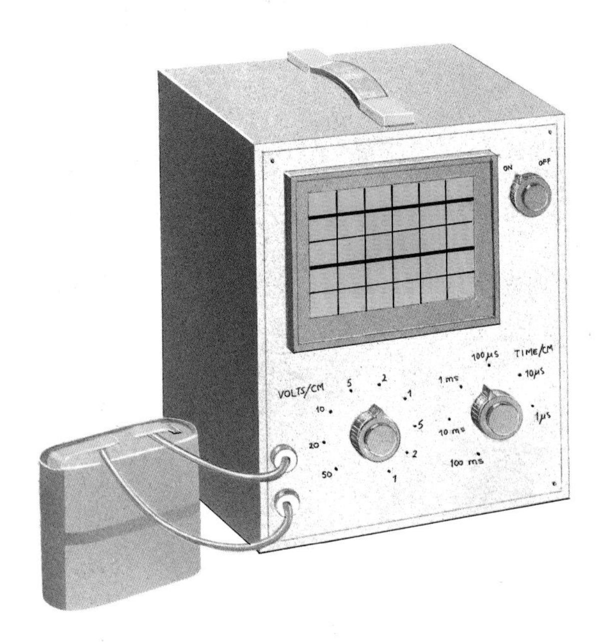

A battery gives a constant voltage in one direction only. It provides a direct current (d.c.). In this example:

voltage = 2 volts/cm × 2 cm
= 4 volts.

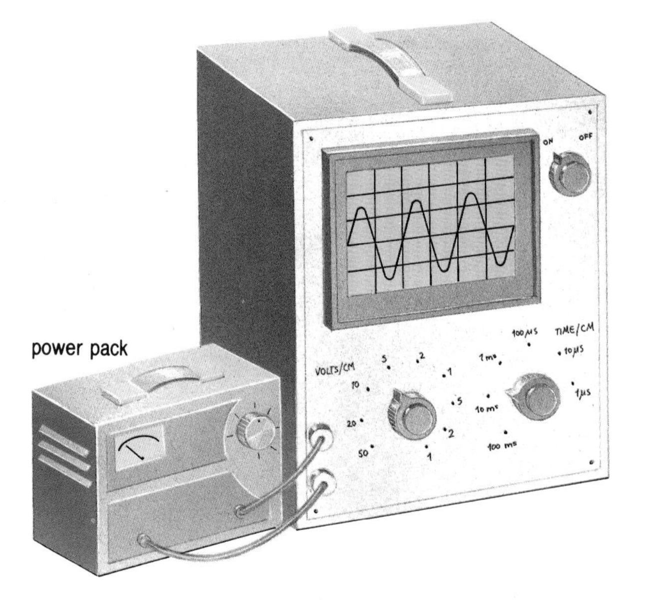

Mains voltage continually changes between positive and negative. It provides alternating current (a.c.). In this example:

peak voltage = 5 volts/cm × 1.5 cm
= 7.5 volts

time between voltage peaks = 10 ms/cm × 2 cm
= 20 ms

20 ms × 50 = 1 second

Thus, mains frequency = 50Hz

An alternating voltage is continually changing. A lamp connected first to a 6 V battery and then to the 6 V a.c. terminals of a power pack will have about the same brightness in each case. But from the oscilloscope trace, the power pack supply appears to vary from +9 to −9 volts! By convention it has been agreed that the a.c. value quoted will produce the **same heating effect** as a d.c. supply of that voltage. A mains voltage of 240 V a.c., for example, produces the same heating effect as 240 V d.c. For this to happen the alternating voltage actually changes from +340 V to −340 V.

From a.c. to d.c.

Many components work equally well on d.c. and a.c. but two do not: diodes and capacitors.

1 A **diode** acts like a 'one-way street' for current. For historical reasons, the arrow of the diode symbol points in the positive charge direction, so electrons can only flow against the arrow.

2 A **capacitor** acts like a reservoir in a water pumping system. Bursts of water come into the reservoir, but a smooth flow can be taken from it. A capacitor stores electrical charge and is said to **smooth** the output from a power supply.

So, with a diode you can make two-way a.c. into one-way d.c. And with a capacitor, the rises and falls of a.c. can be smoothed out. With both, you can produce a fairly steady d.c. flow.

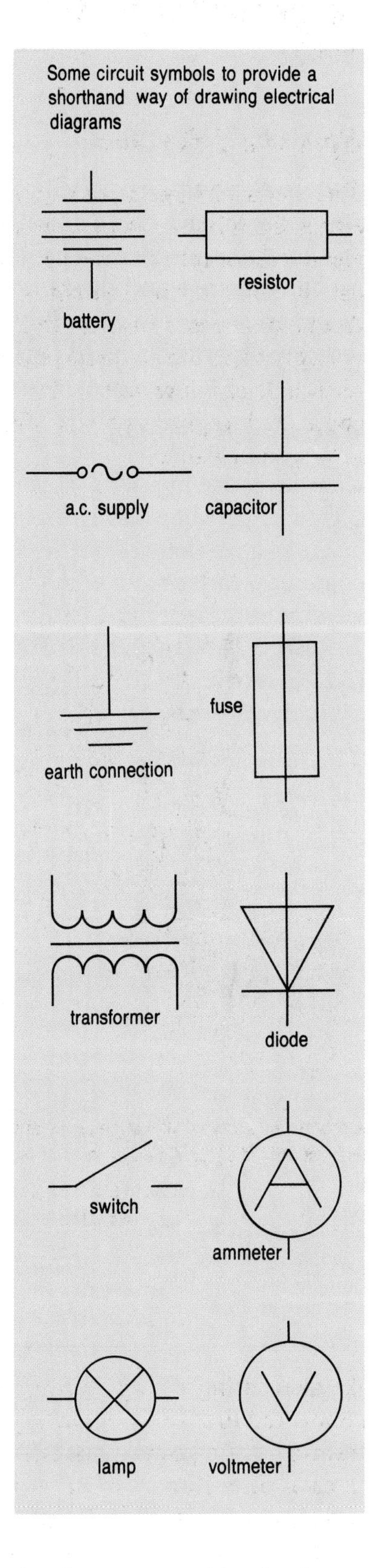

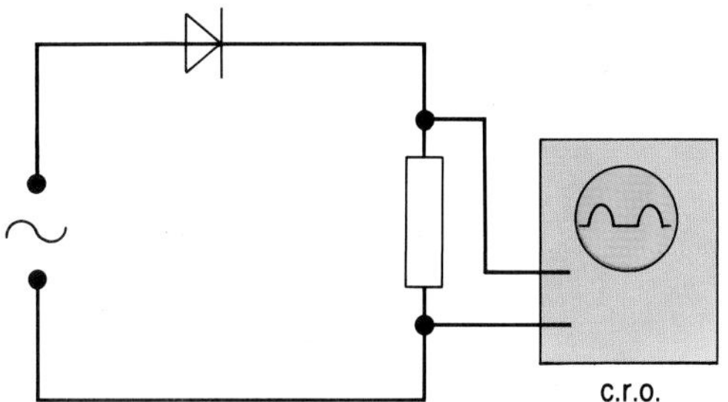

a.c. tries to go back and forth. With a diode in the circuit, current can go one way only. The diode has now changed a.c. into d.c. This process is called rectification.

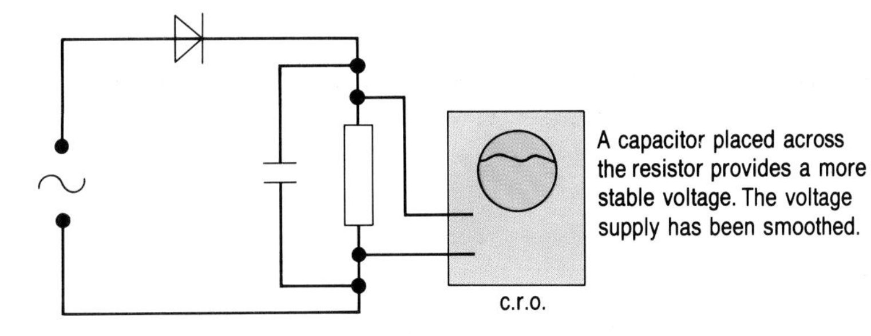

A capacitor placed across the resistor provides a more stable voltage. The voltage supply has been smoothed.

›› *Look inside a battery charger or train transformer and draw the electrical circuit seen. Be certain to disconnect from the mains supply before opening up the equipment.*

1 What is meant by the term rectification?

2 Give one reason for including a capacitor in an electrical circuit?

Resistance put to good use

Resistors recognised

Electronic equipment often contains components which have three or four coloured stripes or bands around them. These components are called **resistors** and they are said to have **resistance.** The value of this resistance is indicated by the coloured bands.

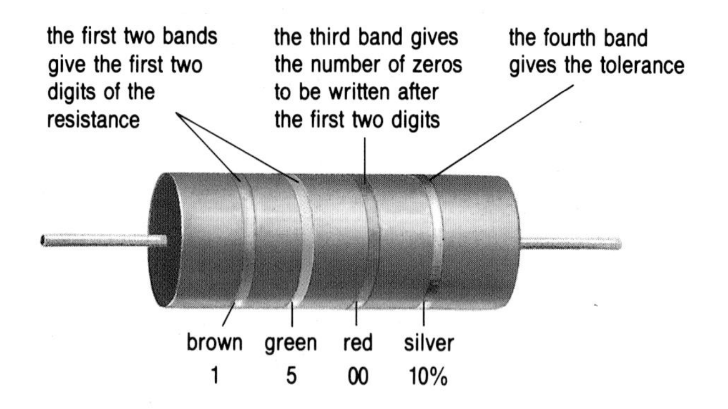

Resistance = 1500 Ohms
10% of 1500 ohms = 150 ohms
Thus, allowing for the tolerance of this resistor, its resistance is within the range of 1350 to 1650 ohms.

Resistance defined

Charges move in a wire if there is a voltage across it. Just how easily that movement can occur in a particular wire – its ability to conduct – varies from wire to wire. Normally we do not measure the ability to conduct. Instead we measure the **difficulty** charges have in moving! The term used to describe this difficulty is **resistance**.

The resistance R for a conductor is defined as follows:

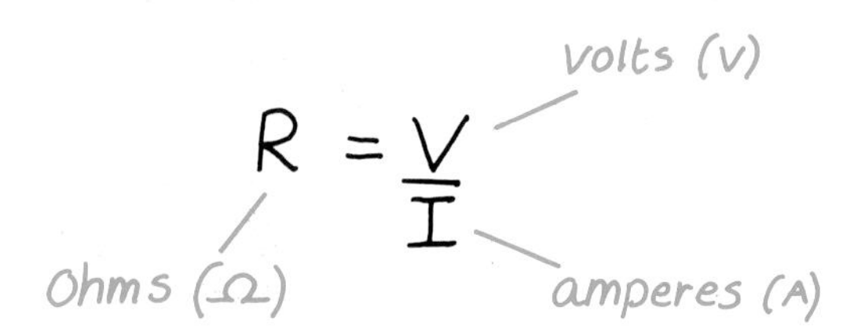

Bands 1, 2 and 3		Band 4 – tolerance	
Black	0		
Brown	1		
Red	2	No colour	20%
Orange	3		
Yellow	4	Silver	10%
Green	5		
Blue	6	Gold	5%
Violet	7		
Grey	8		
White	9		

Resistor colour codes

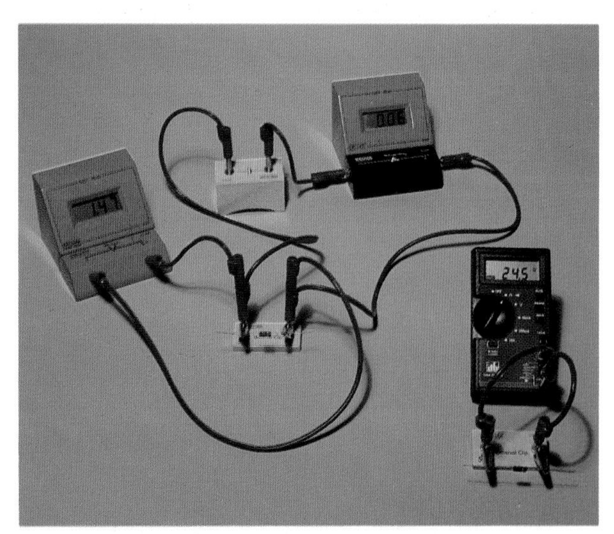

Copy the caption and complete it using the meter readings in the photograph.

R = $\frac{\ldots}{\ldots}$ R (multimeter)

= – – – – ohms = – – – – ohms

Meters, lamps and all the other components we use have some resistance. The value of this resistance can be found experimentally in two ways:

1 The current and voltage are measured and the resistance calculated from the definition of resistance: $R = \frac{V}{I}$; or

2 A multimeter is used as an ohmmeter to give a direct resistance measurement.

Resistance used

1 Resistors are used to reduce the size of currents. The relationship above shows that, for a particular voltage, the greater the resistance the smaller the current. In the example illustrated here a resistor is connected in series with a lamp so that the current can be reduced to the required value.

2 Resistors are used to reduce the size of voltages. By joining two resistors in series we can select **part** of the supply voltage for use in another circuit, e.g. an electronic device. The use of two resistors to divide the input voltage into two parts is called a **voltage divider**.

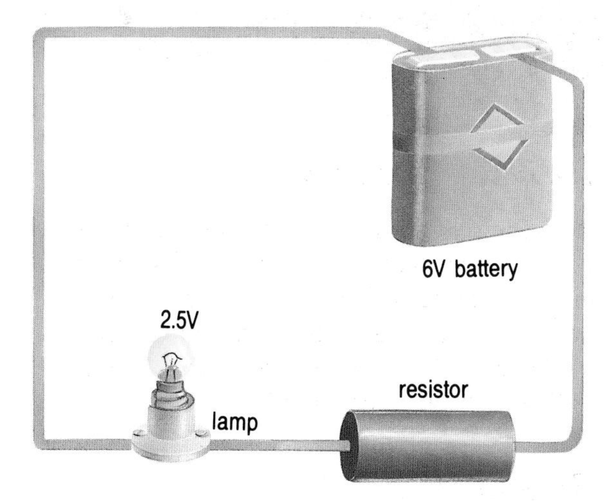

If this lamp were to be connected directly to the battery it would be very bright and quickly burn out. A resistor added in series reduces the current and makes the lamp less bright.

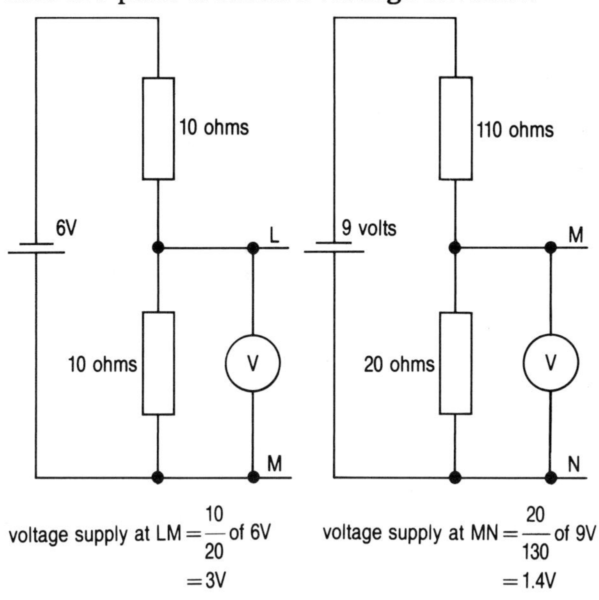

voltage supply at LM $= \frac{10}{20}$ of 6V $= 3V$

voltage supply at MN $= \frac{20}{130}$ of 9V $= 1.4V$

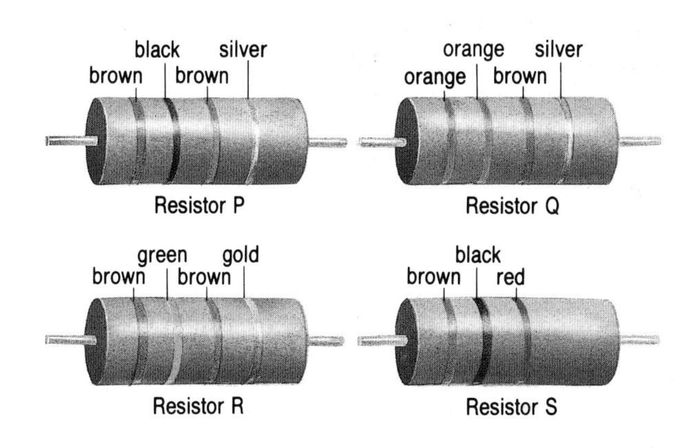

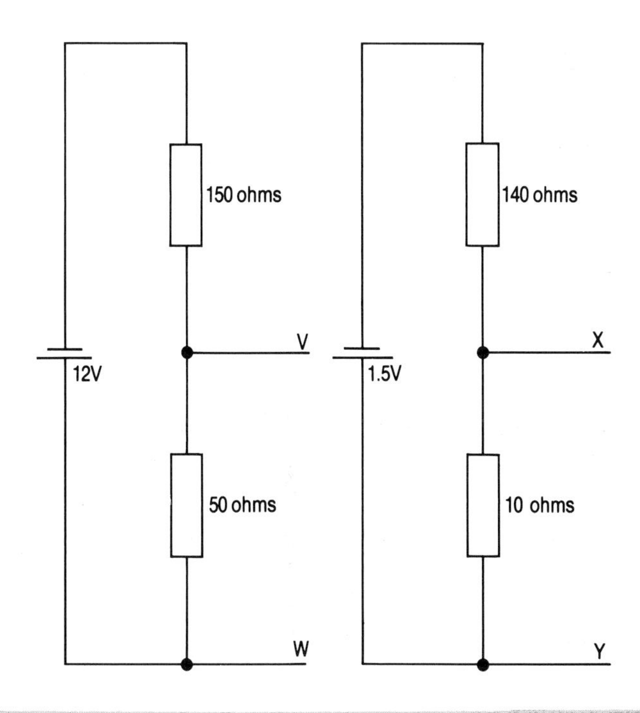

1 For each of the resistors P, Q, R and S shown, determine:
a) the value of its resistance;
b) the range of possible resistance values for that tolerance; and
c) the current which should pass (assuming the quoted resistance is correct) if that resistor were connected to a 10 V supply.

2 a) Calculate the voltage across the wires VW;
b) calculate the voltage across the wires XY;
c) using any of the resistors on this page, draw a circuit diagram to provide 0.5 V from a 6 V battery.

3 A voltage divider circuit consists of two resistors with resistances X and Y. If the input is I and the output has to be O, find an expression for the output O in terms of X, Y and I.

Variable resistors

Room to move

The physical size of a resistor affects the value of its resistance. Just as you find it easiest to travel on wide roads so charges move more easily through a thick wire than through a thin wire. It is easier, too, if the charges don't have a long way to go, and if the material is a good conductor which easily releases electrons from the atoms. The resistance of wire depends upon:

1 the length of wire;

2 the thickness of wire; and

3 the choice of conducting material.

Resistance is greatest for a long and thin wire. When a component allows the length of wire in a circuit to be altered, the component is called a variable resistor.

A short, wide passage gives less resistance to movement than a long, narrow passage.

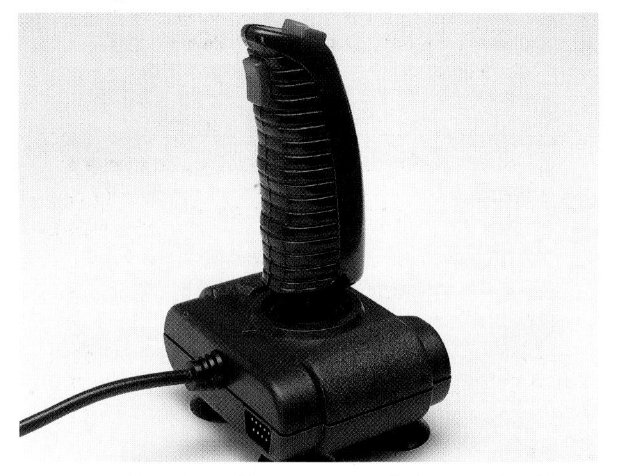

The side-to-side and to-and-fro movements of a computer joystick alter the effective lengths of resistance wires in the two directions.

The length of wire is altered by twisting the knob. This type of variable resistor is used as a volume control in radios and also to make a potential divider (potentiometer).

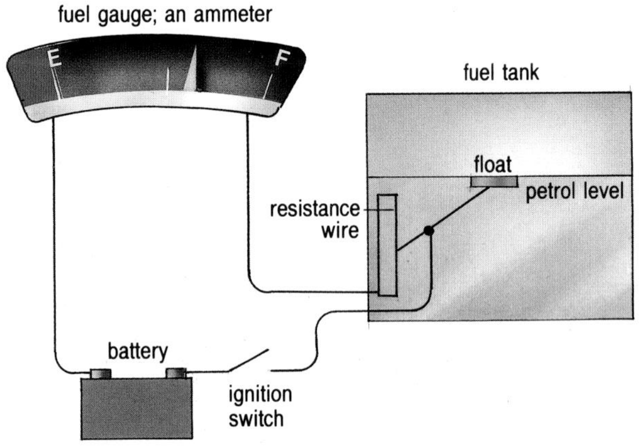

A petrol gauge. A needle fixed to a variable resistor 'floats' on top of the petrol. Variations in petrol levels alter the circuit resistance by changing the length of wire through which the charges move. Resistance changes cause different currents and therefore different fuel gauge (ammeter) readings.

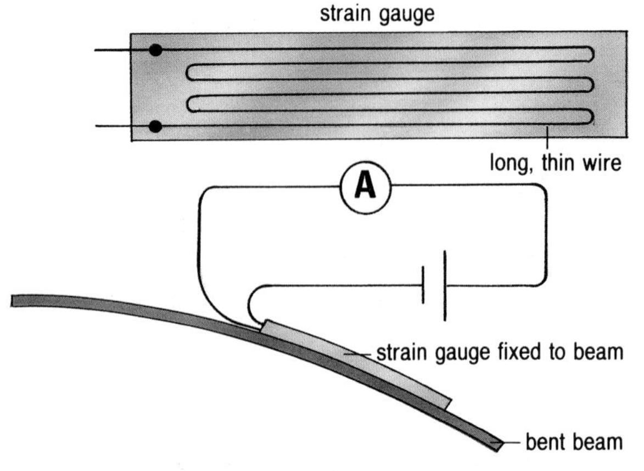

Forces and pressures can be determined from changes in resistance of a wire. In this 'strain gauge' the wire becomes longer and thinner as the beam bends. More resistance and therefore less current means that the beam has bent further.

Temperature and resistance

It was the German scientist Georg Ohm who discovered that for metallic conductors, kept at constant temperature, the resistance stayed the same no matter what the current was in the conductor.

Constant resistance and constant current are often desirable, but as you can see from the selection of variable resistors there are many occasions when it is helpful for resistance to change.

Making the temperature change is another way of altering the resistance of a component. Although the resistance of metals changes only a little with temperature, the resistance of some non-metal (semi-conducting) components – such as thermistors – changes dramatically with temperature.

A thermistor as a thermostat. The change in current caused by resistance changes may be enough to switch other electrical components on or off.

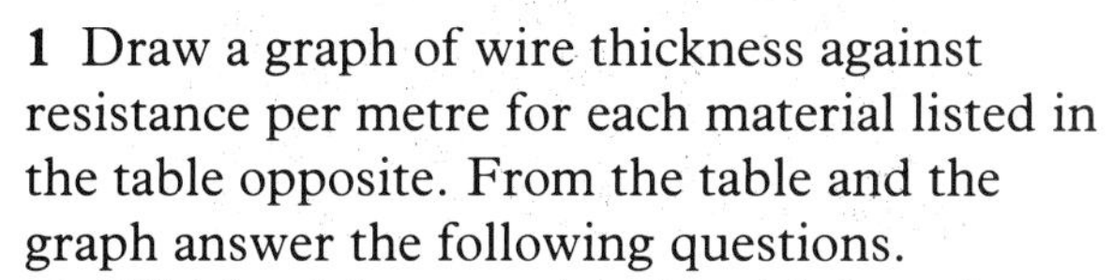

1 Draw a graph of wire thickness against resistance per metre for each material listed in the table opposite. From the table and the graph answer the following questions.
a) Which of the materials listed is best for conducting electricity?
b) What thickness of i) constantan and ii) nickel-chromium would have a resistance of 100 ohms per metre?
c) What thickness of i) constantan and ii) copper would have a resistance of 20 ohms per metre?
d) What thickness of i) nickel-chromium and ii) constantan would carry a current of 2 amperes with a voltage of 240 volts across 1 metre of wire?

Wire thickness (cm)	**Resistance (ohms per metre)**		
	copper	constantan	nickel-chromium
0.038	0.16	4	9
0.027	0.3	8	17
0.019	0.6	16	35
0.010	2.0	60	127
0.006	6.0	165	350
0.003	34.0	950	2030

2 Look at the graphs of resistance against temperature on the right.
a) Which material does not change resistance when its temperature changes?
b) What is the thermistor resistance at a temperature of 50°C?

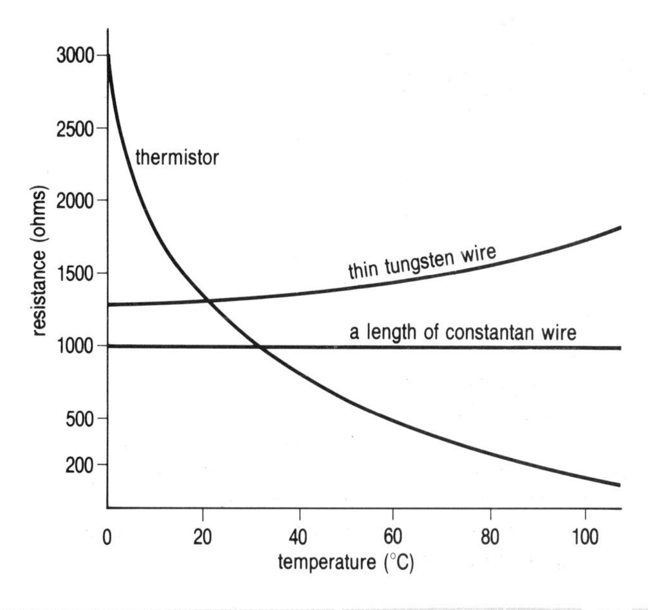

›› *Find out the meaning of the word 'calibrate'. For temperatures between 0°C and 100°C, calibrate an ammeter which is connected to a thermistor and a battery.*

Energy and power

Reading the meter

Electricity provides us with energy. Fires, vacuum cleaners and all the other labour saving devices examined at the beginning of this chapter use electrical energy. We, the consumer, have to pay for that energy. Just as sugar is sold by the kilogram and petrol by the litre, electricity is sold by the 'unit'. The 'unit' is an amount of electrical energy.

The electricity meter in your house records the number of 'units' used. The price paid for each 'unit' depends upon the tariff you have chosen to be the most suitable for your needs.

Power

The 'unit' is also called the **kilowatt hour** (kWh). It is the amount of energy used by a 1 kW appliance working for 1 hour or indeed by a 2 kW appliance working for half an hour.

The power of an appliance is measured in **kilowatts**, and is related to energy by the formula:

$$\text{energy} = \text{power} \times \text{time}$$

energy	power	time
kWh	kW	h

So, power is the rate at which electrical energy is transformed into other energies. An appliance of large power uses up energy more quickly, and so costs more to run than does an appliance of smaller power. Heaters and cookers, for example, with their large power values cost more to run than lamps, radios and televisions.

The domestic tariff

This is the standard method of payment based on the normal domestic meter.

Off peak tariff

At certain times (e.g. through the night) the demand for electricity is much less, but the power stations are still working. To encourage people to use more electricity at 'off-peak' periods a second meter can be fitted to your house. This meter measures 'cut-price' electricity for such things as night storage heaters.

White meter tariff

With this meter all the energy you use is much cheaper at certain times of the day and night. At peak times however, all the electricity you use will be slightly more expensive than the normal domestic tariff.

The joule and watt

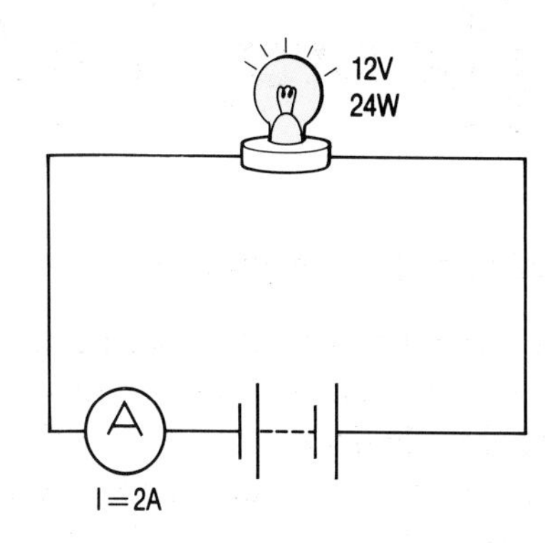

For small amounts of energy it is easier to use the watt as the unit of power and the joule as the unit of energy. These are part of the international system of units (S.I.).

The same formula holds with these alternative units:

energy = power x time

joules (J) — watts (W) — seconds (s)

A 100 watt lamp operating for 5 seconds will use 500 joules of energy.

Current and power

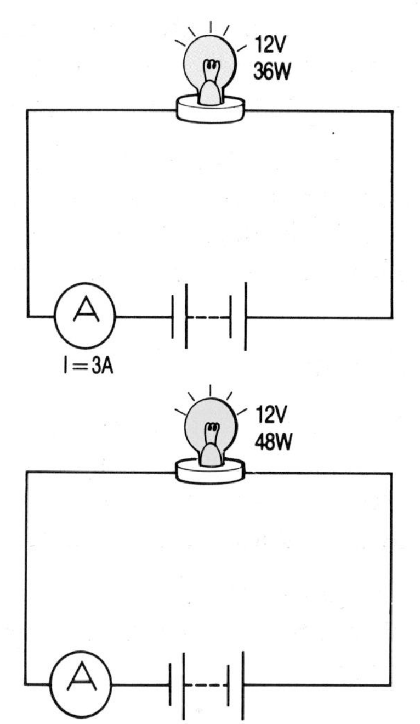

In each example the current depends upon the voltage and the power rating.
power = voltage × current

Appliances of high power use electrical energy very quickly. To do so, they need large currents. They should therefore be fitted with thicker flex and larger fuses. The voltage and power ratings marked on domestic appliances enable you to calculate the current.

power = voltage x current

watts (W) — volts (V) — amperes (A)

Appliance	Power rating (watts)	Voltage (volts)	Current (I) (amperes)	Resistance (R) (ohms)	I^2R
car headlamp	36	12			
toaster	960	240			
kettle	2400	240			
electric drill	240	240			

1 How many 'units' (kWh) are used when a:

a) 1 kW iron is switched on for 2 hours;

b) 3 kW fire is switched on for 45 minutes;

c) 100 W lamp is on for 8 hours?

2 How many joules are used when a 30 W heater is operated for **a)** 20 s **b)** 100 s?

3 What is the power of a lamp which uses 2000 J of energy in 50 seconds?

4 Copy and complete the table above.

5 What is the connection between the power rating and I^2R? Use the equations $P = VI$ and $V = IR$ to explain that connection.

6 Convert power in kW into watts and time in hours into seconds to find the number of joules in 1 kWh?

›› *When some high power appliances (e.g. heaters or cookers) are being used at home read your electricity meter twice with a two hour gap between readings. How many units have you used in that time? Find the tariff being used and then calculate the cost of the electricity being supplied during that two hour period.*

Light up your life

The dawning of the day

The electric lamp is a relatively modern invention. For thousands of years, people burned various fuels to make light. Flaming torches, wax candles, oil and paraffin lamps and finally gas lamps all had their part to play.

In 1810, Humphry Davy demonstrated the first lamp to be operated by electricity. However, his arc lights were never popular. They were too bright, they were noisy and their carbon rods had to be changed regularly. It was 1879 before Joseph Swann in England and then Thomas Edison in America produced the first electric filament lamps. Modern versions of these filament lamps can now be found in virtually every home in this country.

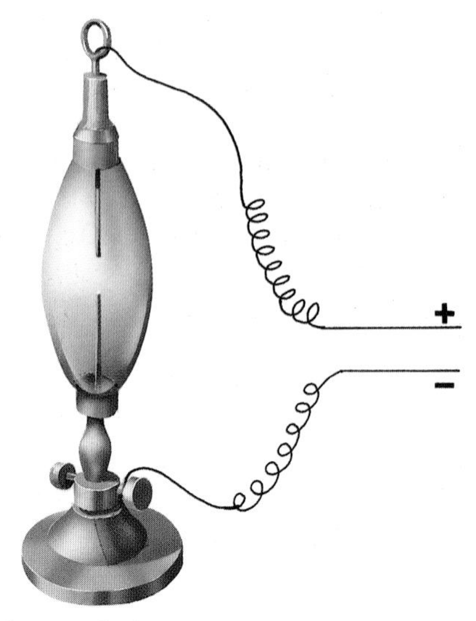

The Davy arc lamp. Davy used a large battery to provide a continuous spark or 'arc' between carbon rods.

Tungsten filament lamps

A filament is a very thin coil of wire of high resistance. It is usually made from tungsten, is thinner than a human hair and when unwound from its coil, is over a metre long. It is so thin that the movement of charges through the wire heats it to temperatures above 2500°C.

Tungsten is used because its melting point is amongst the highest of all metals. Unfortunately at such high temperatures tungsten oxidises and evaporates into the air. To reduce these problems, the filament is surrounded, in a glass bulb, with a mixture of the unreactive gases argon and nitrogen. Even so, evaporation slowly occurs. The filament gets even thinner and eventually breaks. Tungsten filament bulbs waste an awful lot of energy as heat rather than light. In fact only about 10% of the electrical energy supplied is converted to light. But at least filament lamps are fairly cheap!

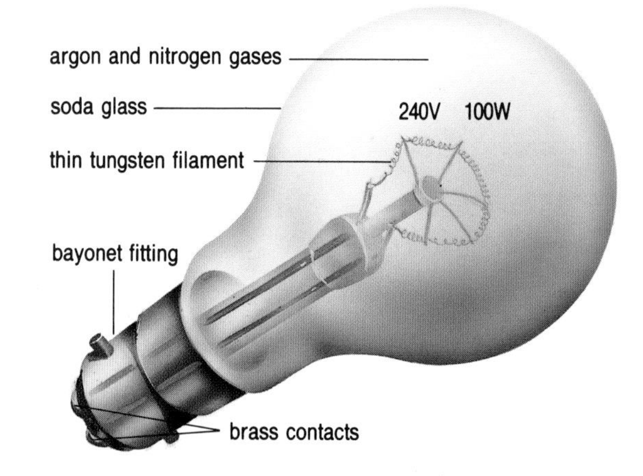

A filament lamp.

A quartz-iodine light bulb gives a very bright light. Larger currents and higher temperatures are possible because iodine prevents the evaporation of tungsten. This is a more expensive filament bulb and is used in slide projectors.

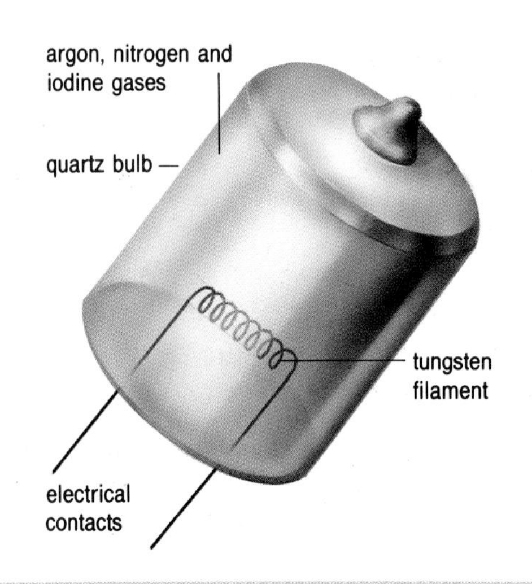

Fluorescent lamps

Gases can also conduct electricity. The current makes the gas, at certain pressures, glow with a colour which depends upon the type of gas in the tube. These gas filled tubes are called discharge tubes, and the ordinary household fluorescent lamp is one of the most common.

The fluorescent tube contains argon and mercury gases. When there is a current in it, moving electrons collide with the mercury particles and ultraviolet radiation is given off. Ultraviolet radiation is similar to light but its frequency is too high for our eyes to detect. Phosphor powder on the inside of the tube absorbs this ultraviolet radiation and gives out light. This process is called fluorescence.

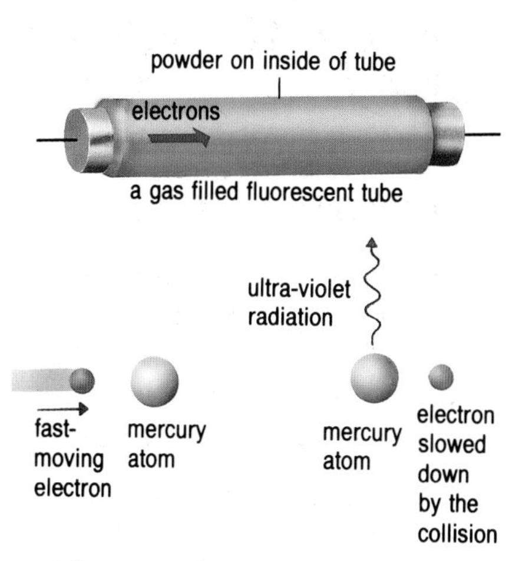

A fluorescent tube.

Don't be fooled by the shape of a lamp! Thin strip lights often used in cabinets have a long thin wire along the length of the tube. They are filament lamps. Similarly compact fluorescent tubes, looking very similar to ordinary light bulbs, can fit into standard bayonet sockets.

Light not heat
Fluorescent tubes convert four times more energy into light than do filament lamps.

Unpopular?
Big and ugly. Fluorescent tubes are sometimes unpopular because of their appearance and bright light.

Last longer
Fluorescent tubes have no filaments to burn out. They last longer than filament lamps.

Starter and choke
A high voltage is needed to start electrons moving through a gas. Once started, the electrons move very easily and need to be controlled. The extra cost of a starter coil and choke is necessary before the tube is part of a useful fluorescent lamp.

Danger
The varying a.c. supply makes fluorescent tubes brighten and darken every 0.01s. This can make machinery appear stationary. No such problem exists for filament lamps because the wire remains white-hot whilst the current changes occur.

A rounded tube with starter and choke, all enclosed in a glass envelope provide a compact fluorescent lamp.

1 What special property makes tungsten suitable for use as a lamp filament?

2 a) What is the current in lamps rated at (i) 240 V 60 W and (ii) 240 V 40 W respectively?
b) From these figures calculate the resistance of the filament wire in each case?

3 What happens to the thickness of a filament wire as the lamp gets older? What effect will this have on **a)** the resistance and **b)** the temperature of the filament?

›› *Use an ohmmeter to measure the resistance of the filaments in the lamps of question 2. Explain any difference between these measured answers and the theoretical answers of question 2.*

Keeping warm around the house

The choice is yours

Electric heaters transform electrical energy into heat when there is a current in a length of resistance wire called the **element**, which is usually wound in the shape of a coil. The shape of the heater and the methods used to transfer the heat from the element to the room can vary greatly. You have to decide on the most suitable type for your needs.

The operating current and the resistance of the element can always be calculated from the quoted power and voltage ratings – provided you remember the formulae $P = VI$ and $V = IR$.

Radiation Infrared rays, sometimes called radiant heat, are waves which behave like light and radio waves. They travel very quickly (3×10^8 m/s) in all directions and can be focused and reflected by shiny metal reflectors. Very hot objects give off a lot of infrared **radiation**. When this radiation is absorbed by something its temperature rises.

Convection When air is heated it expands and becomes less dense. The colder denser air above it is pulled down by gravity and the hot air rises. This transfer of heat by a moving fluid such as air is called **convection**. It occurs in all liquids and gases but cannot occur in solids because the particles in a solid cannot 'change places'.

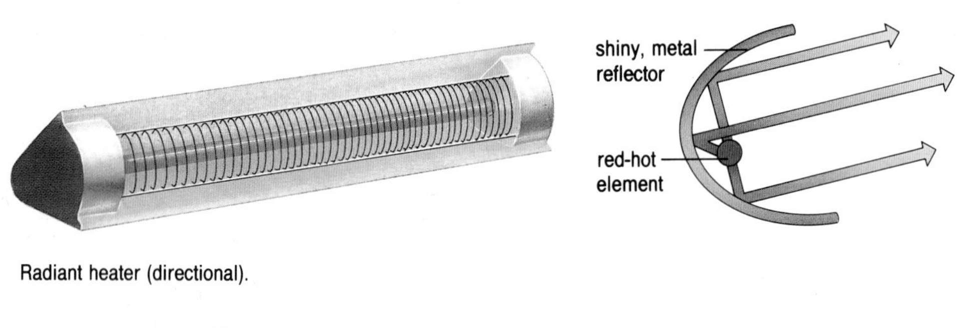

Radiant heater (directional).

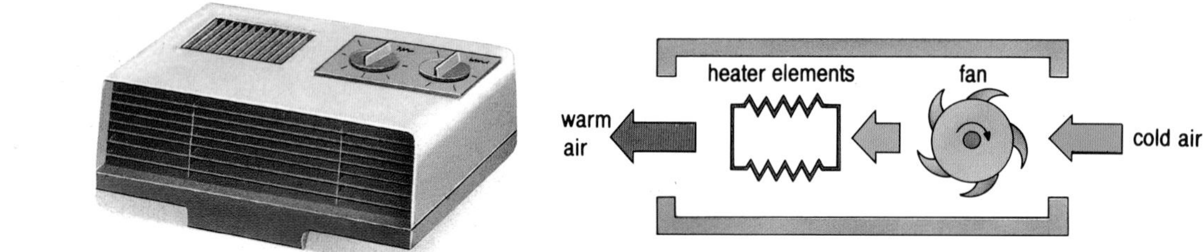

Fan heater (quick acting).

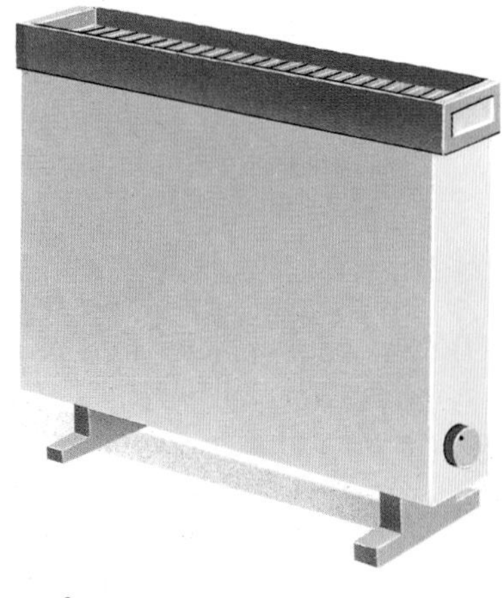

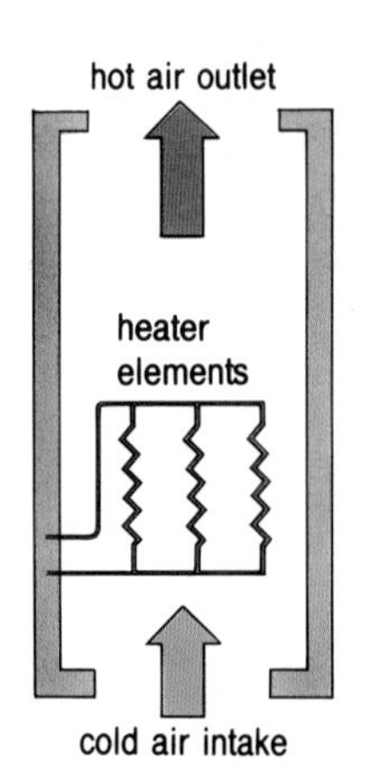

Convector heater (background effect).

Joined together

Components are joined in series and/or parallel for many practical purposes. The voltage divider is an example of resistors joined in series to produce a lower voltage.

In contrast, a 2-bar convector heater has the elements wired in parallel. This enables you to have either one or two bars switched on, depending upon how much heat you want.

Series circuits

In a series circuit, components are joined one after the other.

There is only one possible path for the charges to move around the circuit. The rules to remember are:

1 The current is the same at all points in a series circuit.

2 The voltage of the source is divided. Across resistors, it is shared in proportion to the resistance of each resistor.

Parallel circuits

In a parallel circuit components are joined side by side.

There is more than one path for charges to move around the circuit. The rules to remember are:

1 The voltage is the same across all the components in parallel.

2 The current divides. Some charges move around each of the possible paths. The current is shared and the largest current flows through the smallest resistance, such that the bigger the resistance, the smaller the current. Current is shared in inverse proportion to the size of the resistors.

1 Make a table to list the advantages and disadvantages of any two different types of heater.

2 Assume there is no resistance in the connecting wires of the electric fire drawn here. When both switches are closed:

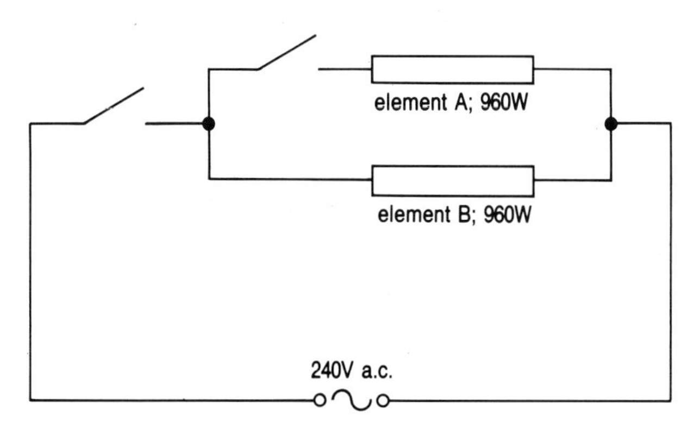

a) what is the voltage across element A;
b) what is the current in element A;
c) what is the resistance of element A;
d) what current is taken from the supply?

3 Draw a circuit diagram for a 2-bar radiant heater with flame-effect light. The requirements are:
a) when the fire is connected to the mains the flame effect light must go on; and
b) both bars have their own switch and can be switched on or off separately from each other.
Check your drawing by wiring up such a model fire using a lamp, small resistors, switches and power pack.

4 What are **a)** the resistance values of the unknown resistors and **b)** the meter readings in the diagrams below?

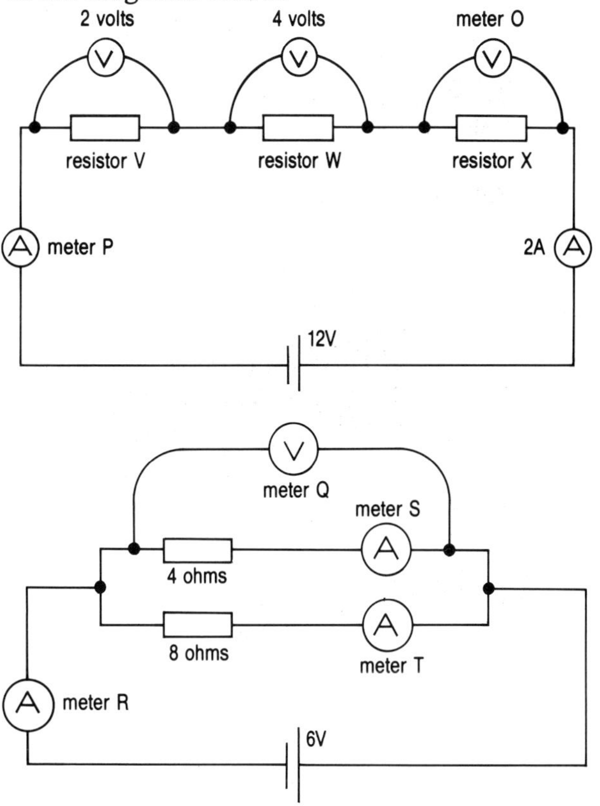

5 You are given two identical heating elements, a single pole switch, a double pole switch and some wire. Show how you would wire these components so that you have an electric heater which can be switched to 'low', 'medium' and 'high' power.

Lighting in car and home

See and be seen

Cars are crammed with lights! Headlights, sidelights, brake lights, number plate lights and flashing indicator lights are all required by law. Add to these fog lights, spot lights and reversing lights, and a car can almost compete with Blackpool illuminations. But these lights help us to see and be seen by others. They also communicate messages to road users.

This page shows a simplified version of a car wiring diagram.

Car lighting unit.

left headlight
right headlight
left sidelight
right sidelight
battery
S_1
S_2
dashboard light
car body
1 2 3 4 5 6 7 8 9 10 11 12 13 14 15

1 By stating which wires will carry current, explain which switch, or switches, turn on: **a)** the sidelights; **b)** the headlights; **c)** the dashboard light; and, **d)** the rear lights.

2 Which lamp or lamps will go out if there is a break in:
a) wire 1; **b)** wire 6; **c)** wire 8?

3 Explain whether or not any other lamps still operate if the dashboard lamp breaks.

4 Draw a circuit diagram to show how:
a) a two way switch switched to the left operates the left indicators (front and rear). Switched to the right it operates the right indicators. In both cases, the ignition switch must be closed;
b) rear brakelamps operate when both the ignition switch *and* the brake switch are closed. (The brake switch closes automatically when you press the brake pedal.)

Looping-in lamps

The simplest possible lighting circuit is a lamp in series with its own fuse and switch. However, one light is not enough for a house! Many are needed. By connecting the lamps in parallel each can be turned on and off separately. **Looping-in** provides such a technique. The connecting boxes where the lamps are fixed to the ceiling are called ceiling roses. The mains connections are made from one ceiling rose to the next.

Safety regulations permit a maximum of ten 100 W lamps in each circuit fitted with a 5 A fuse. An earth wire (not shown) must be connected to each ceiling rose and switch but the lamp holders are not connected to an earth wire.

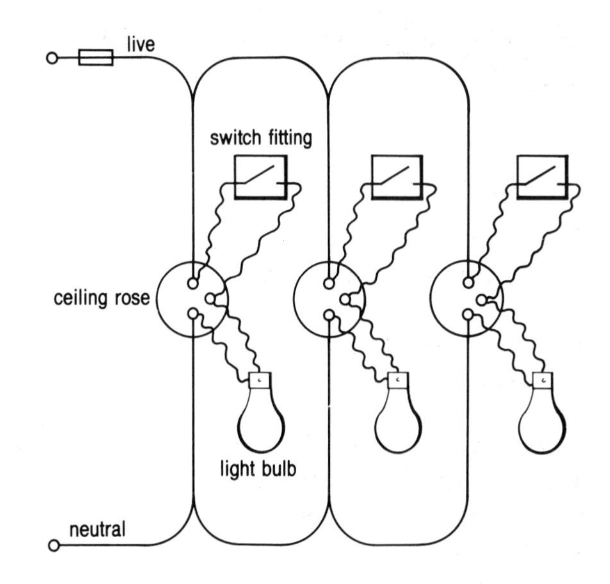

Looping-in lamps connected in parallel. Each lamp is controlled by its own switch. The wires are joined at the place where the wires pass through the ceiling.

Dim the lights

Sometimes a light is too bright. A dimmer switch allows the light to be adjusted to any required brightness, up to the full power of the lamp.

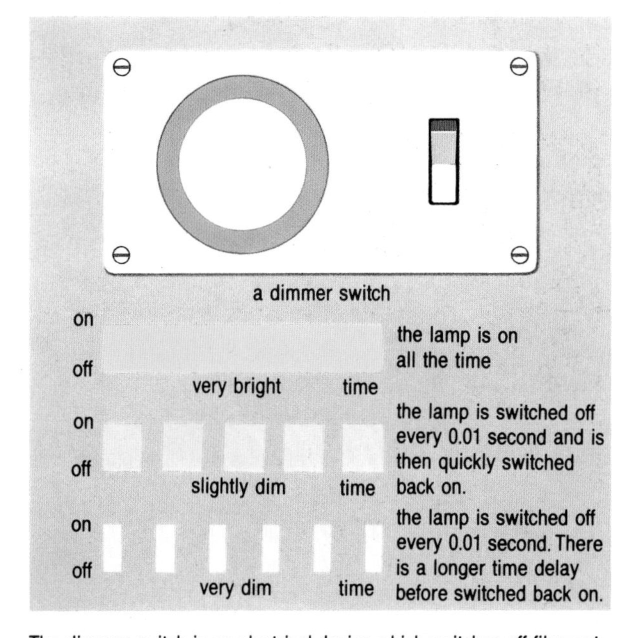

The dimmer switch is an electrical device which switches off filament lamps for very short times. It is put in place of the normal switch. Fluorescent lamps are more difficult to start and should not be used with dimmer switches.

5 What is the total current in a circuit if all ten 100 W, 240 V lamps are switched on at the same time? Will a 5 amp fuse in the main circuit blow?

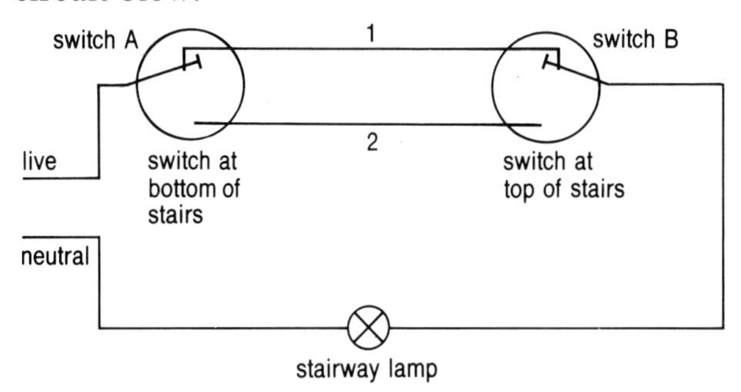

6 Copy the diagram and complete the table.

position of switch A	position of switch B	lamp lit or not lit
1	1	
1	2	
2	1	
2	2	

›› *Count how many light bulbs you can find in a family car and explain why each light is needed?*

Design and draw a circuit which would allow you to switch a single lamp on or off from three different places. Why is it an advantage to be able to switch a particular lamp on and off from three different places? Give an example of where you might want to do this.

Fault finding

The continuity tester

Wires and electrical components often break and tests are needed to find out what has gone wrong. In one such test we look for a complete low resistance conducting path for electric charges. This is called a **test for continuity**, and the test instrument is called a continuity tester. You can make a simple continuity tester from a battery and a small bulb or ammeter connected in series with the circuit to be tested.

Of course, continuity isn't always what we want: it depends upon what is being tested. An unbroken fuse or wire must be continuous. A correctly fitted earth wire should have continuity between the earth pin of the plug and the appliance casing. In contrast, continuity between the live and earth pins of the plug to an iron would prove that there was a dangerous short circuit in the iron which had to be repaired.

Testing is more complicated in circuits which include diodes. If all is well, there should be continuity when the battery is connected in one direction but no continuity when it is reversed. With such circuits remember, therefore, to test for continuity twice across each diode.

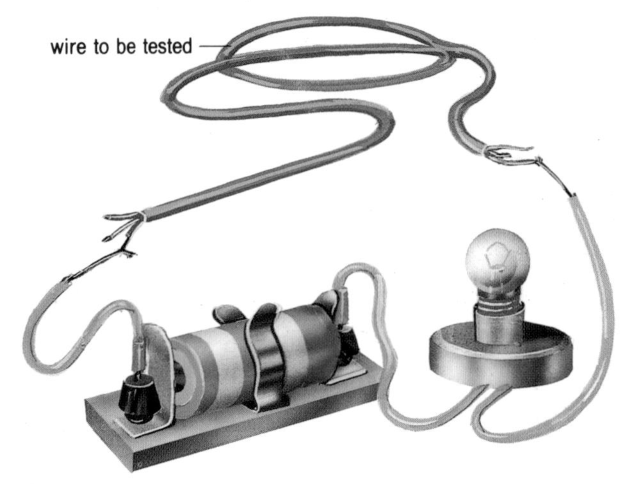

A continuity tester is connected across the ends of the wire, component or appliance to be tested. If the lamp lights there is continuity; if the lamp does not light the conducting path is incomplete.

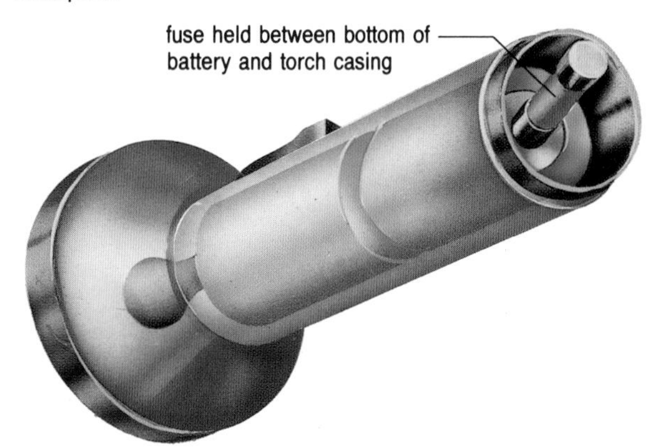

Using your torch as a continuity tester. If the fuse is unbroken it completes the electric circuit and the lamp lights.

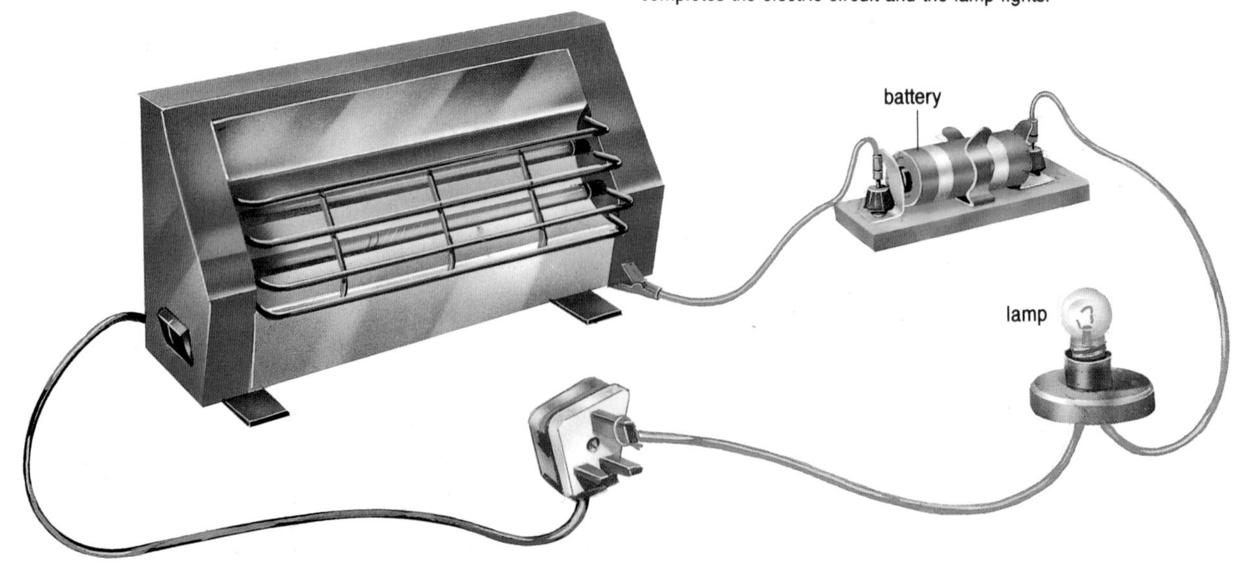

A continuity tester (battery and lamp) used to check the earthing of an electric fire.

Finding the circuit resistance

Continuity testers are designed to show that a circuit has a low resistance. They are not suitable for circuits which contain high resistance components. A multimeter switched to its resistance range would be a far better test instrument for such a circuit.

The measurement of circuit resistance can often provide useful information, for example:

1 the resistance in a bell circuit which contains a transformer winding;

2 the resistance of a bicycle dynamo circuit;

3 the resistance between the household earth wire and the soil. For safety reasons this should be very small.

In each case, faults can be located and problems isolated by such measurements.

Resistors in series and in parallel

Combining resistors one after the other is equivalent to having one length of wire after another. Increasing the length of wire increases resistance.

Combining resistors in parallel is equivalent to putting strands of wire side by side. The total thickness of the combined strands is more than the thickness of a single strand. Increasing the thickness of wire reduces resistance.

Series circuit: the total resistance is more than any of the individual resistors.

$R_{TOTAL} = R_1 + R_2 + \ldots$

Parallel circuit: the total resistance is less than any of the individual resistors.

$$\frac{1}{R_{TOTAL}} = \frac{1}{R_1} + \frac{1}{R_2} + \ldots$$

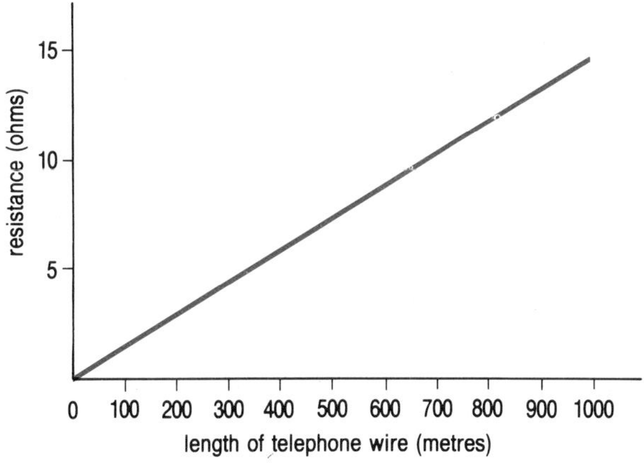

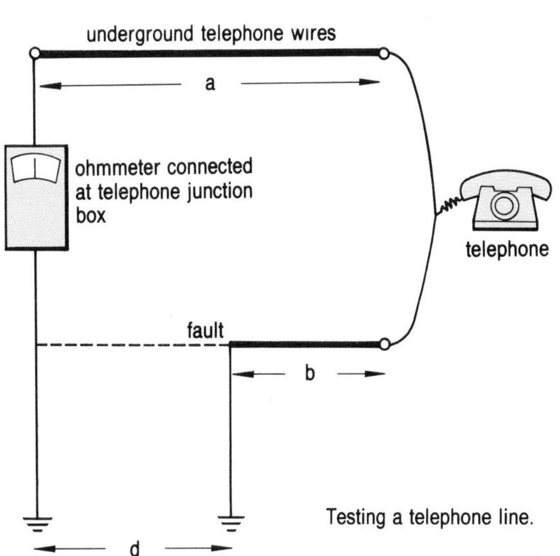

Testing a telephone line.

1 What is the combined resistance of two 50 ohm resistors when connected **a)** in series and **b)** in parallel?

2 What is the combined resistance of a 6 ohm and a 3 ohm resistor in parallel?

3 What possible resistances can be made by combining four 10 ohm resistors?

4 From the graph shown, what length of telephone wire has a resistance of 8 ohms? If the distance from the junction box to the telephone is 350 m in the circuit below and a resistance of 8 ohms is measured, how far away is the fault from the junction box? What assumption are you making?

›› *Put two carbon rods into soil. Measure the resistance between the rods (the earth resistance) with the soil* ***a)*** *dry and* ***b)*** *damp.*

Why is it easier to get a good earth in marshy ground than it is in sandy ground?

Behind the wall

The ring circuit

Our homes are fitted with a large number of power sockets. Each one of them must be capable of supplying up to 13A of current to an appliance. Separate heavy wires from the fuse box to each socket would be expensive, but a clever arrangement normally fitted in most houses allows wiring to be in parallel but still keeps the total length of wire short. It is called a **ring circuit**.

The ring circuit cable consists of three separate conductors called live, neutral and earth.

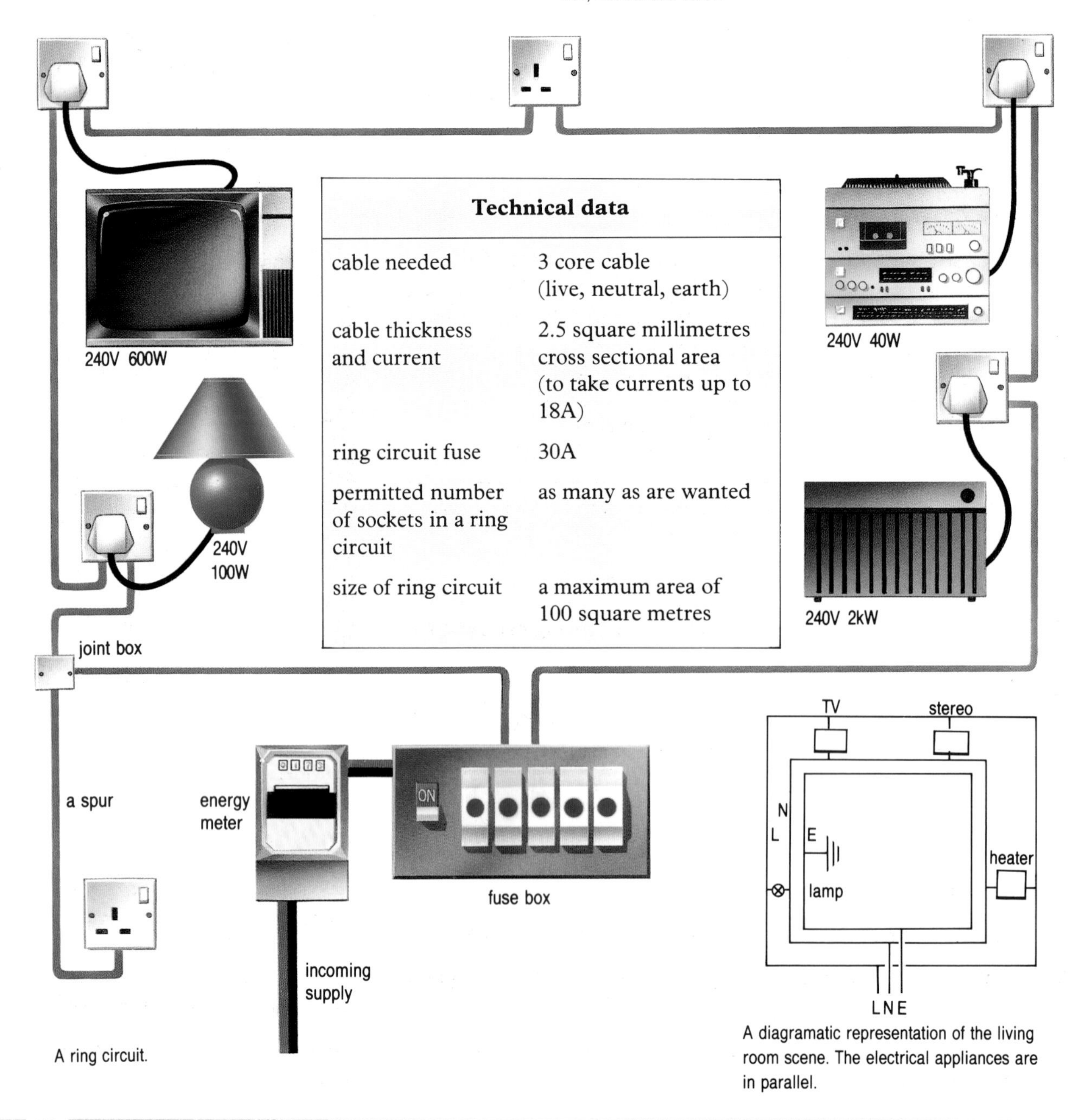

Technical data

cable needed	3 core cable (live, neutral, earth)
cable thickness and current	2.5 square millimetres cross sectional area (to take currents up to 18A)
ring circuit fuse	30A
permitted number of sockets in a ring circuit	as many as are wanted
size of ring circuit	a maximum area of 100 square metres

A ring circuit.

A diagramatic representation of the living room scene. The electrical appliances are in parallel.

The choice of cables

The wiring used for the ring circuit is made from three solid copper conductors which are difficult to bend. These three-core inflexible wires are called **cables**. The cables are thicker than those used for lighting but are still only thick enough to carry currents up to 18 A without overheating. However, as there are two paths from each appliance to the fuse box in a ring circuit a safe total current of about 36 A is possible. Bigger currents would make the cables too hot and there would be a risk of fire.

Fuses are fitted in the mains fuse box to protect the cables behind the wall and reduce the risk of fire. Typical fuse values are 5 A for lighting, 30 A for the ring circuit, 15 A for immersion heaters and 45 A for electric cookers.

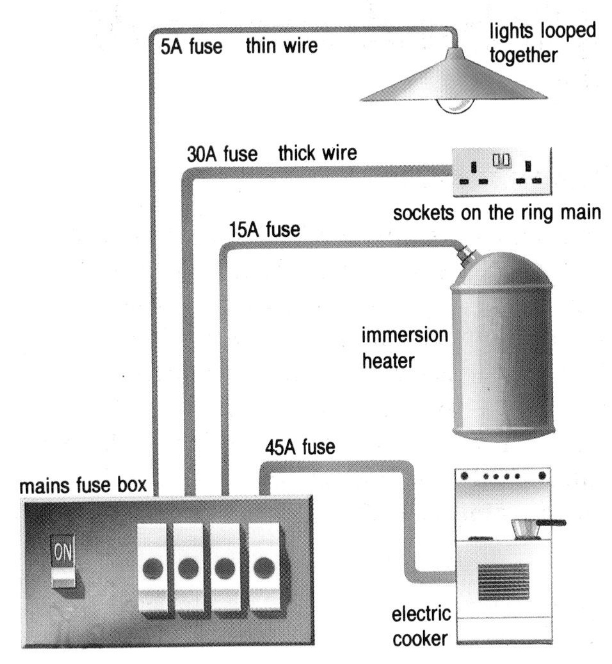

The kitchen is a place for high-power electric appliances. The tendency is to wire these separately to their own fuse in the fuse box.

Circuit breakers

Some houses have miniature circuit breakers in place of the fuse box. Circuit breakers are automatic switches which switch off and break the circuit when there is too much current. Once the cause of the large current has been found and repaired the circuit breaker is simply switched back on again.

Advantages of using circuit breakers	Disadvantages of using circuit breakers
They switch off very quickly when the current is too large (This gives better protection to the cables)	They are more expensive than fuses
They can be quickly switched back on and reused	They have moving parts and so need to be regularly tested
For some the maximum current (above which they switch off) cannot be changed They are tamper proof	Maximum current can be affected by the temperature of the surroundings

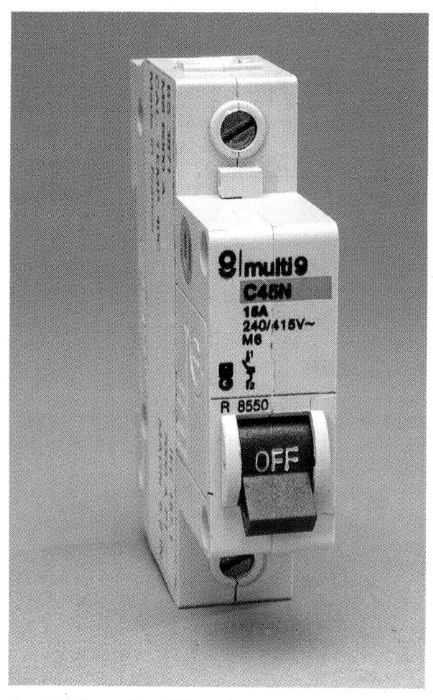

A miniature circuit breaker.

1 Why is the ring circuit method of wiring used?

2 What is the difference between a cable and a flexible cord?

3 Calculate the current in the ring circuit on the opposite page when all the appliances are switched on.

4 Why must an electric cooker not be connected to a ring circuit power socket?

›› *Ask your parents to show you the fuse box at home. How many fuses do you have, what are their current ratings and what circuit does each protect?*

Find out how a circuit breaker switches off the current.

Electromagnetism

The magnetism of current

Many well known scientists such as Ampere, Faraday, Oersted and Edison investigated the magnetic fields produced by an electric current in a wire. In one experiment a small plotting compass is used to investigate the direction of the magnetic field around a wire. When the current is switched on, the compass needle turns until it points in the direction of the magnetic field at that point.

If a long wire is wound into a coil (or solenoid) the combined effect of all the turns of wire is to produce a field which is similar to the field of a permanent bar magnet.

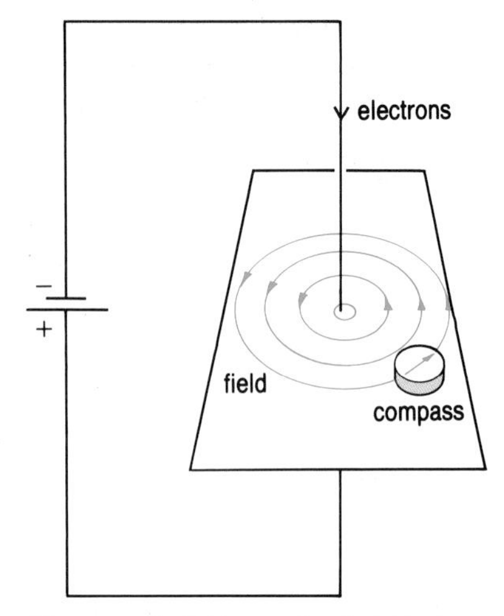

The magnetic field around a straight wire. Notice that the field is circular and the field direction reverses when the current direction reverses.

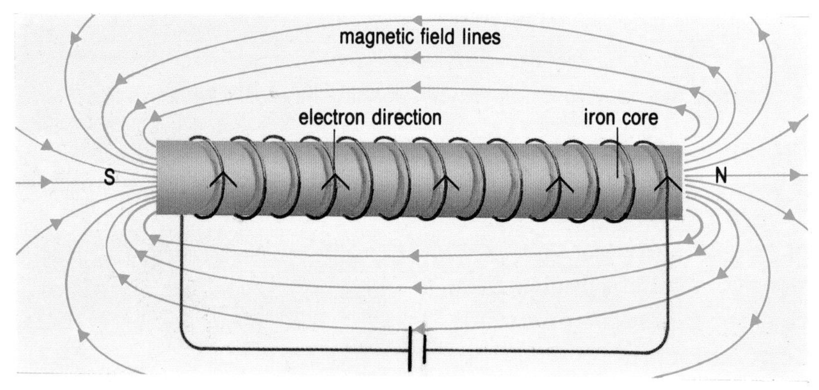

The magnetic field around a solenoid is similar to that of a bar magnet.

To increase the field strength use:

1 a larger current in the solenoid;

2 more turns of wire; and

3 a soft iron core inside the solenoid.

When the current is switched off, the soft iron core is no longer magnetized and so the magnetic field around the solenoid is lost. A solenoid with an iron core is called an **electromagnet**.

An electromagnet. The very strong magnetic field is used to lift heavy ferrous metals and machinery. Electromagnets are often used in place of permanent magnets because: 1) the magnetic field can be made very strong; 2) the magnetic field can be switched off.

Magnetic fields

A freely hanging magnet rotates to align itself with the Earth. The end of the magnet which faces the Earth's North Pole is called the north end or north pole of the magnet. The other end is the south. A compass needle which we use to find our direction is just a small magnet.

Electricity has two types of charges which attract or repel each other. To describe the action of forces at a distance, electric fields were introduced. Similarly for magnets.

Each magnet has a north and a south end.
Two north ends repel each other.
Two south ends repel each other.
One north and one south end attract each other.
Either end of a magnet attracts ferrous (iron) materials but has no effect on materials such as copper, brass and most non-metals.

To describe how the magnetic forces act from a distance, physicists use the picture of a **magnetic field**. The direction of the field at any point is defined as the direction in which the north end of a compass is made to face.

The reed relay

Just as in a relay race at an athletics meeting, a baton is passed from one team runner to another, so in electrical circuits, a **relay** allows a signal to be passed from one circuit to another. The relay is a switch in one circuit which is operated by a current in another control circuit.

Relays have been around for a long time, but one recent development is the reed relay which is smaller and more reliable than some earlier relays and operates as follows.

1 Current in the solenoid produces a magnetic field similar to that of a bar magnet.

2 The field magnetizes the iron reeds and one contact becomes a north pole, the other a south pole.

3 Unlike poles attract and so the contacts close. The reed contacts form the switch in the second circuit. When the contacts close this circuit is complete.

4 When the control current is switched off, the reeds lose their magnetism and spring back into place thus opening the contacts.

The direction of current in the solenoid doesn't matter. The reeds will always have opposite magnetic ends when magnetized and will therefore always attract each other. Reed relays will operate with either a.c. or d.c. supplied to the solenoid.

Reasons for using a relay:

- one circuit can switch other circuits;
- a safe low voltage can switch a dangerous high voltage;
- a very small current can control a large current;
- relays can be designed to delay their switching action;
- relays can be operated with a permanent magnet or with a solenoid supplied with a.c. or d.c.

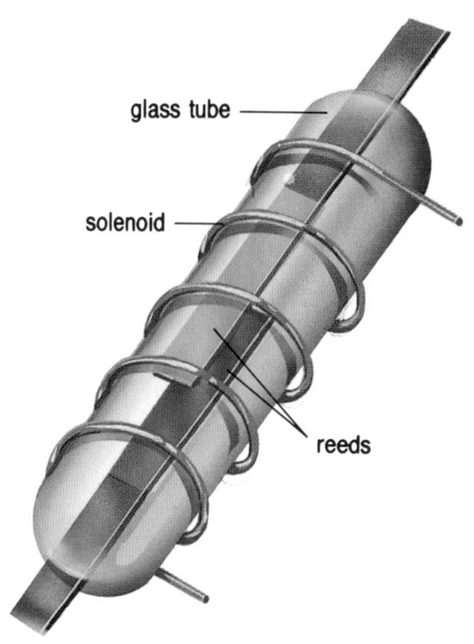

The reed relay. The switch consists of two metal contacts (called reeds) made of iron. The tips are gold or silver plated for good conduction and are housed in a glass envelope filled with an unreactive gas. The envelope excludes dust and prevents corrosion. A solenoid surrounds the contacts. Relays can make or break a circuit. The one shown here will make (complete) a circuit when there is a current in the solenoid.

Reed relays act as safety switches on the door of a microwave oven. With the door open, the reeds must be in the open position so that the microwave heater circuit cannot be turned on. With the door closed, the magnets close the reed contacts so that the microwave heater can be turned on.

1 What are the advantages and disadvantages of using an electromagnet as opposed to a permanent bar magnet?

2 Explain whether the reed relay would operate from an a.c. supply.

›› *Using a reed switch and a permanent magnet, design a circuit which would set off an alarm if a window were opened.*

Design a reed relay which will break the circuit when there is current in the solenoid.

On the move

Electric motors

Electric motors use electrical energy to do useful work. They range from small motors in food mixers, vacuum cleaners, drills and lawnmowers through larger motors in washing machines to very large motors that drive electric trains and heavy machinery. In each case, the magnetic fields produced by current-carrying wires help to provide the force needed for movement. Different designs of motors can make the motor rotate or move in a straight line.

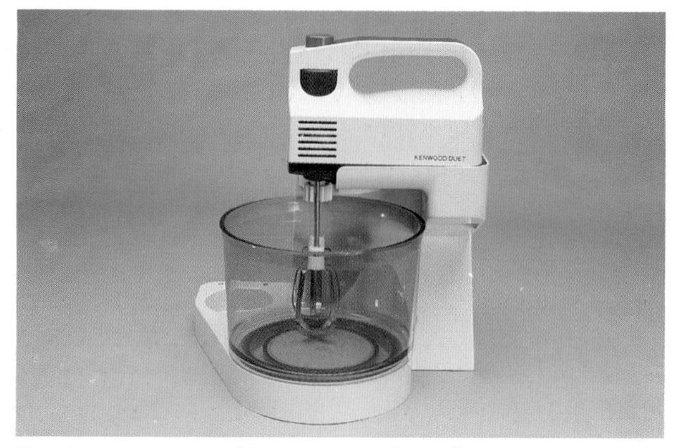

Electric motors are used in many domestic appliances.

Look at the simple d.c. motor shown here.

1 It has a coil of wire which is free to rotate in a magnetic field.

2 When the coil is in the horizontal position there is a current in it and so each side of the coil experiences a force.

3 As the currents in the two sides are in opposite directions, the forces acting on them are in opposite directions. One side is forced up and the other down. The coil tends to rotate.

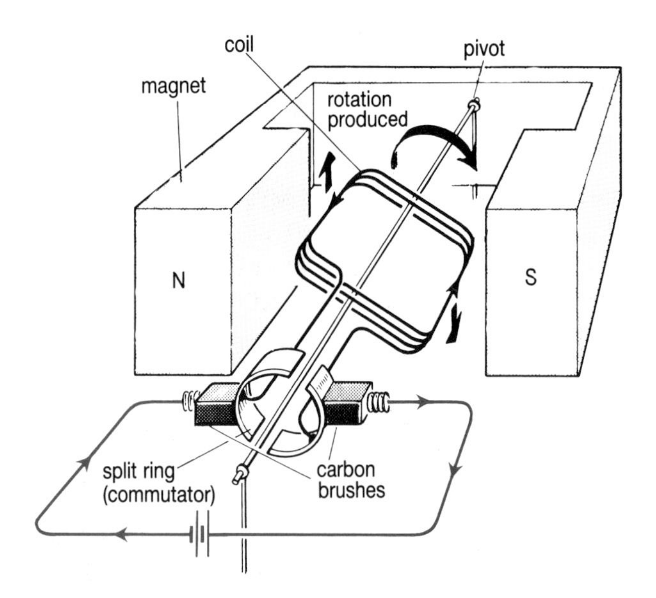

4 When the coil has turned to the vertical position, its sides can go up and down no further. At this point the **split ring commutator** reverses the current in the coil. This changes the direction in which the forces act. The side which was going up is now forced downwards and the side which was going down is forced upwards. So, the coil continues to rotate. This reversal process is repeated every half revolution.

5 The current has reached the coil through **brushes** which rub against the commutator. So, the stationary wires to the battery and the moving wires in the coil are always part of one complete electric circuit.

London Underground trains use large electric motors.

Commercial motors

Commercial motors are much more complicated than the simple version described above. The single rotating coil is usually replaced by several coils called **armature windings**. They are wound on thin **iron strips (laminations)** to produce a series of magnetic poles. The armature windings are connected to the many strips of copper called segments which make up the commutator. With this type of **segmented commutator** and multicoil armature, a much smoother continuous rotation is possible.

The permanent magnet of the simple motor is replaced by an electromagnet called the **field magnet**. The armature windings and field windings can be connected in series or in parallel. Thus, one electrical supply provides the energy for both sets of coils.

Electromagnets are preferred because:

1 They can produce very strong magnetic fields. More powerful motors can be built than is possible with permanent magnets.

2 The same motor will operate on a.c. as well as d.c. Reversing the current reverses both the magnetic field due to the armature and that due to the field windings and so rotation continues in the same direction.

simple motor	commercial motor
permanent magnet	electromagnet (the field magnet)
one coil	multicoil windings
one core	thin strips of iron (laminations)
split ring commutator (for d.c.)	segmented commutator (a.c. and d.c.)
wiring direct to the brushes	armature and field windings in series or parallel

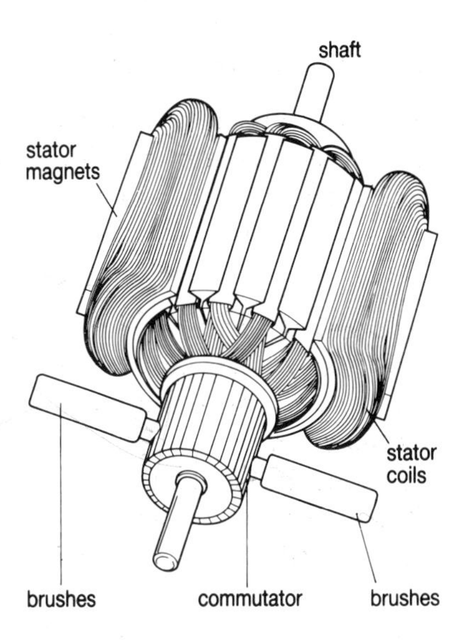

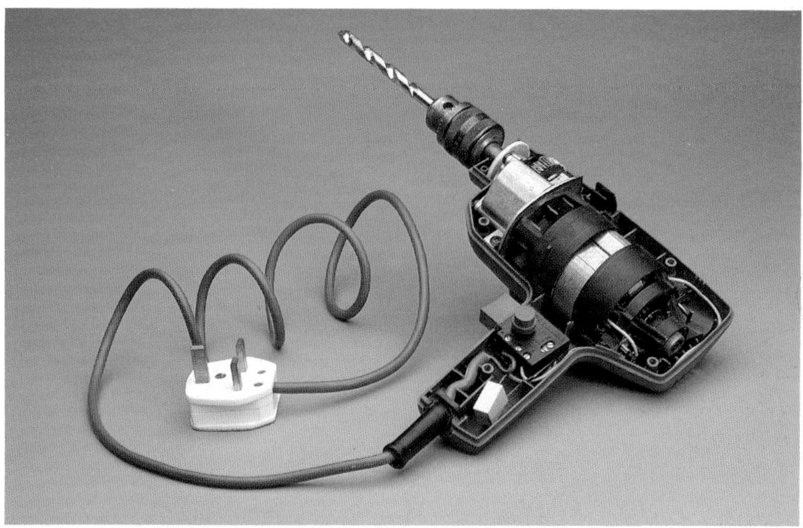
An electric motor is used to drive this drill.

1 What is the purpose of **a)** the brushes and **b)** the commutator in a motor?

2 Why is a multicoil armature used in commercial motors?

3 What are the advantages of using field windings in a motor?

›› *Connect an ammeter in series with a 'motor and flywheel' in a d.c. circuit. What happens to the current as the motor speeds up? Can you suggest a reason for this?*

Did you know?

The current in a motor is greatest at low speeds. Large motors have resistors fitted to prevent the coils from overheating at low speeds. Passengers in Underground and other electric trains can sometimes hear 'clicks' as the train gains speed. The clicks are switches operating to cut out these resistors.

HEALTH PHYSICS

3

Body temperature

Stethoscope sounds

Silent ultrasound

A pain in the ear

See you!

Eye problems

Lenses and fibres

Lasers in medicine

X-rays

Using the electromagnetic spectrum

Cell-killing radiation

Detecting radiation

Taking care with radiation

Helping the disabled

Body temperature

Clinical thermometers

If you're healthy your temperature is likely to be between 36° and 37°C. Your temperature can be checked with a digital thermometer or with a standard clinical thermometer in which liquid mercury expands when it is heated. Unlike ordinary room thermometers, however, a clinical thermometer reads only the maximum temperature reached. Here's what happens.

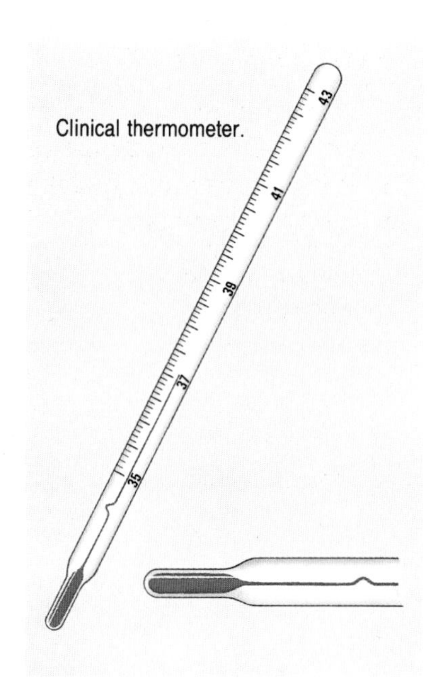

Clinical thermometer.

To find your temperature the thermometer bulb is placed under your tongue and left for a few minutes. As the mercury expands it pushes its way past a kink in the tube and finally stops rising. The thermometer is then removed from your mouth. The narrow kink in the tube causes the mercury thread to break as it cools and contracts. This leaves a short thread of mercury which indicates the maximum temperature. To reset a clinical thermometer it has to be shaken fairly strongly.

Infection causes the body to heat up more than usual. If your body temperature is a few degrees above 37°C – the normal body temperature – you're ill!

Don't throw out the baby ...

The temperature of a baby's bath water should be kept close to its body temperature. Often toys with built in *liquid crystal* thermometers are used to check the temperature of the water.

If you place the fish thermometer in the water a number showing the temperature will appear. The baby will enjoy playing with the fish and, by leaving the thermometer in the bath, you will know when the water becomes cool.

Measuring body temperature.

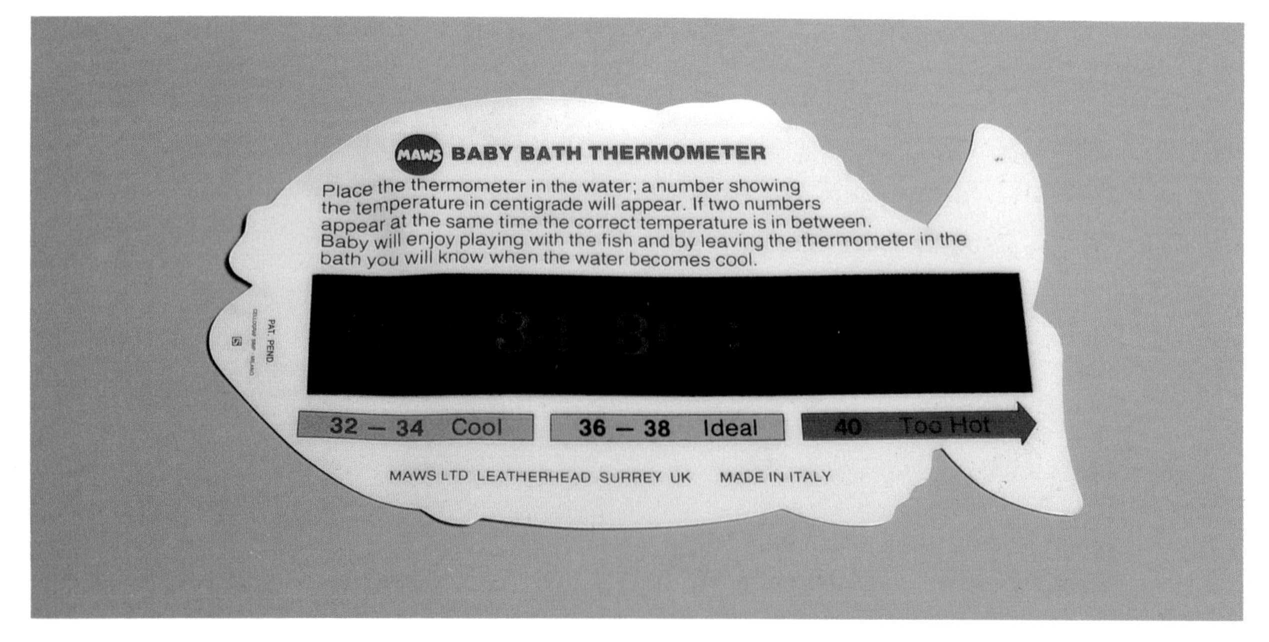

Hypothermia

Too cold for comfort.

Normal 'body' temperature (37°C) is the temperature of the inside organs such as the heart, lungs and brain. Elsewhere, the temperature is lower for example the skin has an average temperature of about 33°C.

A lowering of body temperature is known as **hypothermia** and can be caused by exposure to cold and damp conditions. The table summarises typical body behaviour during various stages of hypothermia.

Temperature inside the human body	Condition of person
37°C	Normal
34°C	Shivering, narrowing of blood vessels, reduced heart rate
33°C	Slow responses. Loss of memory
30°C	Temperature regulating system fails. Sleepiness
28°C	Loss of consciousness. Breathing slows down. Irregular heart beats
about 26–28°C	Death

Too hot for comfort

Too hot for comfort.

There is a gradual worsening of the way the body works if it is exposed to *high* temperatures for too long. Hot surroundings, fever or vigorous exercise can make the body temperature rise. The result is that the blood vessels get wider, the heart rate rises and the blood flow to the brain is reduced. This may result in unconsciousness. When the body temperature rises to about 41°C, the nervous system starts to deteriorate and convulsions begin. Finally death occurs between 43°C and 45°C.

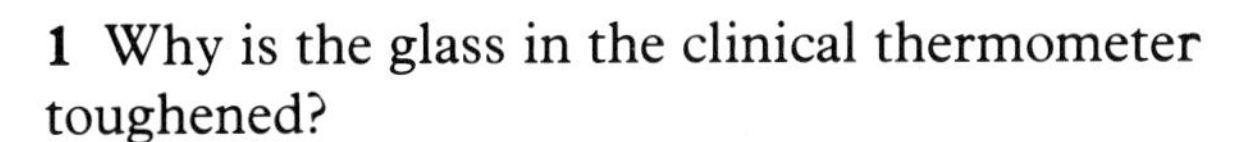

1 Why is the glass in the clinical thermometer toughened?

2 What temperature is shown on the clinical thermometer on the opposite page?

3 The clinical thermometer glass tube has a special shape.

a) Why is the tube shaped to act as a lens?
b) Why does it have a white background?

4 Use the temperature chart to answer these questions.
a) What was the patient's highest temperature?
b) When was an ice bath used to cool him?
c) For how many hours was his temperature above normal?

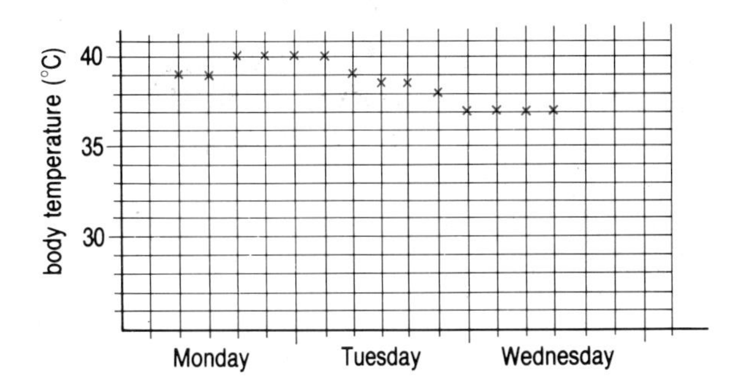

Stethoscope sounds

How the stethoscope was invented

Until the early years of the nineteenth century a doctor listened to the sounds coming from inside his patient by placing his ear directly on the patient's chest or back.

In 1818. R.T.H. Laennec complained that this method 'is always inconvenient, both to the physician and the patient; in the case of females it is not only indelicate, but often impracticable'.

Once, when Laennec was examining a girl with symptoms of heart trouble, he decided to try out a new method of examination. He rolled several pieces of paper into a cylinder and held one end against his ear and put the other end to the girl's chest above her heart. The results were excellent.

Laennec had invented a simple instrument that allowed him to hear sounds from inside his patients – a sort of 'hearing aid.' His success encouraged him to improve the instrument. He developed a hollow wooden cylinder 30 cm long with an inner diameter of about 1 cm and an outer diameter of about 7.5 cm. The stethoscope used today is based on this instrument.

Before the stethoscope was invented.

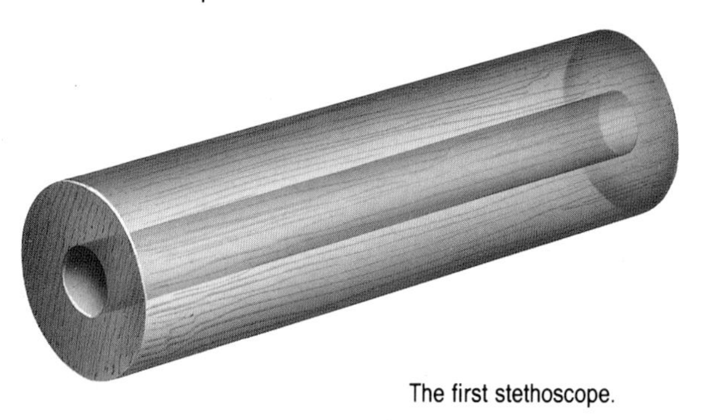

The first stethoscope.

The modern stethoscope

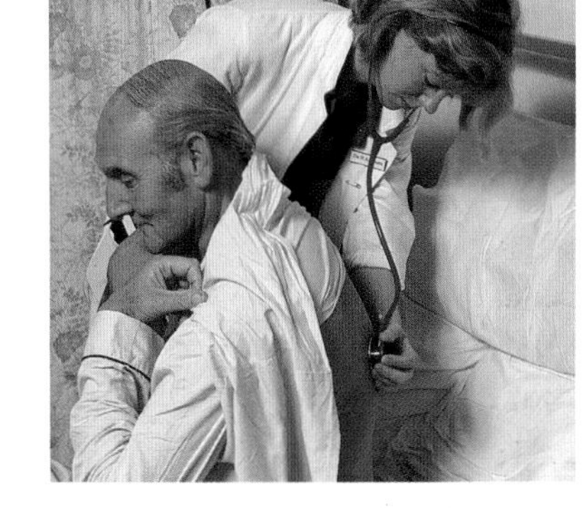

The stethoscope allows a doctor to hear sounds made inside the body. In particular sounds from the heart and lungs can be useful in the diagnosis of various diseases. At the end of the stethoscope tube there are two 'bells', one closed (for higher frequency sounds) and one open.

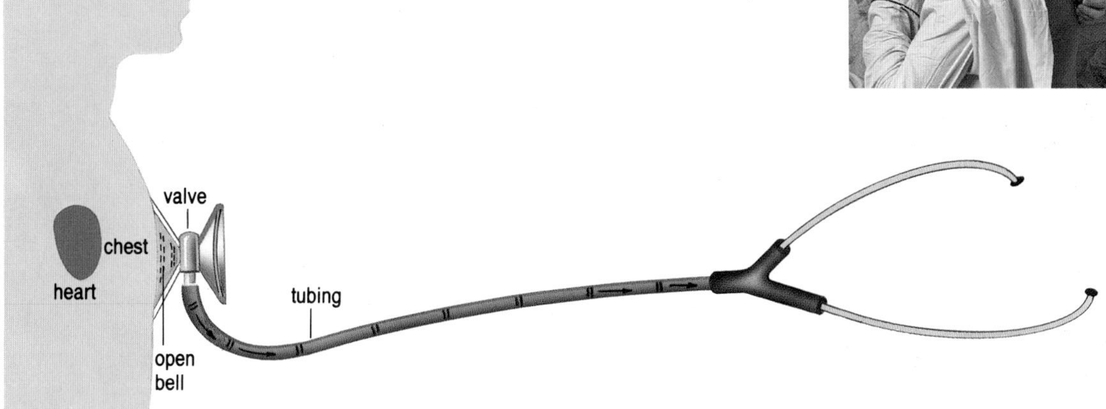

The closed bell has a diaphragm with a high natural frequency to 'tune out' unwanted low frequency sounds and is used for listening to lung sounds of higher frequency than heart sounds. When using the open bell the skin acts as a diaphragm with a natural vibrating frequency. The tighter the skin the higher the natural frequency. To produce the loudest sounds the frequency of the sound should equal the skin's natural frequency. This produces resonance. The doctor varies the pressure of the bell on the skin to tune the skin frequency to the heartbeat frequency causing the skin to vibrate and produce the loudest sound. The doctor's ear-drum is vibrated by changes in air pressure in the tube. To cause enough change the bell's volume should be a minimum.

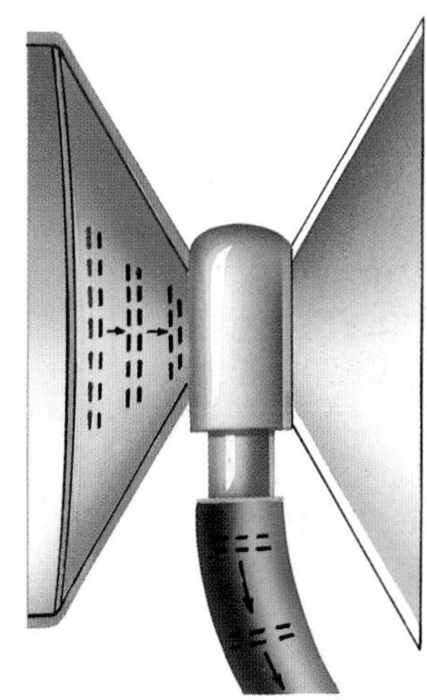

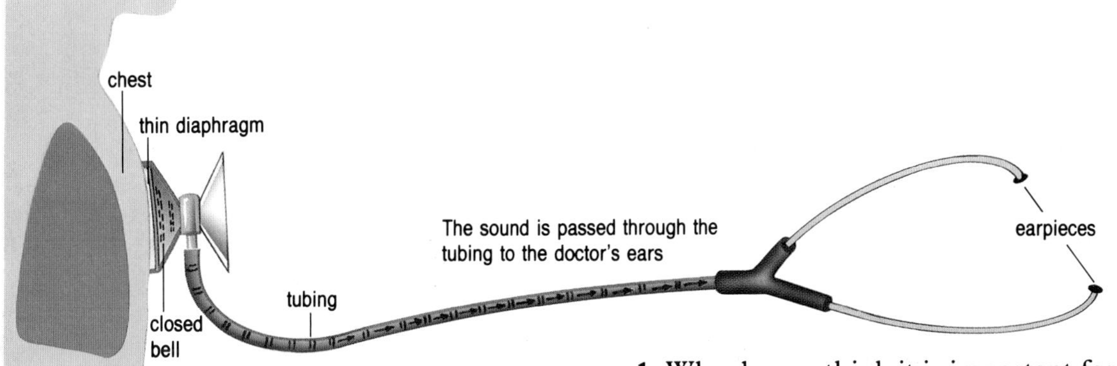

›› *Take a piece of wood and place one end against your ear. Scratch the other end with a pin. Describe the level of noise heard. Try this again using different materials in place of wood, and then hollow cylinders of various sizes. Write a report of your results.*

1 Why do you think it is important for the earpieces to fit snugly into a doctor's ears?

2 If a patient's heart rate is 70 beats per minute, to what frequency should a doctor adjust the patient's skin under the stethoscope bell to pick up maximum sound?
What is this frequency called?

Silent ultrasound

Echoes

If you clap your hands some distance from a wall you will hear an echo. This is caused by the reflected sound. If you know the speed of sound in air and you measure the time delay between the clap and the echo you can calculate the distance to the wall.

This technique is used in Sonar. This is a system for finding the depth of the sea. A boat sends out bursts of 'sound' and picks up the reflected pulse.

For this purpose it is necessary to produce a narrow directed beam of waves. But this is possible only if the source is large compared to the wavelength used. As ordinary sounds have wavelengths of a few centimetres or more the beam generator would have to be quite large. By using very high frequencies, beyond the range of human hearing, the wavelengths can be short. The generators – or transducers as they are called – can therefore be very small.

Sound with a frequency higher than the threshold of human hearing is called ultrasound.

Ultrasound scanning in medicine

Ultrasound, with a frequency of several megahertz, is used to form images of unborn babies. A very narrow beam of ultrasound is fired in short pulses through the body of the mother in direction OA. The spot on the oscilloscope moves down the screen in the same direction as the beam inside the mother. Echoes are produced as the waves pass from one type of tissue to another, eg. from muscle to bone. When echoes are received, the spot on the screen glows brightly. The beam is then sent out along other directions (OB, OC etc) and a series of bright spots produced. These eventually build up an image of the unborn baby.

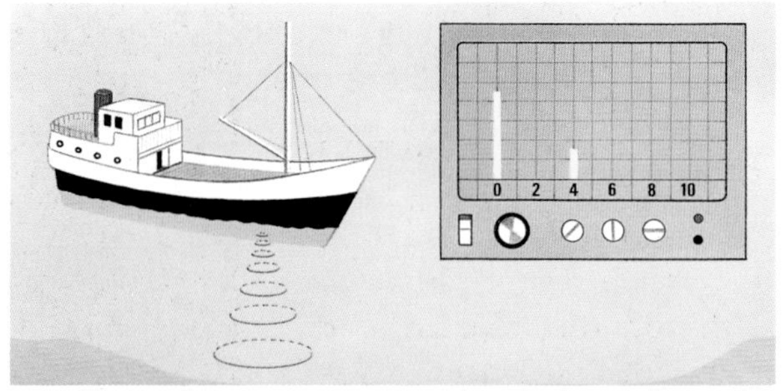

Sonar. The scale shows time in milliseconds.

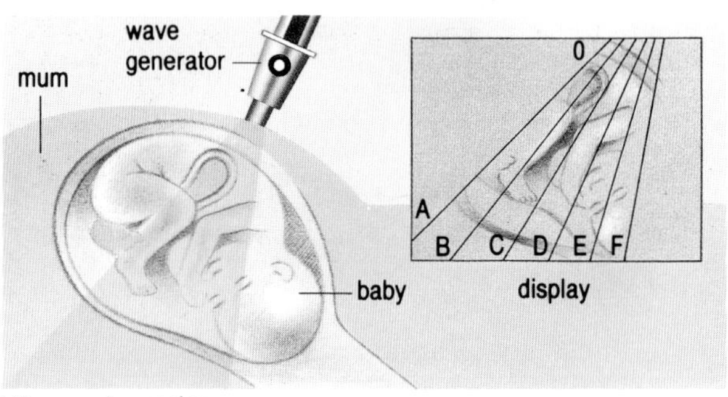

Ultrasound scanning.

1 a) From the information given in the diagram (top right) find how long it took for ultrasound to travel to the sea bed and back.
b) The speed of sound in water is 1500 m/s; calculate the depth of water below the bottom of the boat.
c) By drawing oscilloscope screens show the pattern produced by deeper and shallower water.
d) What is the wavelength of 10 MHz ultrasound in water? (Assume its speed to be 1500 m/s.)

2 a) At a frequency of 1 MHz how long would it take to transmit 100 waves?
b) If the frequency is raised to 1.1 MHz, what will the new transmission time be for 100 waves?

3 a) If a pulse is sent out every 0.001s how many pulses are sent out in 0.1s?
b) The generator moves through a small angle between sending out pulses. If the generator moves through 50° in 0.1s, what is the angle between the pulses?
c) If this swing by the generator is called a scan – how many scans are produced each second?

First snapshot for the album

This is a picture made with ultrasound. It shows the face of a healthy baby three months before its birth. By this stage the unborn baby can blink its eyes and suck its thumb. In this case a computer has been used to give a much clearer photograph from the reflected pulses. This ultrasonic body-scanning technique does no damage to the mother or the unborn child.

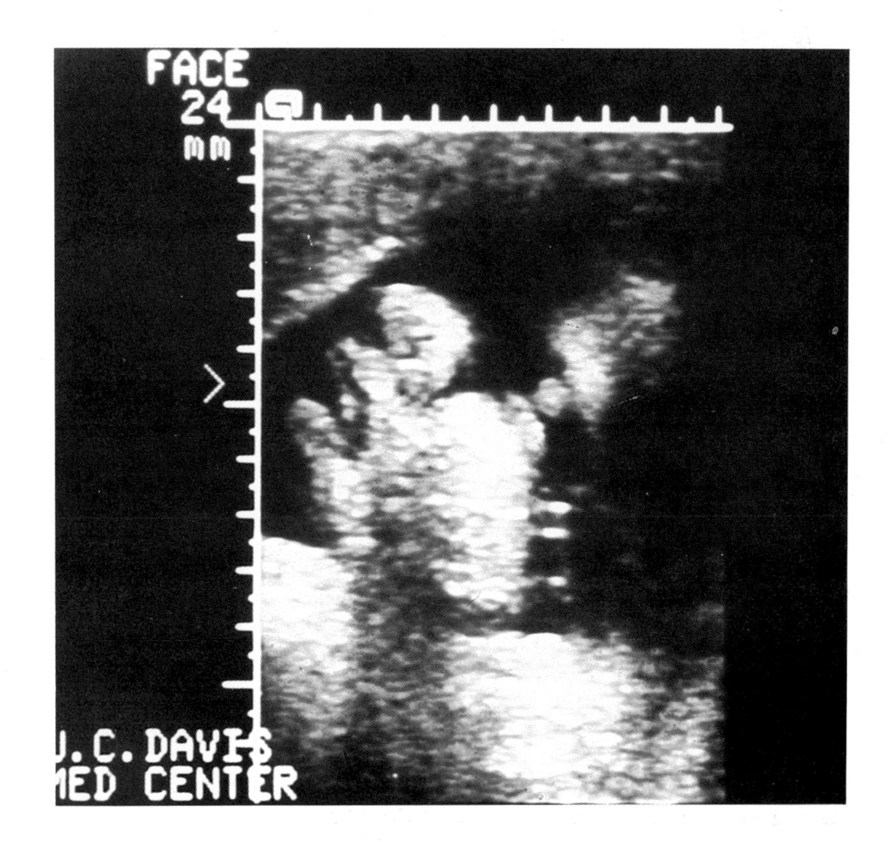

Saving a life

This photograph is a scan taken at the 28th week of pregnancy. Sometimes such a scan can prove a lifesaver. If for example the scan showed that the baby's mouth was wide open and its tongue was stretched forward there would be a real problem. This may result from a large growth in the neck which would stop the baby from swallowing fluids. If a scan showed up such a problem it would give a warning to the surgeon. At birth the baby would turn blue, unable to breathe, but the specialist would go into action and open up the breathing passage blocked by the growth. Thus a life could be saved by ultrasound.

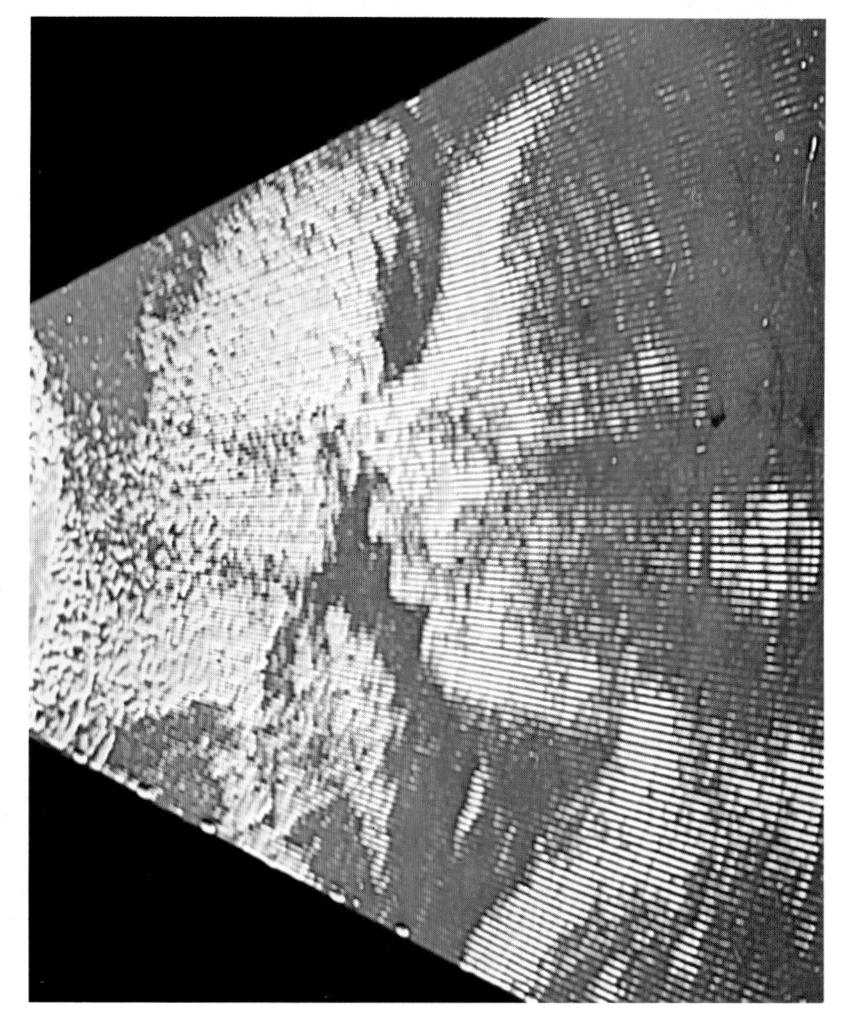

›› *Investigate the reflection and absorption of ultrasound using several different surfaces.*

A pain in the ear

Loudness

Too loud for comfort.

Some sounds can be painful. Try standing near a road drill or a jet plane taking off!

Our ears can detect very small amounts of sound power – less than a millionth of a watt. A loudspeaker producing just 10 watts of sound power would be deafening.

The decibel scale

Sound level is measured in decibels. The zero point on this scale is the smallest change in air pressure that the ear can detect. It is *not* 'no sound' any more than 0°C is absolute zero. The table on the right allows you to compare the sound levels of some familiar noises.

Source of noise	Sound level in decibels
Silence threshold	0
Whisper	20
Background noise at home	40
Normal talking	60
Vacuum cleaner at 3 m	70
Door slamming	80
Lorry passing	90
Pneumatic drill at 5 m	100
Disco, 1 m from loudspeaker	120

Sound level meters

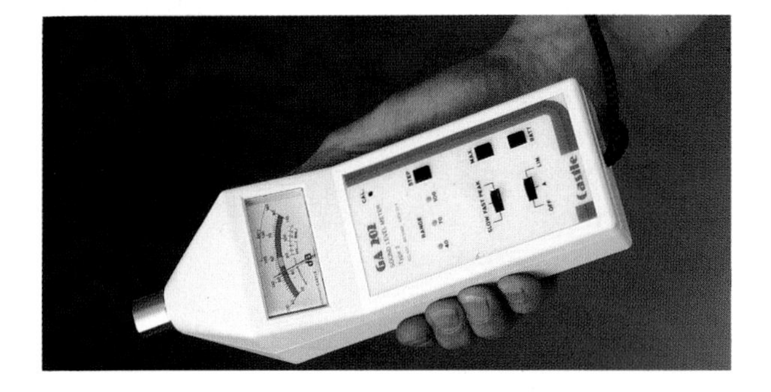

This is a kind of 'electronic ear' consisting of a microphone, amplifier and output meter. The amplifier responds to frequencies in a similar way to our ears – it is most sensitive to frequencies around 1–2 kHz.

The scale of the output meter is marked in decibels.

Noise exposure

Measurement of sound levels in a factory.

The risk of damage to hearing depends not only on loudness but also on the length of time a person is exposed to the noise.

Short term sound levels above 140 decibels may give the sensation of pain and cause temporary and sometimes permanent damage to the ear. Exposure to 90 decibel sound levels for a person's whole working life is likely to cause a permanent hearing loss of around 25%.

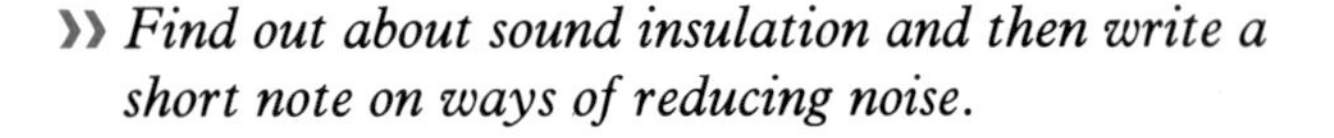

›› *Find out about sound insulation and then write a short note on ways of reducing noise.*

Hearing problems

A recent survey has shown that almost 20% of our population has a hearing loss. Total deafness is usually caused by a defect at birth, by disease or by accidental damage to the ear mechanism. Hearing loss varies from total deafness to the slight loss which often results from growing old. A young person with good hearing can detect frequencies from about 20 Hz to 20 000 Hz.

A hearing test

An instrument called an **audiometer** is used to test a person's hearing loss. It is similar to a signal generator with a pure sine wave signal feeding a pair of headphones. The signal is often called a pure tone as it contains no harmonics (ie. multiples of the stated frequency). The output control of the audiometer is however calibrated in a peculiar way. It indicates, in decibels, output sound levels which are greater than those needed for a person with normal hearing.

So, for a person with normal hearing, the output should read 0 decibels for all output frequencies. A reading of 30 dB indicates a 30 dB hearing loss.

During the test the patient listens to the signal through the headphones. The operator (audiologist) then picks a particular frequency and increases the output until the note can just be heard by the patient. The frequency and output levels are noted and the results plotted on a graph, called an **audiogram.**

The audiogram shown is for a patient with a 20 decibel hearing loss up to 500 Hz and then severe high frequency hearing loss above about 2000 Hz.

A hearing test being carried out.

The hearing aid

Hearing can be improved by using a hearing aid. Such aids consist of a small microphone, amplifier, battery and earphone.

Frequency response

There is more to helping hearing-impaired people than just making sounds louder. Hearing aids have tone (frequency) controls which alter the amplified frequencies to suit the particular need. The tone controls adjust the amplifier so that it boosts sounds over a particular frequency range.

A person with a hearing loss mainly at higher frequencies would adjust the tone control to 'cut' low frequencies and to 'boost' higher frequencies. This is like setting your hi-fi treble control to its maximum position.

›› *Use a signal generator with a pure tone (sine wave) output and a loudspeaker to find the maximum frequency you can hear.*

1 What are the advantages of being able to measure loudness?

2 Hi-fi manufacturers often claim that their equipment gives an even output from 15 Hz to 30 000 Hz. Comment on the worth of this claim for different age groups.

3 Draw a graph of sound output level against frequency for a hearing aid to help someone who has hearing loss particularly at high frequencies.

See you!

The human eye

The eye is our main channel of communication with the outside world. It lets us grasp, hit or touch things 'out there' and prevents our being grasped, hit or touched by things that threaten us.

The way the human eye works is not fully understood but some of its optical principles are similar to those of the lens camera. In the eye the cornea and lens combine to focus a picture (image) on a screen called the **retina**. The retina consists of about 100 million tiny nerve endings called rods and cones which respond to light falling on them. The rods are sensitive to small amounts of light. The cones are responsible for our colour vision and for forming sharp images.

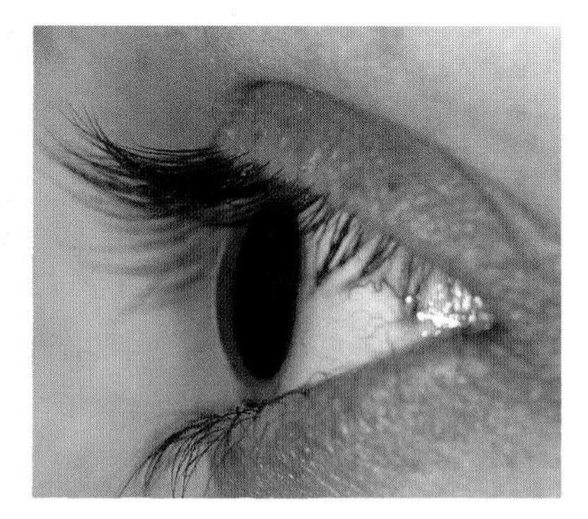

The outward curve of the cornea can be seen here.

When light passes from one material to another it changes speed and wavelength. The frequency stays the same. This is called **refraction.** This is what causes the bending of light by a lens.

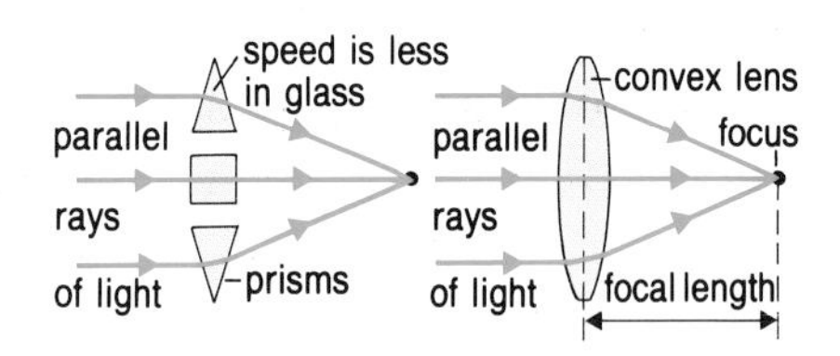

Forming an image

The front part of the eye, the **cornea,** is transparent. Behind the cornea there is a clear liquid called the aqueous humour. Together the cornea and the aqueous humour form a 'fixed focus' lens. Most of the light bending (refraction) takes place at the front surface on the cornea. However, in order to allow us to focus on things near and far, an adjustable lens is also needed.

Each 'point of light' from the object forms a 'point of light' on the retina. These points of light will form an image which is upside-down and smaller than the original object. The tiny nerve endings affected by the image send a pattern of signals to the brain. The brain then takes this pattern and forms an impression of the real size, distance and position of the object.

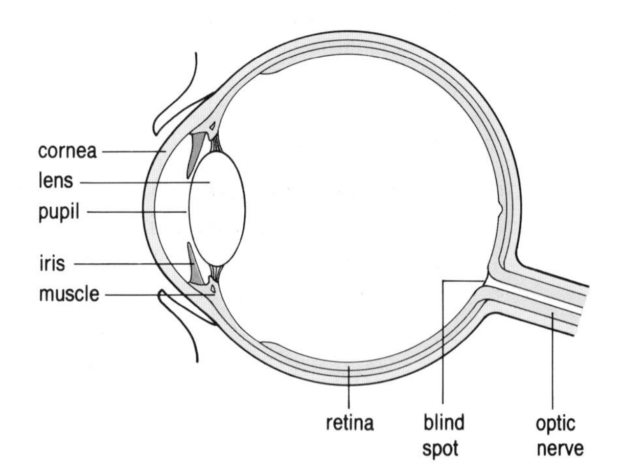

Section through a human eye.

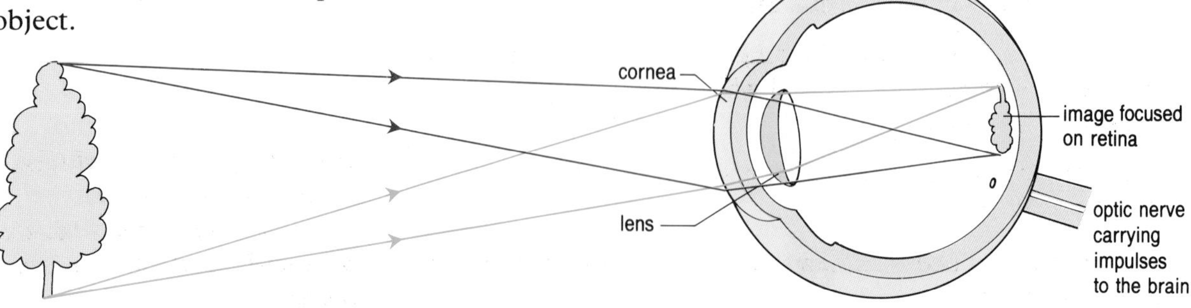

Focusing the light

The focal length of your eye lens can be changed by the ring of muscle round it. The ability to change the focal length is called **accommodation**.

When you focus on something close to you, the eye lens bulges. The distance from your eye to the nearest thing on which you can focus is called, appropriately, your **near point**.

When your eyes are resting the muscles make the lenses thin – so that your eyes are focused on far away things.

You focus, then, by changing the 'bulginess' of your eye lenses.

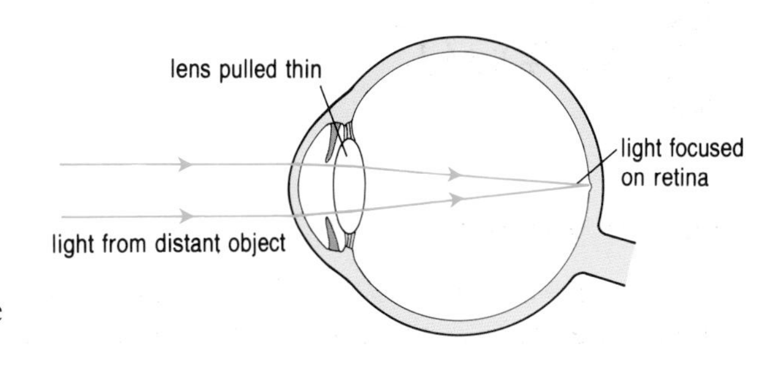

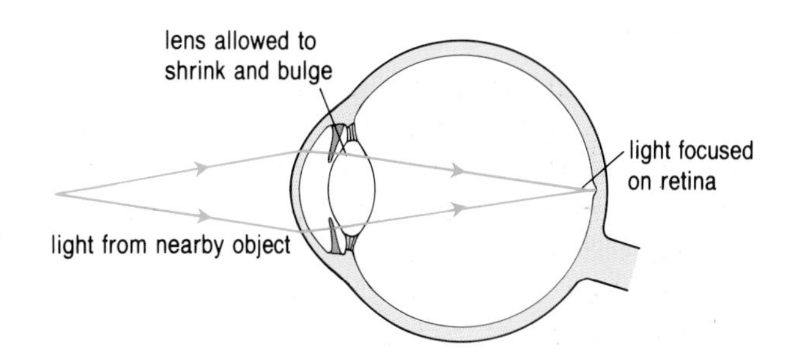

Control of brightness

The **iris** is an adjustable ring with a central hole, the **pupil**, through which light enters the eye. The size of this hole and the sensitivity of the retina change automatically with the amount of light coming into the eye. The iris adjusts almost instantaneously. However the changes which take place in the rods and cones of the retina take several minutes. If, for example, you go into a dark cinema after being in sunlight it takes some time before your eyes adjust to the different light level.

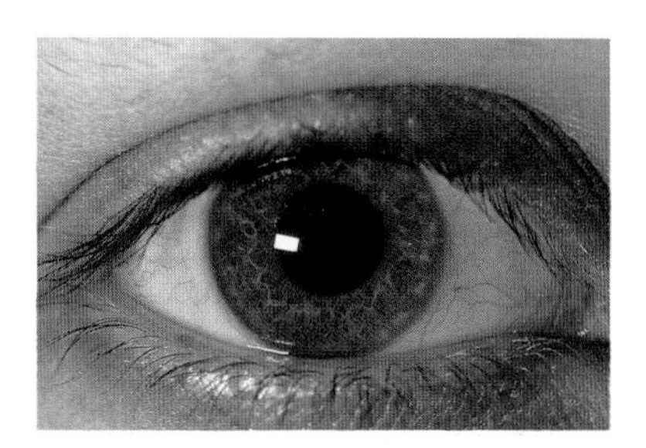

Eye in normal light. Pupil open.

Persistence of vision

As we have seen, images are formed on the retina which then transmits signals to the brain through the optic nerves. But we continue to 'see' these images for at least a $\frac{1}{20}$ of a second after the light has disappeared. This is why movie films and television pictures produce what looks like continuous movement from a series of still pictures. This effect is called **persistence of vision**.

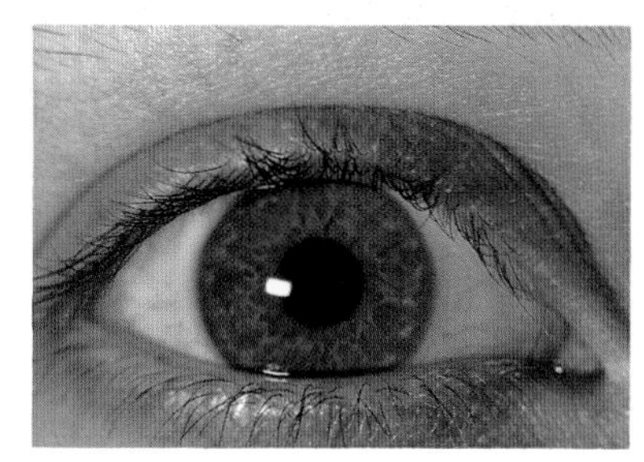

Eye in bright light. Pupil closed.

1 What is the 'screen' of the eye called?

2 Draw diagrams of the eye showing the lens shape when it is looking at something far away and when it is looking at something close to the eye.

3 Draw pictures of the iris and pupil size when in
a) bright light **b)** dull light.

4 Draw a diagram showing how a convex lens forms an image of a distant object.

›› *Find, roughly, the focal lengths of several convex lenses. Comment on how these lengths vary with the thickness of the lenses.*

Eye problems

Normal sight

You can see an object clearly when the light reflected from it is focused sharply on the retina of each eye. Someone with normal vision can focus sharply on nearby and distant objects.

normal sight

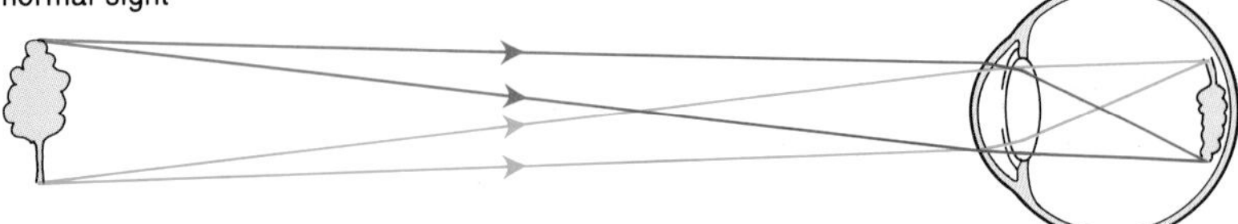

Short sight

If the length of the eye is longer than it should be, the image falling on the retina will be out of focus and blurred. This is called short sight, because it is possible for someone with this defect to focus sharply only on things which are a *short* distance away.

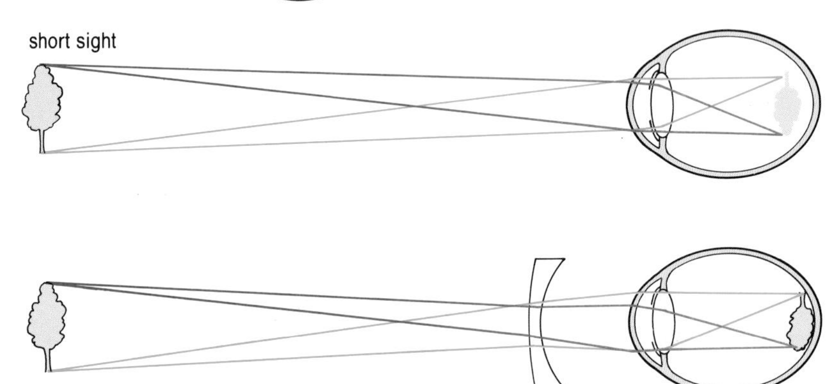

Short sight can be corrected by wearing diverging (concave) lenses.

Long sight

If the sharp focus position falls behind the retina the condition is called long sight, because people with this defect can focus sharply only on things which are a *long* distance away from them. But light coming from a nearby object, such as a book will produce a blurred image on the retina. Long sight happens when the retina is too close to the front of the eye.

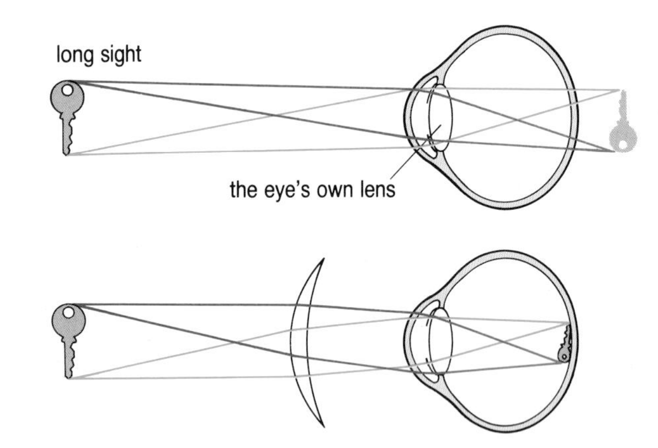

Long sight can be corrected by wearing converging (convex) lenses.

Power of a lens

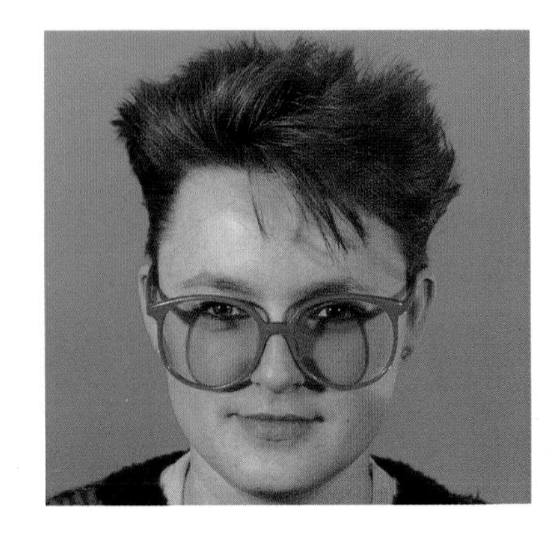

Opticians speak about the *power* of a lens. A powerful lens is a 'strong' bulging lens. The power of a convex lens is positive and the power of a concave lens negative.

Power is measured by taking the reciprocal of the focal length (expressed in metres):

$$\text{power} = \frac{1}{\text{focal length}}$$

Lazy eyes

An area known as the 'yellow spot' is situated on the retina. It has a diameter of about 1 mm with a small depression (the fovea) at its centre. When light is focused on this area a very clear picture is formed.

The eye patch covers the good eye.

A child develops a 'lazy eye' if he or she does not learn to focus on this yellow spot during the first few years of life. The eye muscles should always turn the eyeball so that the image falls on the 'yellow spot' area. This gives a picture with fine detail.

When the image seen by *one* eye is out of focus and blurred it is ignored by the brain. This leads to one lazy eye. Wearing a pair of glasses to make the pictures seen by both eyes equally clear sometimes corrects this problem.

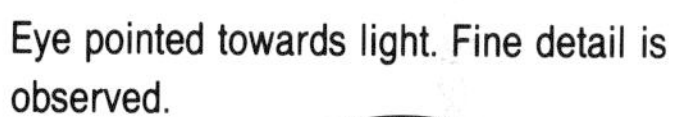

Eye pointed towards light. Fine detail is observed.

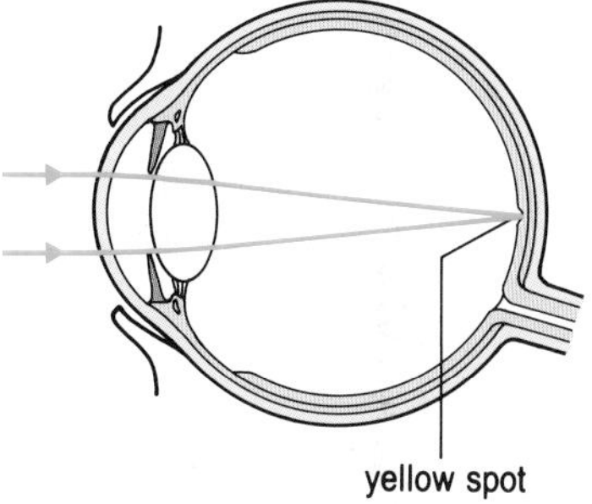

For the brain to be able to produce clear pictures, this condition must be detected early – certainly well before the child learns to read. Muscular training, involving patching the 'good' eye, is often used to correct a lazy eye.

Lazy eye not pointed towards light. Fine detail is not observed.

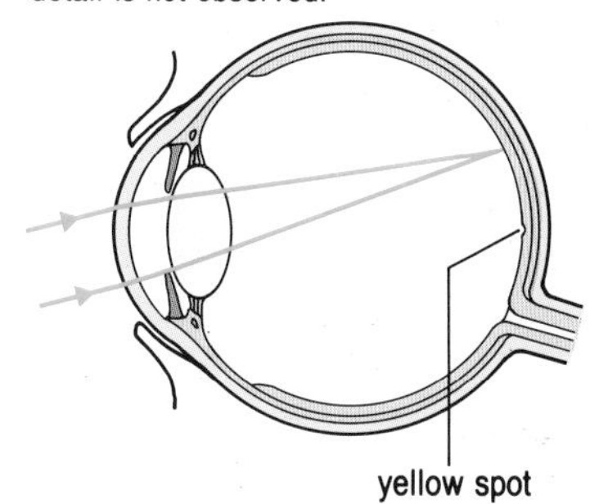

Colour blindness

Most people are able to recognise colours and very few people are completely colour blind. However it is estimated that about 10% of men have trouble in distinguishing certain colours, particularly red and green. Very few women are colour blind, but men usually inherit it through their mothers. (The mother is not colour blind herself, but her father was!)

If left undetected colour blindness can be hazardous in certain working situations.

1 Many people, as they get older, have to wear spectacles for reading. What sort of lenses will such people need? Explain your answer.

2 Find the focal length of each of the following lenses: $-1.5\,D$; $-3.5\,D$; $+1.8\,D$; $+2.7\,D$.
State which are converging and which are diverging lenses.

3 An electrician who is colour blind might make dangerous mistakes. Explain why this is so.

›› *Devise a test for colour blindness. Choose a range of colours and shades which you think may be effective in helping to decide whether someone is colour blind.*

Lenses and fibres

Contact lenses

When the refraction (bending) of light by the cornea and eye lens is not sufficient to produce a clear image on the retina, additional lenses can help.

Rather than wearing spectacles many people now prefer to use contact lenses. These are lenses placed in direct contact with the eye.

Although they have only become popular recently, the idea of contact lenses was first thought of over 150 years ago.

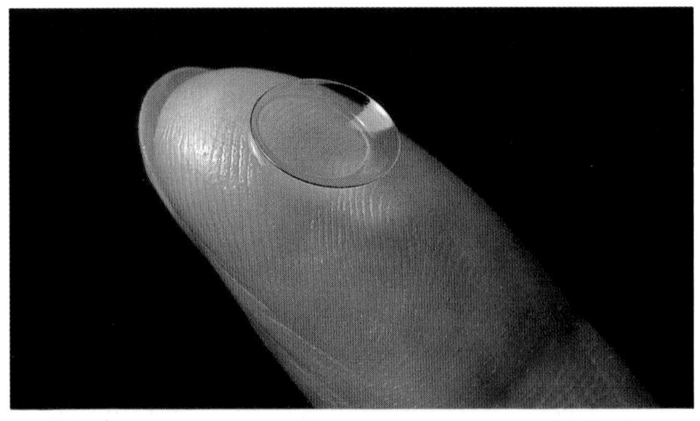

Most contact lenses are about 1 cm diameter. About a million people in Britain wear them. This is one of a number of types in use.

Why use contact lenses?

Contact lenses can give more natural vision. There are three main reasons for this:

1 There is no restriction in your field of view when looking straight ahead – no frames to get in the way.

2 When you turn your eyes, the lenses move with them. You therefore always see through the lenses and not, as with spectacles, past the edges.

3 Objects all appear a more natural size – with spectacles an object is viewed through different sections of the lens.

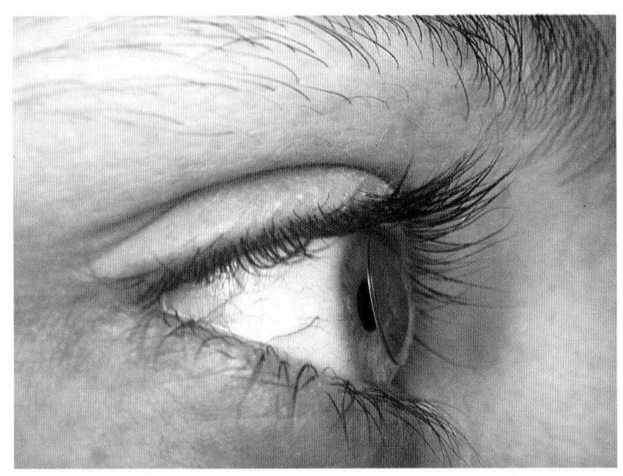

Close-up of eye with contact lens.

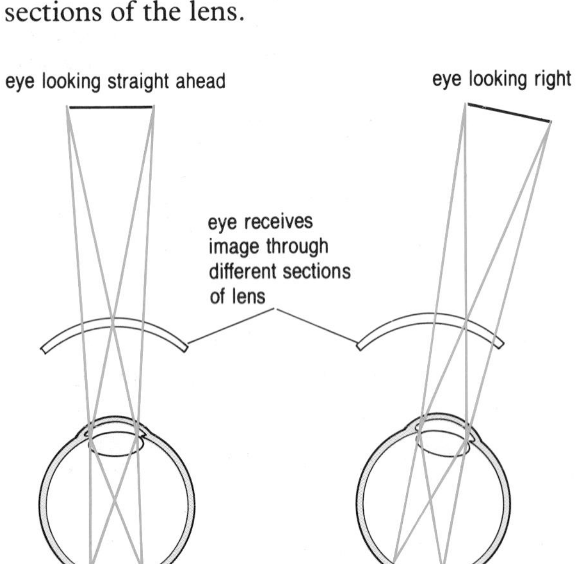

Looking through spectacles.

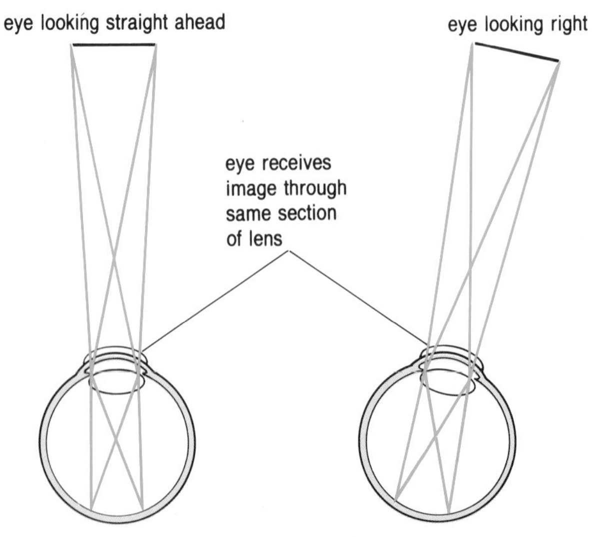

The contact lens, being placed directly on the eye, cuts down on distortion and gives the wearer a much extended field of corrected view.

Refraction and total internal reflection

A ray of light passing from glass to air or from dense glass to less dense glass is refracted at the surface as shown.

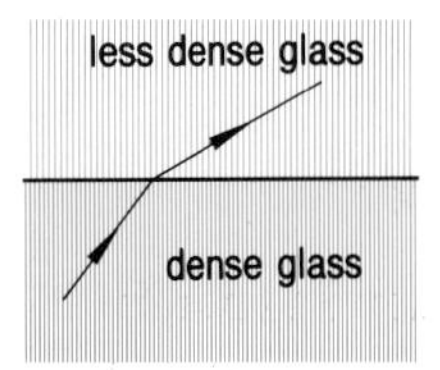

Refraction.

If the angle of incidence (i) is large enough, all the light is reflected at the boundary. This is called **total internal reflection**.

This idea is used to great effect to 'trap' light in a very thin optical fibre made up of a glass core with an outer cladding of less dense material.

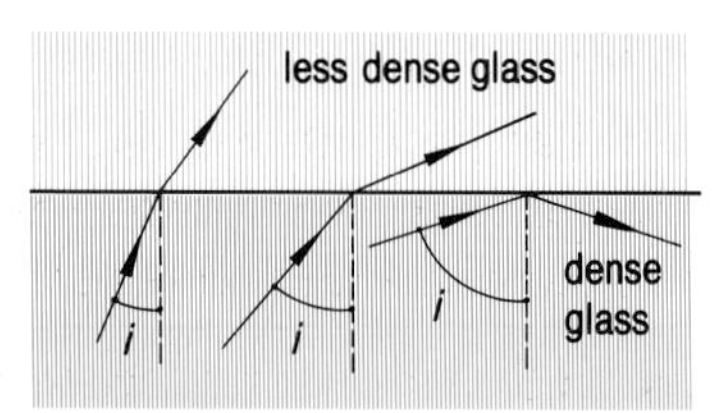

Total internal reflection.

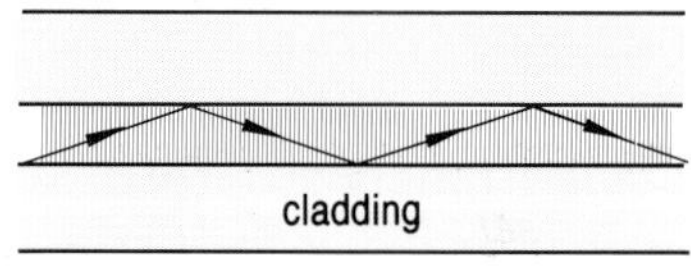

Optical fibre.

The fibrescope

In a fibrescope light can pass from one end of a bundle of optical fibres to the other even when the fibres are bent. As each fibre is coated and lets light pass through it independently of the others it is possible to 'see' through or take photographs through the system. This idea is used to enable doctors to see inside parts of the body eg. the stomach.

The instrument shown is called a **fibrescope** or **endoscope**. It has two fibre optic bundles, the light guide and the image guide, each with thousands of very thin glass fibres.

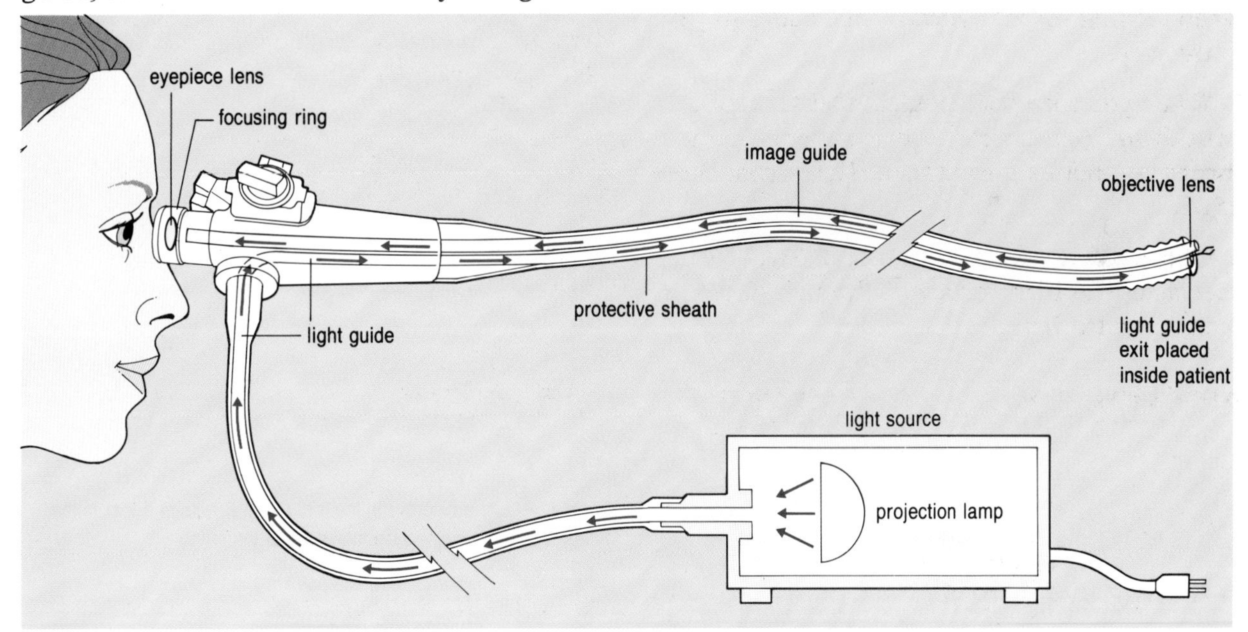

Fibrescope.

1 How does a contact lens alter the shape of the front of the eye? Draw diagrams for long and short sight correction.

2 Why is it important that the fibrescope 'end' is cold.

3 Use diagrams to explain why optical fibres need claddings of less dense materials.

4 Why do you think fibrescopes usually have a controllable bending section near the tip?

›› *Design an experiment to find the minimum angle of incidence,* ***inside*** *a block of glass or perspex, for a ray of light to be totally reflected at the boundary.*

Lasers in medicine

A super light

In medicine the laser is extremely valuable for some types of surgery.

A laser is a light source which emits a narrow beam of light of a single wavelength. However, each of the waves is 'in step' with the others near it. The beam stays narrow over long distances and can be focused to a spot of only a few millionths of a metre in diameter. Because all the energy is concentrated in such a small area the power per unit area is very large.

In medicine lasers are used to deliver energy to human tissue, and so a wavelength is chosen which is strongly absorbed. When the laser beam is directed at the tissue there is a rapid rise in temperature. The amount of damage to the tissue depends on the temperature reached and also how long it is maintained.

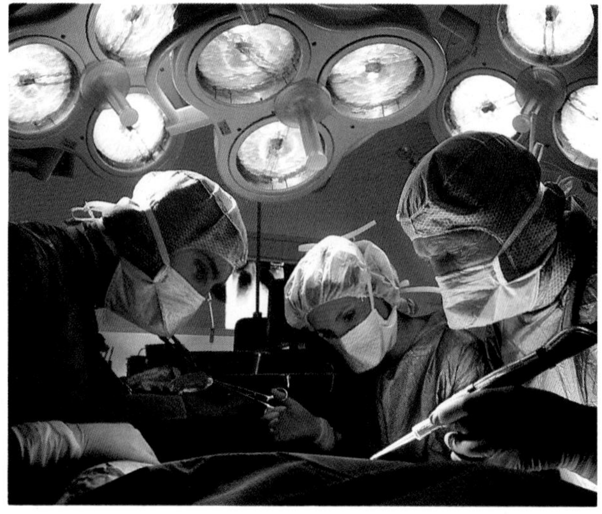

Surgeon at work with a laser.

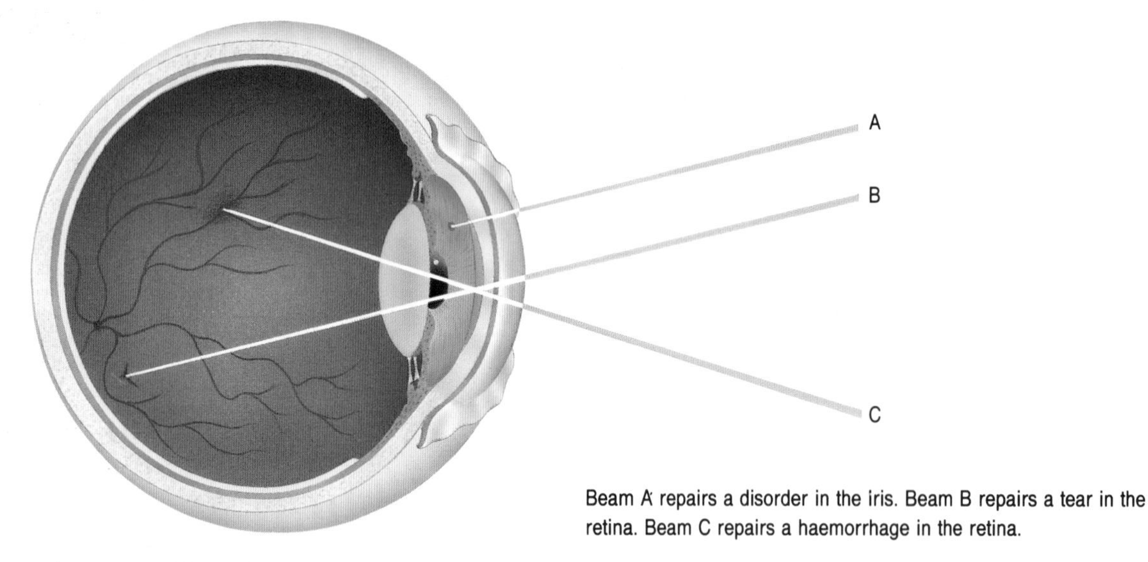

Beam A repairs a disorder in the iris. Beam B repairs a tear in the retina. Beam C repairs a haemorrhage in the retina.

Treating eye disorders

A very important use of lasers in medicine is the treatment of certain eye disorders. In one operation, called an iridectomy, excess fluid build-up in the eye is relieved by burning a hole through which the fluid can escape.

The retina of a diabetic person sometimes does not get enough oxygen from the blood vessels which therefore grow forwards and bleed into the eyes. This reduces vision and blindness can occur. The eye surgeon uses a laser to heat the blood vessels until the blood clots and blocks the vessel. This technique (photocoagulation) is also used to secure detached retinas.

Beams that heal

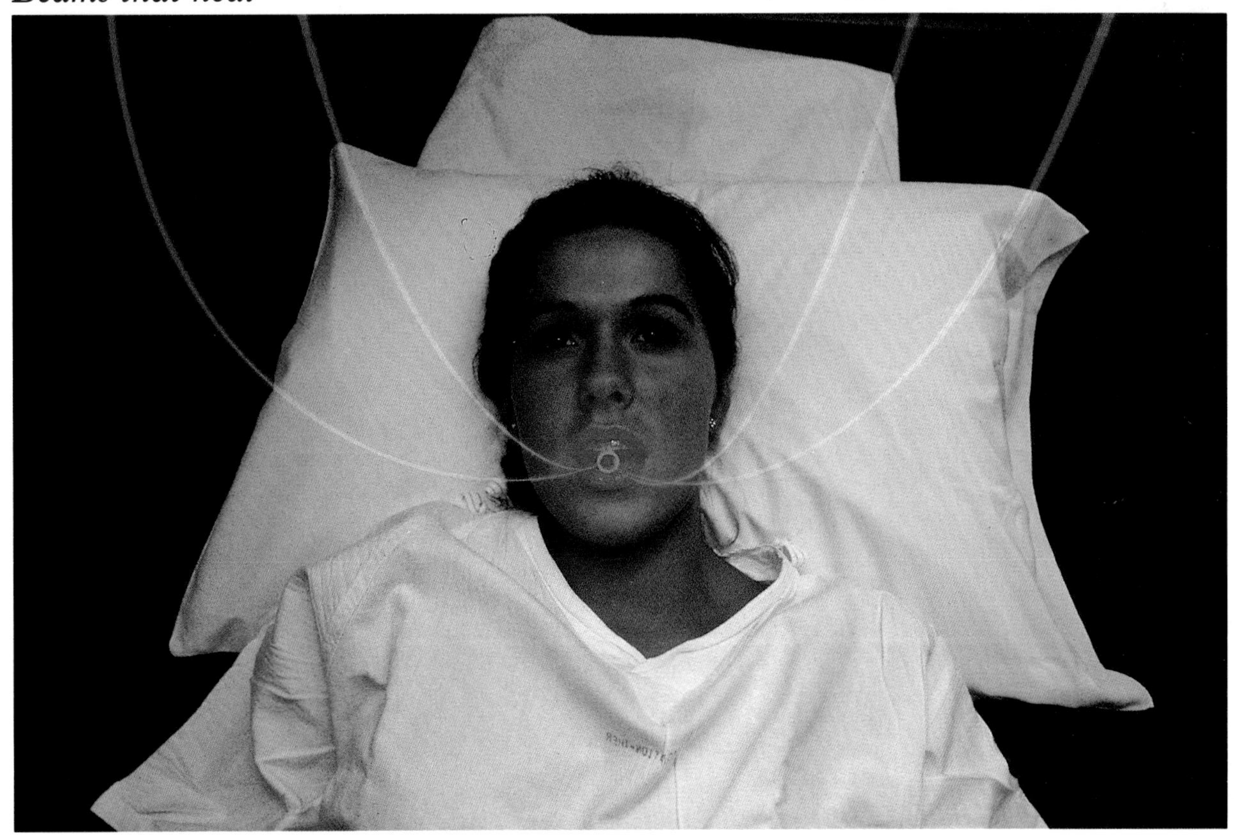

Use of a low power red argon laser, directed through four fibre-optic waveguides to treat a cancerous throat tumour. Unlike other laser devices, the treatment does not involve the use of heat energy generated by the laser but instead the laser light activates a drug, previously injected into the patient, which kills cancerous cells but does no harm to normal cells.

At present laser technology is expensive, but it is likely to be increasingly used in the future as an alternative to the scalpel.

Tumours in the brain and spinal cord are very difficult to operate on, and lasers are now being used to great effect. They allow the surgeon to operate inside the body without using a scalpel. A particularly useful kind of laser is the carbon dioxide laser. As the carbon dioxide laser beam is almost totally absorbed in the first 0.1 mm of human tissue it is used as a 'laser scalpel'. Tumours of the larynx can be vaporised without scarring the vocal cords.

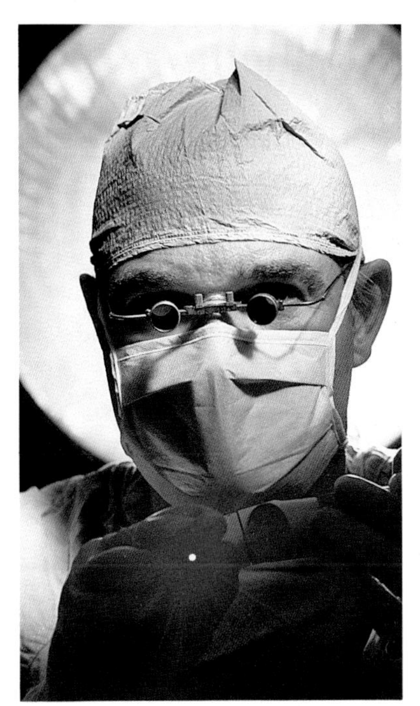

The laser surgeon wears spectacles to protect his eyes from the intense red laser light coming from the end of the operating torch.

1 Why must special protective glasses be worn by laser surgeons?

2 If the total energy of a particular laser pulse was 1 millijoule and it was delivered in 1 microsecond what is the power of the pulse?

3 Why must a surgeon pay particular attention to the time he or she has the laser beam switched on?

4 When fatty cholesterol gathers on the artery walls it can cause heart disease. How might a laser be used to solve this problem?

X-rays

I can see through you

X-rays have been used since 1895 when W.C. Röntgen discovered them by accident. He was experimenting with a high voltage cathode ray tube when he noted that a nearby screen, covered with a barium salt, started to glow. He placed a piece of wood and then a sheet of aluminium between the tube and the screen and it still glowed. He then put his hand between the cathode ray tube and a photographic plate and found that it produced a shadow photograph of the bones in his hand! Röntgen called this new radiation X-rays.

X-rays are electromagnetic waves with a very high frequency and very small wavelength.

X-rays are produced when fast moving electrons strike a metal target. A modern X-ray tube consists of a glass case containing a tungsten filament and a tungsten anode (target). When the filament is heated electrons are given off – a process called thermionic emission. By applying a high voltage between the target anode and the filament the electrons are accelerated towards the target.

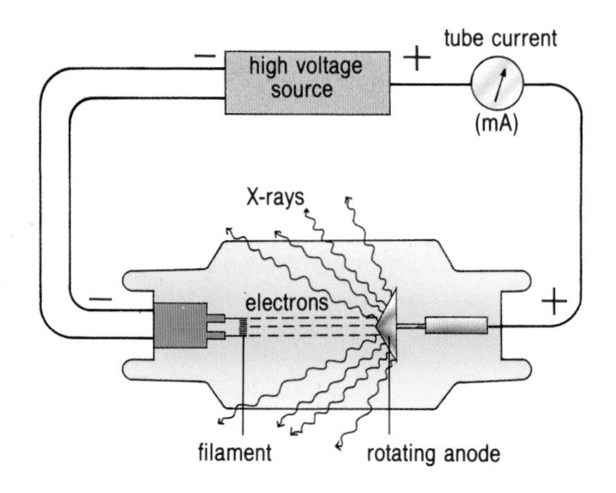

The main parts of an X-ray system.

This beam of electrons is then focused on to the target and when they hit it about 99% of their energy is lost in the form of heat. So only about 1% of the electron energy is changed into X-rays.

The X-rays which are produced at the target are given off in all directions but lead shielding is used to absorb most of them. However, a small hole is left in the shielding so that the X-rays travelling in that direction can escape from the tube.

In modern X-ray tubes the target rotates so that the electron beam hits different parts of it. This spreads the heating effect over a bigger surface area and reduces damage to the metal target.

Mobile X-ray systems, which can be moved around a hospital, still use a fixed anode target.

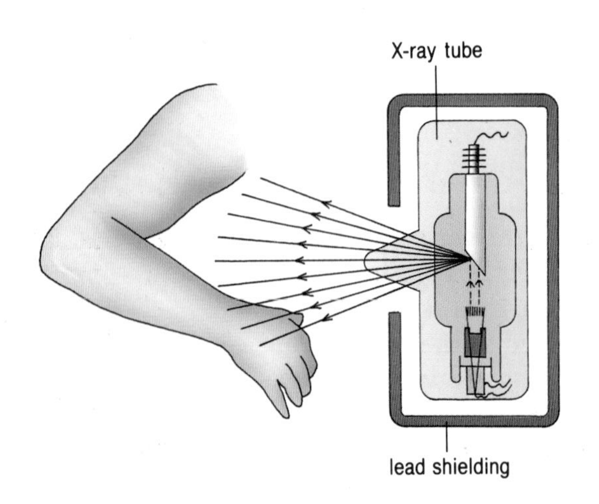

X-ray tubing showing X-rays escaping through a hole in the shielding.

X-rays in medicine

X-rays provide a powerful tool which allows doctors to see inside patients. When the rays are directed towards a patient some will be absorbed and some will pass through. The rays which pass through the patient then strike a piece of photographic film which is kept inside a light-tight box. A chemical change takes place in the film and when it is developed different degrees of blackness can be seen: the greater the radiation absorbed, the blacker the film.

The quantity of X-rays which pass through each part of the patient depends on the amount of bone and flesh in the path of the rays.

Bone absorbs more of the X-rays than the same thickness of flesh, therefore in the developed film the bone structure stands out very clearly.

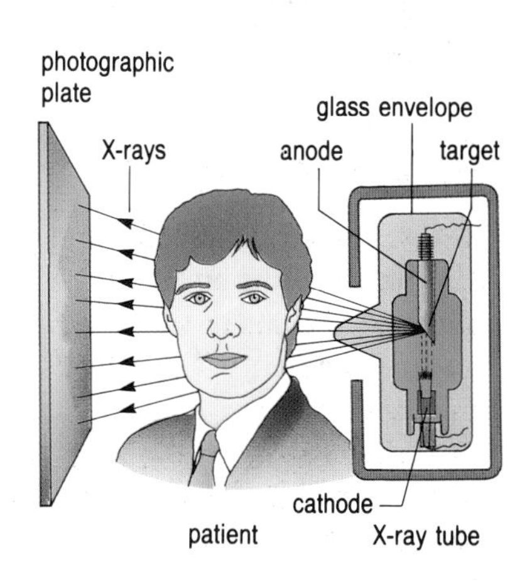

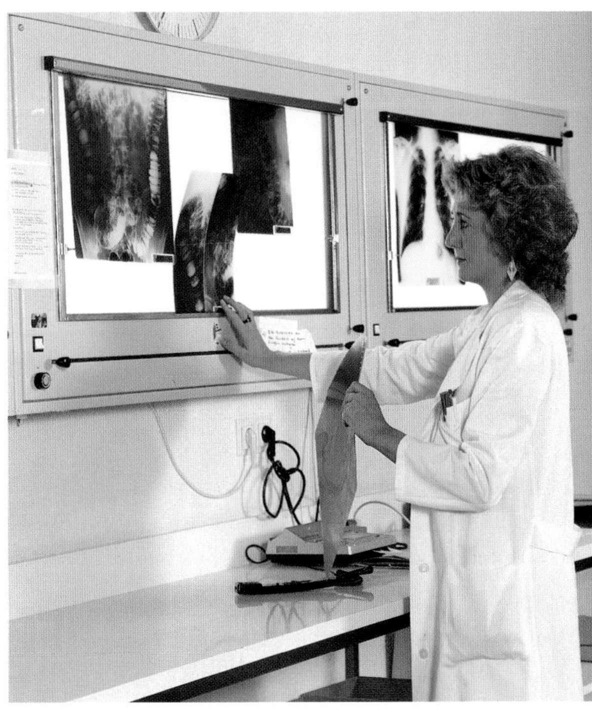

The doctor who specializes in interpreting the X-ray film is called a radiologist.

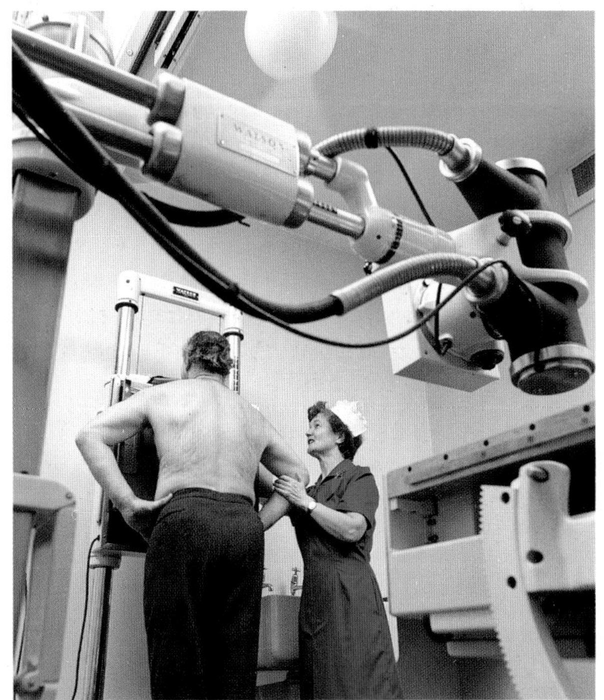

The person employed to operate X-ray equipment is called a radiographer.

More about X-rays

Continuous X-ray pictures can now be obtained by using an image intensifier screen instead of a film. These images are displayed on a TV screen and can also be recorded on video tape or stored in a computer.

1 Why is the X-ray film put in a light-tight container?

2 What advantages can you think of in using continuous X-ray pictures?

3 Explain why the melting point of the metal anode target in an X-ray tube has to be high.

4 If X-rays of wavelength 0.03 nm are used for a particular patient, calculate their frequency. ($c = 3 \times 10^8$ m/s.)

Using the electromagnetic spectrum

Ultraviolet (UV)

When carefully controlled ultraviolet radiation can be used in hospitals to treat skin conditions.

Ultraviolet light can also kill harmful bacteria, which is why hairdressers shine UV lamps on combs and brushes. Such lamps are also used in the air conditioning systems of buildings.

Energy often travels through space in the form of electromagnetic waves. This family of waves includes gamma rays, X-rays, ultraviolet, visible light and infrared. They all travel through space at 300 000 000 m/s. The different members of this family differ in frequency (number of waves per second) and in wavelength.

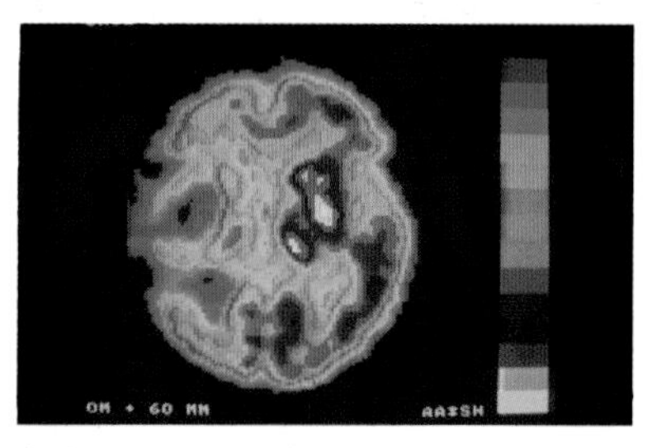

An image of the brain taken with a gamma camera.

Part of the energy given off by the sun, sunbeds and sunlamps is called ultraviolet radiation.

γ-rays

X-rays

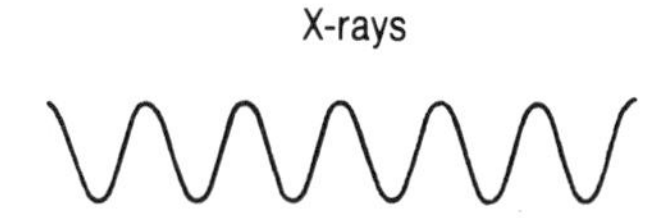

ultra-violet

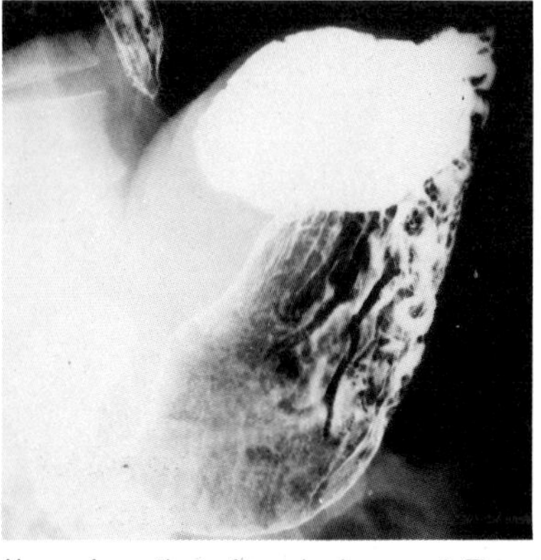

X-ray of a patient after a barium meal. The barium (shown in white), has mostly passed from the stomach into the duodenum.

Inside the CAT

In the CAT (computer axial tomography) scanner the patient lies with his body through the middle of a hoop. An X-ray source on one side of the hoop sends a beam which passes through the body. The beam is picked up by detectors on the opposite side. The source moves around the circle and information from the detectors is fed to a computer. This allows a picture to be built up of the slice of the body under investigation.

The CAT scanner has several important advantages over normal X-ray tubes. It reveals much greater detail and will show up differences between very similar types of tissue. Finally, there is no problem with the parts of the picture being hidden by bone shadows.

Infrared radiation

Our bodies radiate a range of infrared frequencies. The frequency of the radiation depends on the temperature of the part of the body concerned.

Thermography uses electronic equipment to translate this infrared radiation into visible light which can then be photographed. These photographs are called **thermograms.**

A thermogram of a hand is shown here. This technique is used in medicine to detect tumours which are seen to be warmer than healthy tissue.

Physiotherapists use infrared radiation to warm damaged muscles, and so speed up healing.

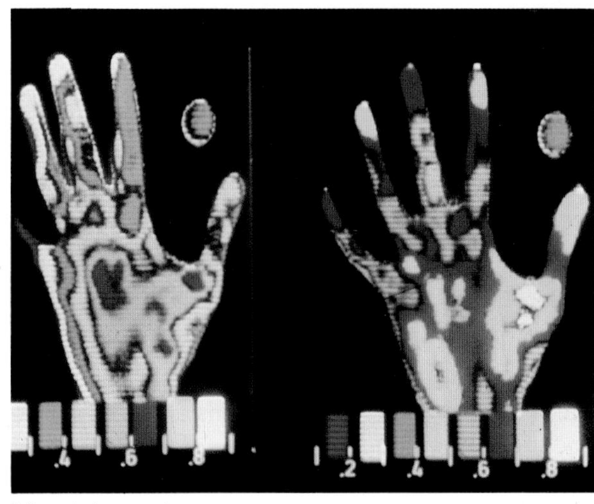

Infrared photographs.

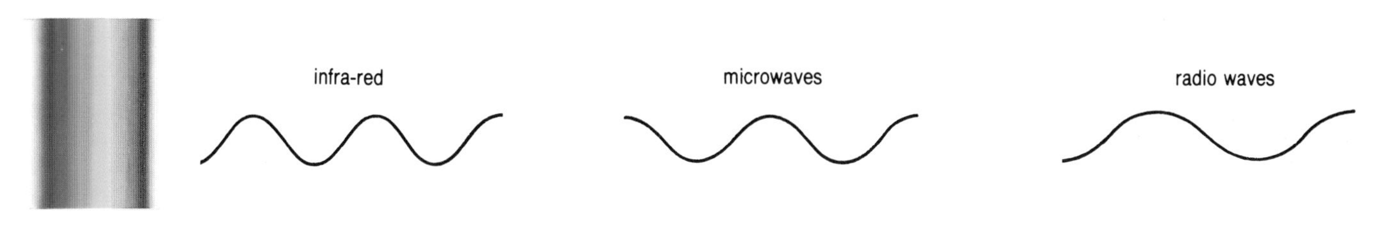

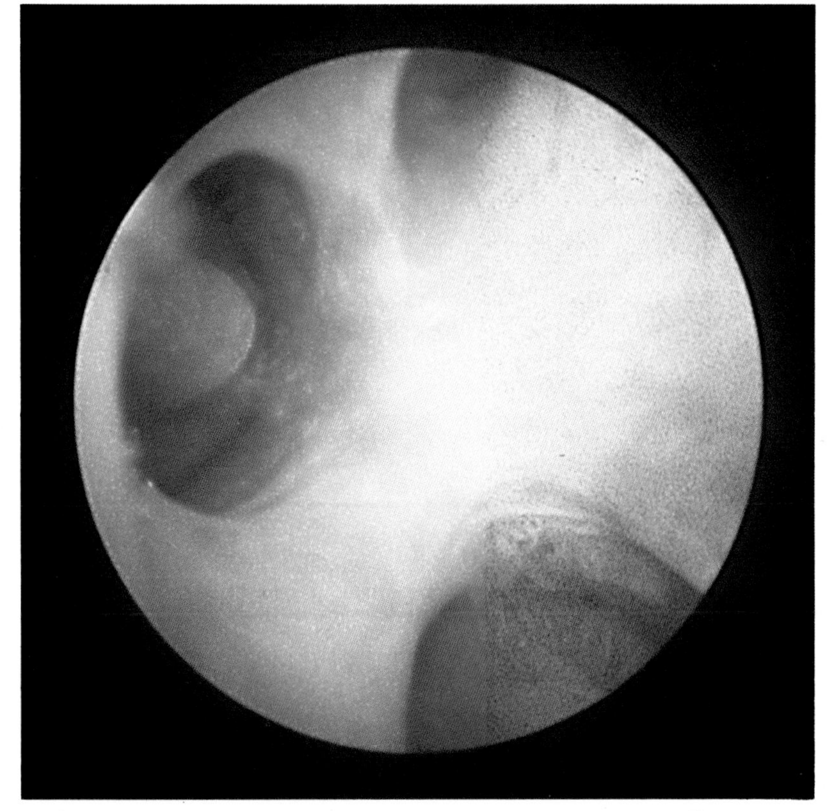

Visible light is used with a fibroscope to look into a kidney. The photograph shows the branching tubes where urine is collected.

1 State two disadvantages of a normal X-ray compared with a CAT scanner.

2 Why are thermograms helpful in finding tumours?

3 Explain why the skin cancer rate increases in people living nearer the equator.

›› *Find out how sun-tan lotions help to protect the body. Are lotions equally effective against ultraviolet and infrared radiation?*

Cell-killing radiation

Radioactivity

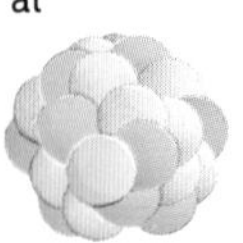

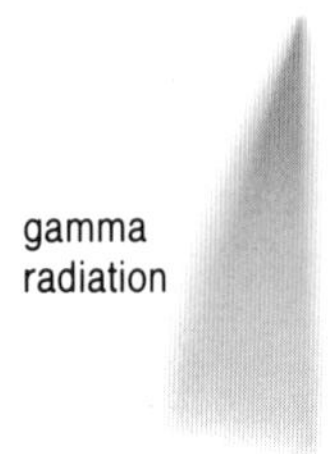

Matter is made of atoms. Atoms contain nuclei, and each nucleus contains protons and neutrons. If a nucleus has the 'right' number of protons and neutrons the atom is stable. A stable atom stays the way it is. Fortunately, most of the things with which we deal are stable.

But if the number of protons and neutrons is 'wrong' the nucleus will be unstable, and will try to change until the number of protons and neutrons is 'right', that is, it will try to become stable. The nucleus can do this in several different ways but in each case it will give off energy in the form of radiation.

Radioisotopes

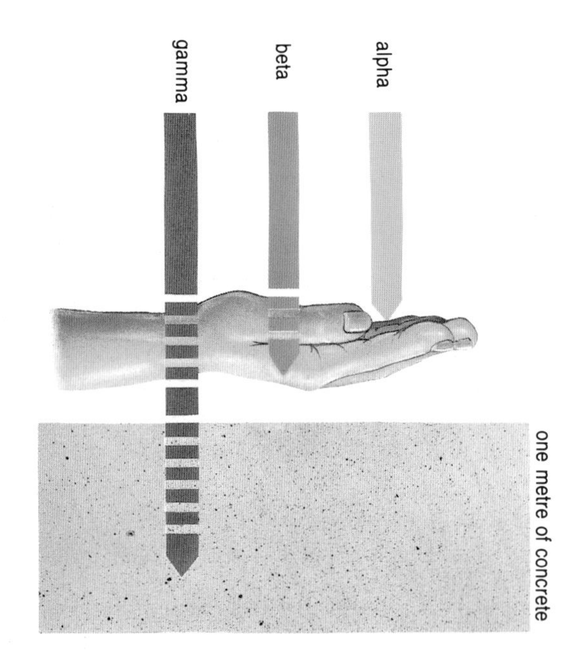

Some elements occur in different forms. These are the same chemically but their nuclei contain different numbers of neutrons. These different forms of the elements are called **isotopes** of the element. Some isotopes are unstable and are called radioisotopes.

An atom is described by the chemical symbol of the element and also a number which represents the sum of the protons and neutrons.

Some carbon atoms have more than six neutrons in their nuclei, eg. C^{13}.

The different versions are known as isotopes of carbon. Carbon has another isotope, C^{14}, which is unstable. It is a radioisotope. The C^{14} nucleus can become stable by giving off a fast moving electron – called a beta particle.

The medical uses of radioactivity

Over the past 50 years, radioisotopes have played an important part in medicine, both for diagnosis and for therapy. Many of these radioisotopes are made artifically for a specific medical application.

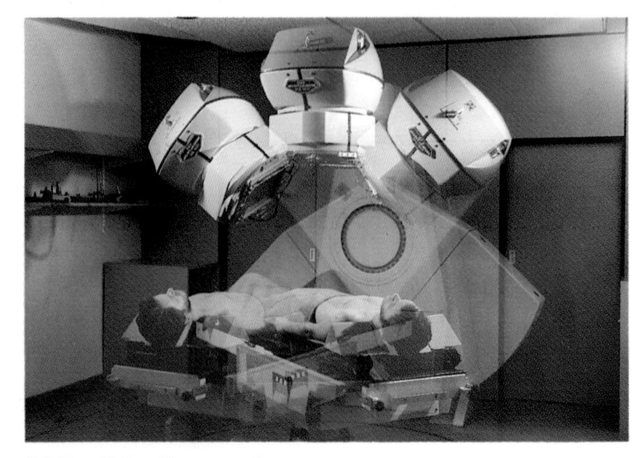

A 3-D radiation therapy unit.

Therapy

For therapy the radioisotope dose tends to be higher and it often acts like a time bomb, detonating only when it reaches its target – a tumour cell.

All living things are made of cells. These are usually only about one-thousandth of a centimetere in diameter. Each cell has an even smaller central nucleus which carries the special code (DNA) of the living organism. This determines what the organism will be like.

Radiation can strike the cell nucleus, and stop the cell from working properly. If radiation damages blood cells, then vomiting, loss of hair, and increasing sensitivity to infection can all occur. High doses of radiation can destroy body cells completely.

Treating cancer

The ability of radiation to destroy human cells is used when a patient has cancer. The amount of gamma radiation used must be just enough to destroy the cancerous cells and not the surrounding healthy tissue. To do this a cobalt-60 source is made to rotate in a circle in such a way that the beam of gamma rays it produces is always directed towards the centre. The patient is then placed so that the tumour is exactly at the centre of this circle. As the source rotates the radiation dose is very great at the centre where it is needed but not sufficiently great elsewhere to do damage.

There are two ways of using radioisotopes in therapy. The cobalt-60 source directs its gamma radiation at the cancerous cells from *outside* the body. The other common method is to attach the radioisotope to another substance which carries it to the correct part of the body – a tumour cell. The radioisotope then acts from within the body.

The three main types of radiation given off are alpha-particles (α), beta-particles (β), and gamma rays (γ).

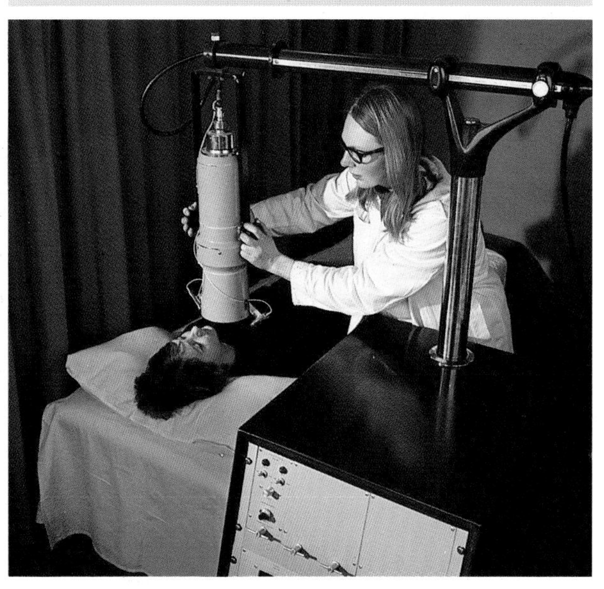

Radiation detector used to check the patient's thyroid gland.

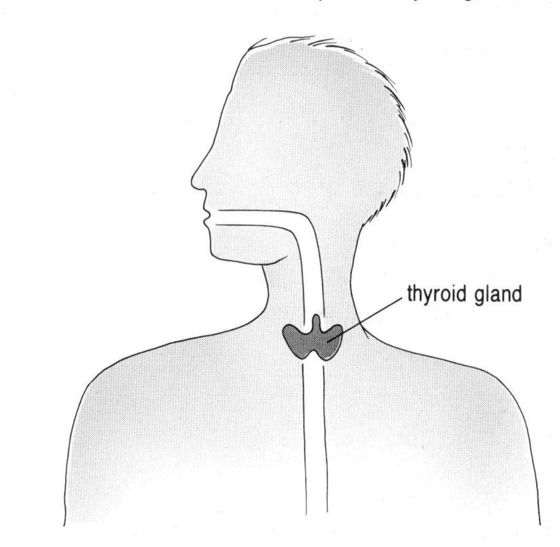

Cancer in the thyroid can be treated with iodine-131. Iodine is absorbed by the thyroid gland.

1 By considering the penetrating power of each form of radiation explain why only *gamma* radiation can be used from outside the body to treat cancerous cells deep inside the body.

2 The radioisotope Iodine-131 is a beta-emitter. It can be used to treat cancer of the thyroid. What method of therapy do you think is used? Explain your answer.

3 Alpha particles are very effective at damaging human tissue. This is due to their large mass and charge. What limits their use?

Detecting radiation

Radioisotopes in diagnosis

When used for diagnosis the radioisotope is attached to a substance which makes its way inside the body to the part being studied. The radioisotope behaves like a transmitter – its radiation passing from the inside to the outside of the body.

Ideally the radioisotope should give off only gamma radiation as alpha and beta particles would damage the body tissue.

The 'half-life' of the radioisotope is also important. This is the time it takes for the radioactivity to fall to half its value.

If the radioisotope has too short a half-life it would be difficult to detect, as the radiation given off would fall rapidly. If the half-life is too long the radioisotope would remain radioactive inside the body for a dangerously long time.

Detection devices such as the gamma camera are so good at detecting even tiny amounts of gamma radiation that only very small amounts of radioisotopes need to put inside the body.

The most popular diagnostic radioisotope is technetium-99, it gives off gamma radiation and is used in bone scanning. The radioisotope is joined to phosphate compounds which home in on the bones. In one year more than 250 000 people in Britain received diagnostic doses of technetium-99.

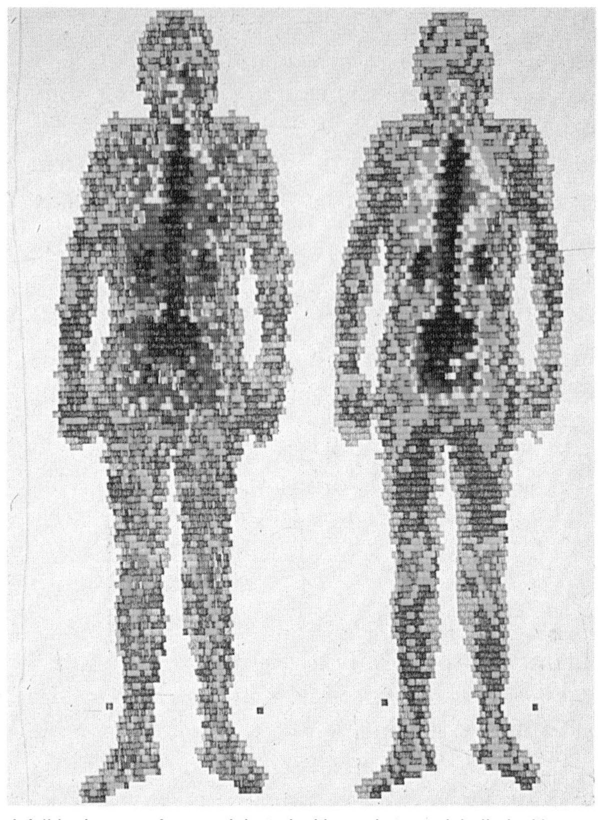

A full-body scan of a man injected with a substance labelled with a gamma-emitting isotope, technetium-99. The substance is concentrated in the bones as indicated by the regions with the highest emission of gamma rays (brown) compared with lower emission (blue) elsewhere.

Safe but strong

Iodine can be made in a number of different isotope forms. Although Iodine-131 is easy to make and is useful for treating tumours of the thyroid it gives off too much radiation – beta and gamma – for safe use in diagnosis.

The UK atomic energy authority's site at Harwell makes a safer radioisotope – Iodine-123 – which is sent weekly to hospitals thoughout Britain. Iodine-123 is a strong source of gamma rays without emitting the more damaging beta particles. It is very useful for obtaining images of the thyroid gland.

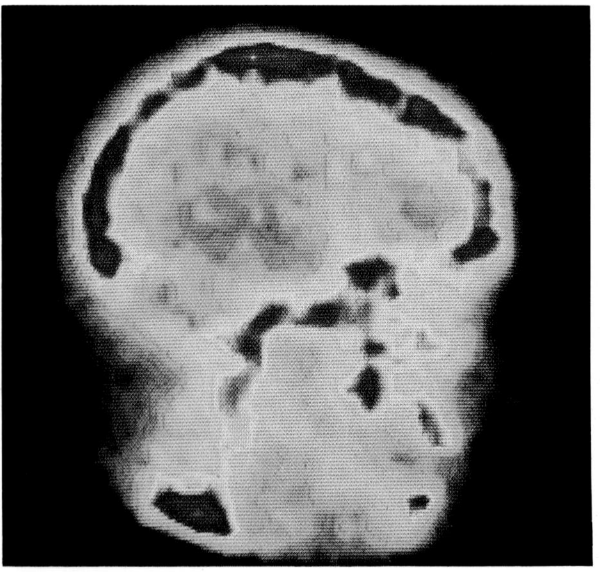

View of skull and jaw-bone of a healthy adult human. The image is obtained using radioactivity

The gamma camera

The main equipment used to detect radiation from inside the patient is called a gamma camera.

Suppose the radioactive material which has been given to the patient has been carried by the blood until it reaches the lungs. We now want to get a picture of the lungs to determine whether any parts do not have a normal blood supply.

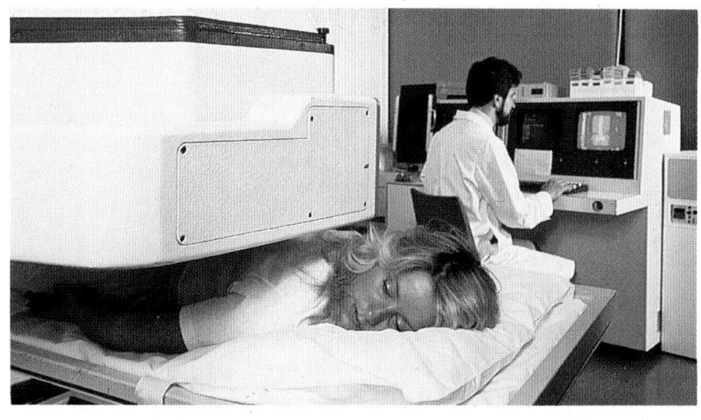

Gamma camera in use.

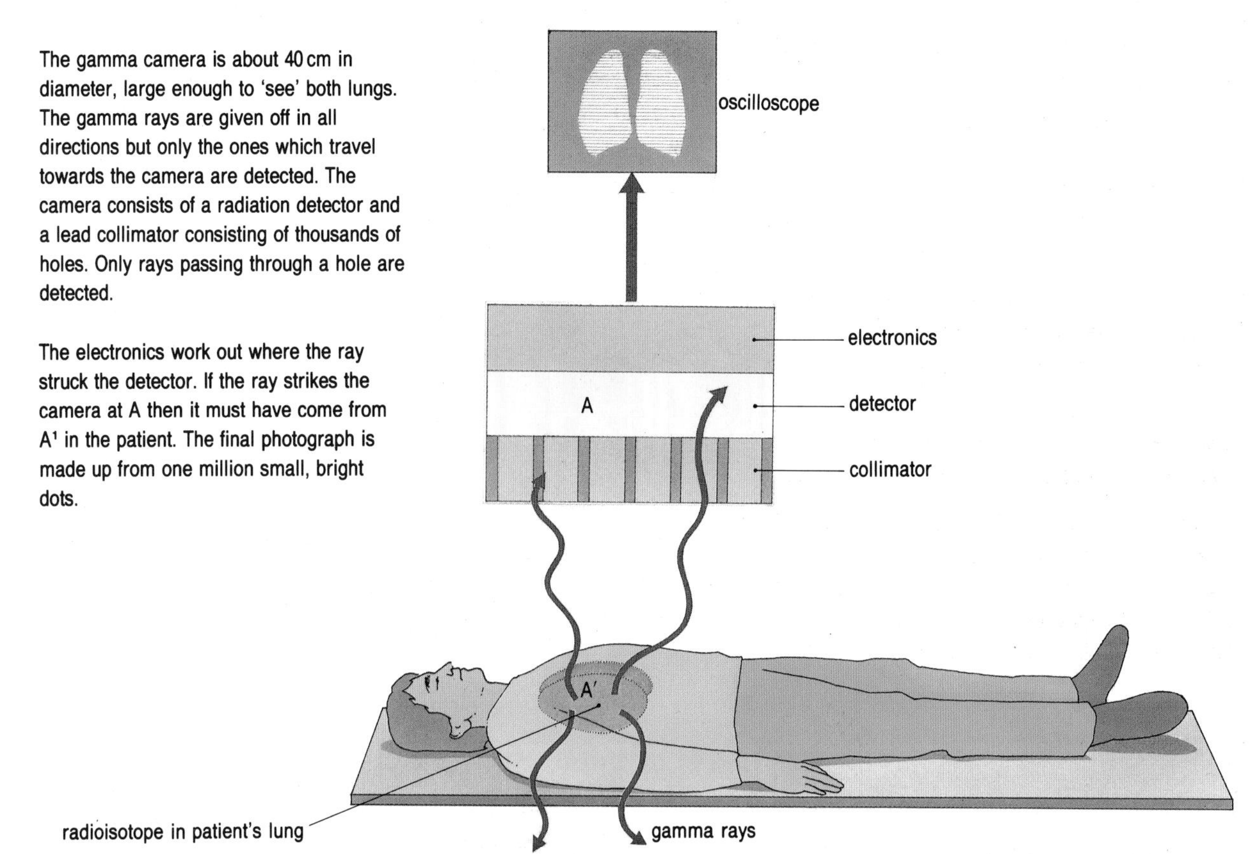

The gamma camera is about 40 cm in diameter, large enough to 'see' both lungs. The gamma rays are given off in all directions but only the ones which travel towards the camera are detected. The camera consists of a radiation detector and a lead collimator consisting of thousands of holes. Only rays passing through a hole are detected.

The electronics work out where the ray struck the detector. If the ray strikes the camera at A then it must have come from A' in the patient. The final photograph is made up from one million small, bright dots.

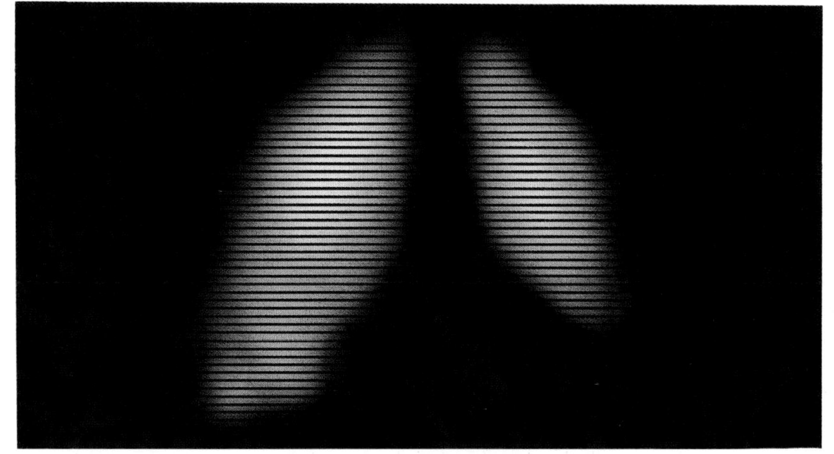

Lung scan taken with a gamma camera. The scan shows that most of the patient's lungs contain the radioisotope but there is an area which does not. This means that there is no blood flow to this part of the lung.

1 Explain why most radioisotopes used in medicine as tracers are chosen because they give off only gamma rays.

2 Why is lead used in the collimator?

3 If there is an accumulation of a radioisotope in one particular part inside the patient, describe how this would show up on the screen?

Taking care with radiation

Checking radiation exposure

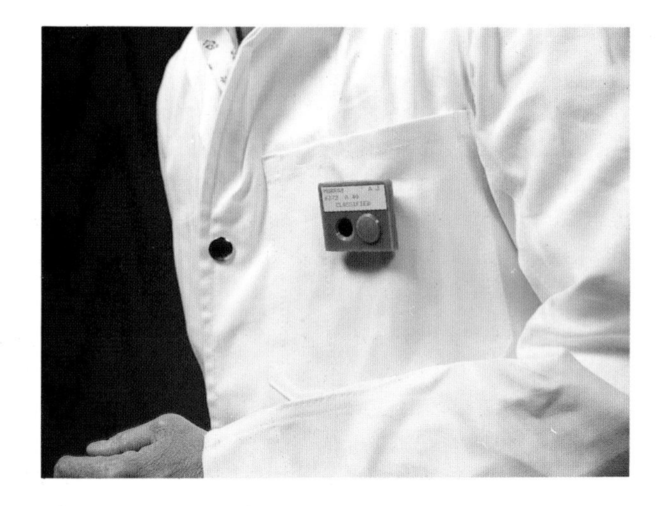

People working with radioactive materials in hospitals are themselves exposed to dangerous radiation and so must be strictly monitored. To do this hospital workers wear badges which detect the radiation. Each badge contains a piece of photographic film which is fogged when it is exposed to radiation. When the film is taken from its holder and developed the amount of blackening gives an indication of the amount of radiation received.

Protecting staff

Radioactivity is put to a wide variety of uses within hospitals and staff using radioactive materials must find ways of avoiding exposure to large doses of radiation. A worker is shown here making up radioactive materials for hospitals.

Debugging

A strong beam of ionizing radiation will kill bacteria and free medical equipment from infection. Gamma rays are used for this purpose with bandages, drugs, scalpels and hypodermic syringes.

1 Explain the safety procedures shown in the photograph.

2 Gamma radiation is used to sterilize articles *after* they have been pre-packed in plastic film.
a) Why is the gamma radiation still effective on these articles?
b) Why are the articles covered in plastic film?

3 Why could syringes not be made of plastic materials before gamma rays were used for sterilization?

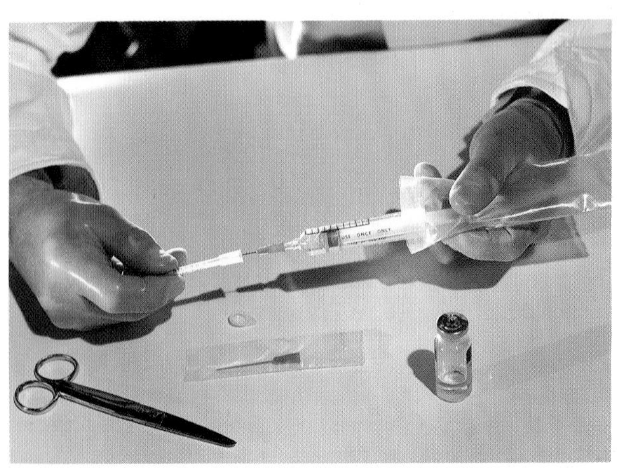

A disposable syringe sterilized by gamma radiation.

Helping the disabled

The powered wheelchair

The application of physics to the design of a wheelchair allows the disabled person to lead a fuller life.

The vehicle is driven by an electric motor which is powered by two 12 V car batteries. They can be fully charged overnight and will then, under ideal conditions, enable the wheelchair to travel up to 12 miles.

By building the wheelchair from light strong alloys the weight is kept to a minimum. The frictional forces are therefore small; the force needed to keep the vehicle moving is also small. The best use can then be made of the energy available from the battery. This is also helped by the microelectronic circuits which control the motor and ensure that it runs efficiently.

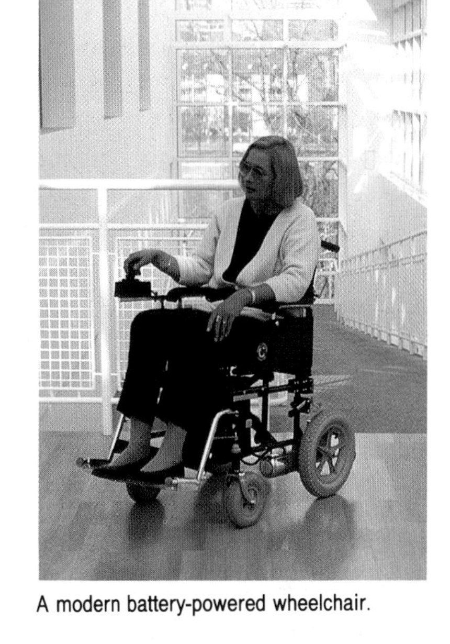

A modern battery-powered wheelchair.

One particular wheelchair has the following specifications.

Technical Data

Motor	: 24 volt permanent magnet DC
Batteries	: 2 × 12 volt lead acid
Brakes	: Dynamic, electromechanical and parking
Gradient	: Maximum 20% (1 in 5)
Range	: 10–12 miles on level ground depending on surface and weight of user
Speed	: High-low speed selection and proportional control in both modes. Maximum speed 3 mph (4.8 km/h)
Battery charger	: Fully automatic; 24 volt, 8 ampere
Battery capacity	: 21 Ah
Current at maximum speed	: 6A

Wheelchair batteries.

1 What energy changes are involved when the batteries are being charged?

2 What energy changes are involved when the wheelchair is being driven?

3 What part does the electric motor play?

4 Assuming the battery can deliver 21 Ah with 6 A discharge how far will the wheelchair travel on the level?

5 What is the power of the motor on the level?

6 How much energy (in kWh) can the battery store?

7 If the current increases to 10 A and the speed drops to 2 mph when the wheelchair is climbing up a 20% incline what is its range? What assumption are you making?

ELECTRONICS

4

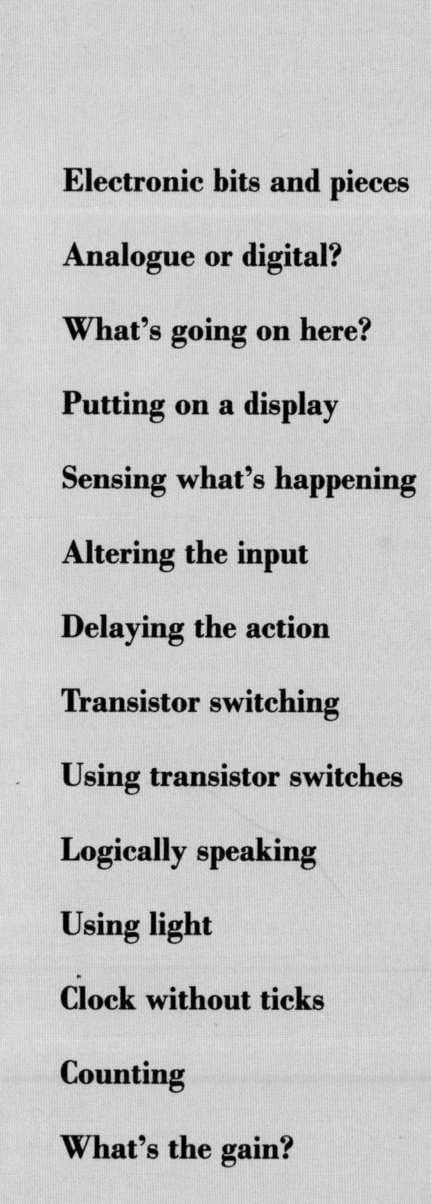

Electronic bits and pieces

Electronics today

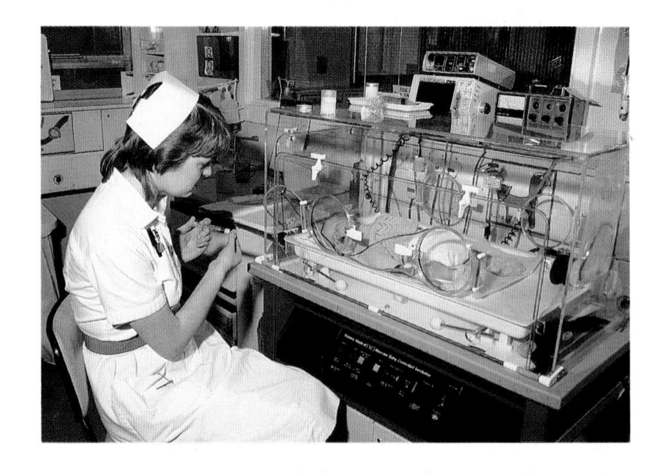

Electronics uses **electrons** to carry out useful tasks which make our lives easier and more comfortable. Electronics affects almost every aspect of our everyday existence. Televisions and telephones; radios and robots; computers and cameras; satellites and stereo; audio and video recorders; word processors and washing machines ... the modern world is full of things which use electronics. The bits and pieces which are put together to form these things are called electronic **components**.

Components

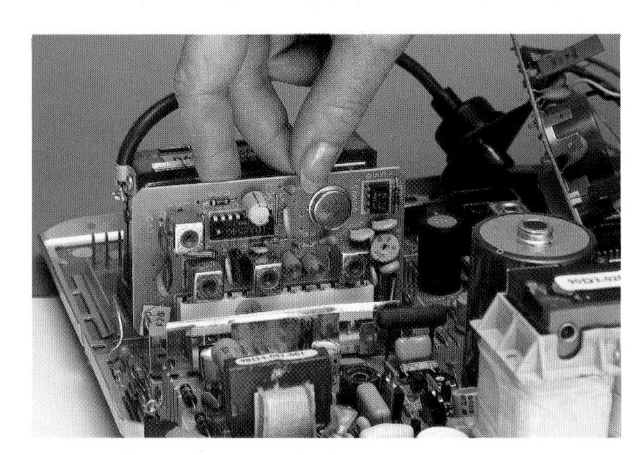

The components through which the electrons travel have been made smaller and smaller as electronics has developed. The T.V. tube is probably one of the largest components we use. The smallest components are the tiny resistors, capacitors and transistors which are inside the 'micro-chips' of modern systems.

Systems

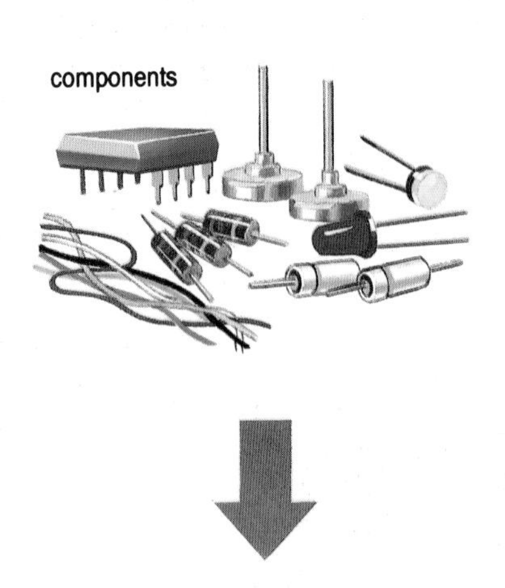

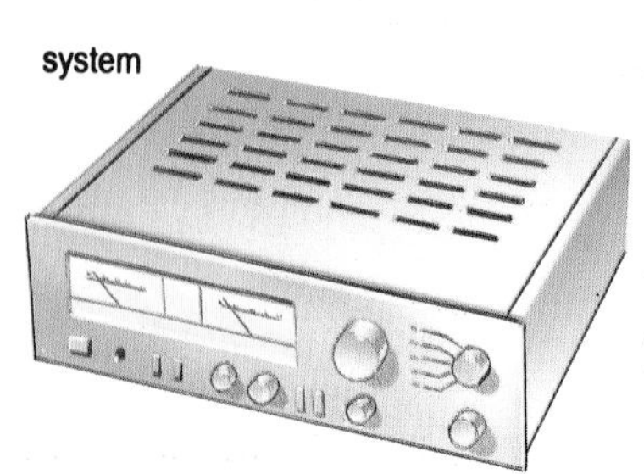

A stereo system, a central heating system, a body system and a transport system are all made up from lots of components. When the components are joined together they form a working **system**. It is much simpler to speak about the system than all the individual components. Imagine describing your central heating system as 'the boiler, chimney, water tank and pipes connected to the radiators which supply heat ...' or your stereo system as 'the box with the transistors, capacitors, resistors, inductors and bits of wire connected to the loudspeakers ...'

Inside most electronic systems are smaller sub-systems (eg. micro-chips) and each of these is made from individual components. Electronics involves the study of individual components as well as the design of systems which use them.

Electronics is a branch of physics and technology which is concerned with the movement of electrons, and with the ways in which this movement can be usefully applied.

Input–Process–Output

It is often easier to think of many quite complicated electronic systems in terms of three basic sections:

- an *input* section which starts the system working;
- a *process* section which changes the input in a way which will produce the desired output;
- an *output* which gives the result we want.

The input section of a stereo system is the pickup cartridge or tape deck. The process section is the amplifier. The output section is the pair of loudspeakers.

Systems can be shown as block diagrams, in which the arrows between the blocks show how information is passed electrically from one section to another.

The pick-up cartridge (input) senses the 'ups' and 'downs' of the record grooves and converts them into very small electrical signals. They are fed to the amplifier (process) which 'processes' them so that they become strong enough to operate a loudspeaker. They are then fed to the loudspeakers (output) where their electrical energy is converted to sound energy.

There are many ways of joining together a variety of input, process and output devices. Electronics engineers use this 'input–process –output' routine when looking for electronic solutions to problems.

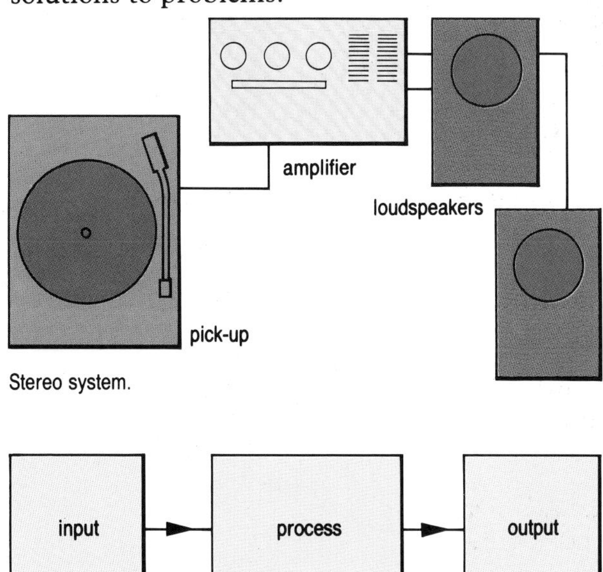

Stereo system.

Midi stereo system.

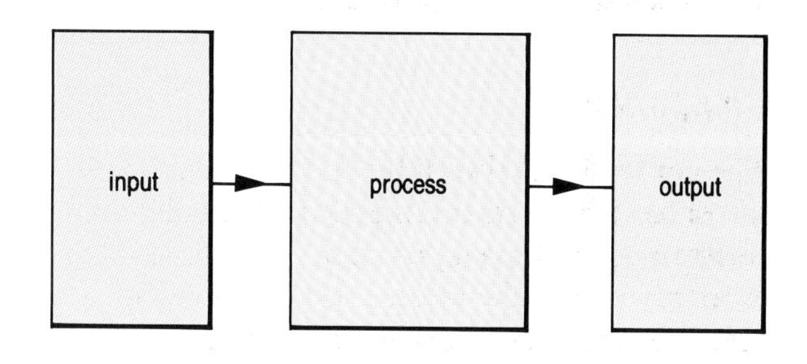

1 Name three electronic components.

2 List five electronic systems.

3 What are the input and output sections of a television set?

4 Name a system which has a microphone input and a loudspeaker output. What process links them?

5 What system uses money at the input and gives coffee at the output?

6 What input is needed for a system which uses a computer to process and gives money at the output?

7 Draw a block diagram of a calculator system.

8 Copy and complete the block diagram for the public announcement system.

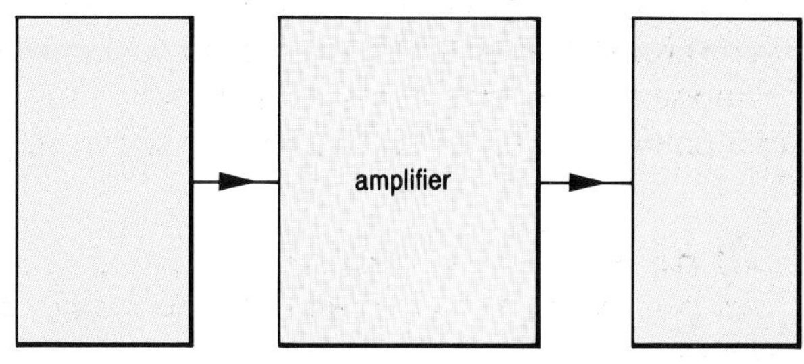

Public announcement system.

Analogue or digital?

Analogue

A wind-up watch shows the time by the position of its hands, which move continuously around the dial. The hands are always on the move, never stopping in any one position.

A mercury thermometer shows the temperature by the position of the liquid, which expands continuously up the narrow tube as the temperature increases.

Both of these systems have a *continuous range* of conditions – they are **analogue**.

Digital

A digital clock shows only certain times. If the time is shown in hours and minutes then the time displayed changes once every minute.

A digital thermometer, reading in degrees, changes its reading in one degree steps, even though the temperature may only have changed by half a degree. A digital balance with a display giving weights to the nearest 10 grams changes in steps of 10 grams. **Digital** information switches directly from one value to the next.

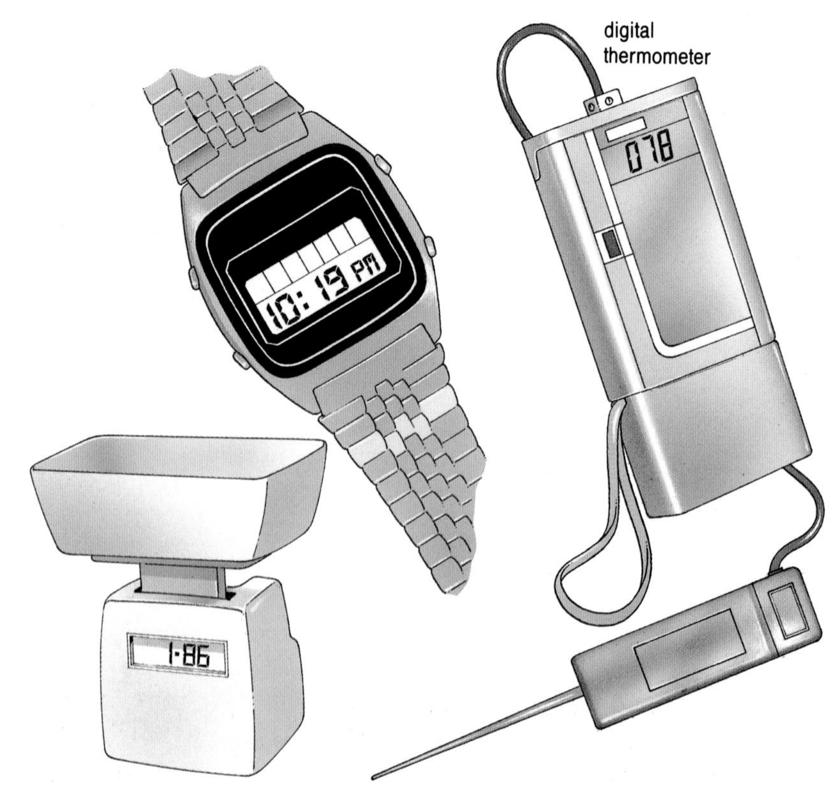

See the difference

The output signal from a telephone mouthpiece can be examined using an oscilloscope. The pattern shows that the signal has changes which are continuous – it is an analogue signal. A computer program tape, played by a cassette recorder, can be examined in the same way. The pattern on the oscilloscope screen has a series of electrical pulses each having the same amplitude. The signal goes between 'high' and 'low' and never shows an in-between value. It is a digital signal.

Most natural things change gradually, and so most physical measurements reflect the analogue properties of the surroundings. The temperature of a bowl of hot soup, for example, does not drop abruptly from one value to the next. Temperature is an analogue quantity.

The input section of an electronic system can, however, use a varying physical quantity such as sound, to produce a varying electrical signal. The process section can then convert these electrical variations into a series of individual values which then appear as a digital signal at the output.

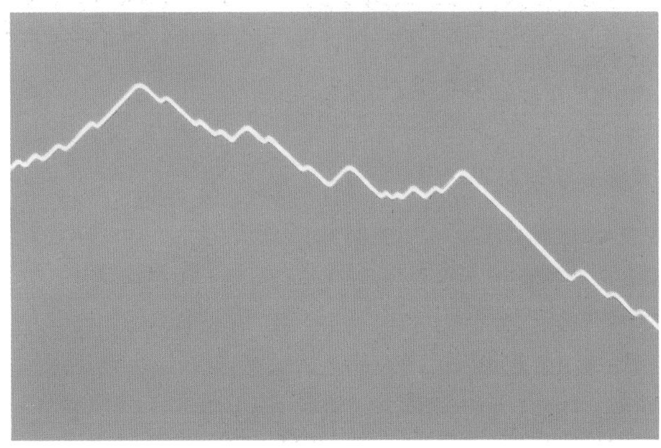

Analogue signal.

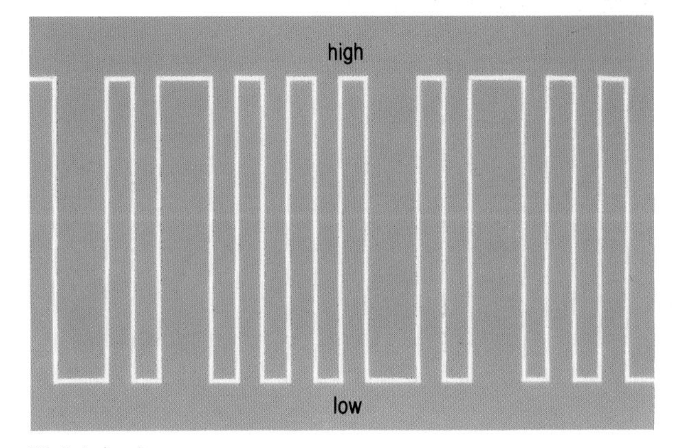

Digital signal.

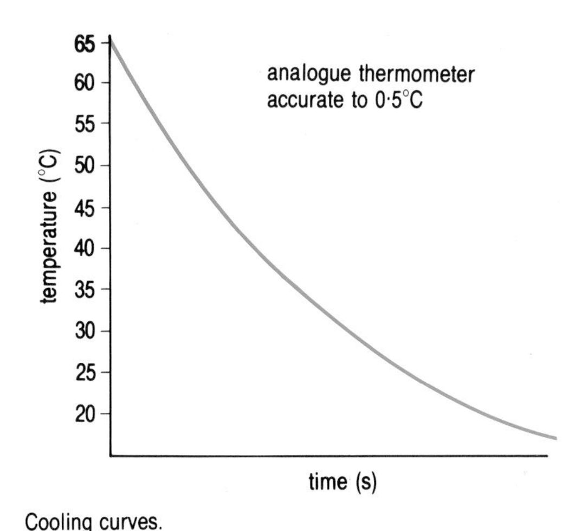

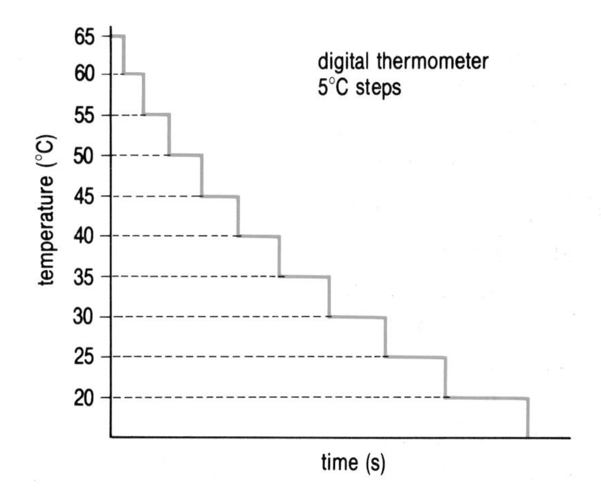

Cooling curves.

›› *Look at the electrical pattern from a computer tape program using a CRO. Compare the type of trace produced when the tape player volume control is 'low' and 'high'.*

1 Sort the following devices into 'analogue' and 'digital': radio; electronic thermometer; computer; cassette recorder; stopwatch.

2 List the advantages and disadvantages of digital and analogue systems.

What's going on here?

Output devices

We can see the effect of an electronic system by looking at its output where an energy conversion usually takes place.

The device which carries out this conversion of energy from one form to another is called a **transducer**.

In an output transducer electrical energy is converted into another more useful form.

The output of the transducer can be in an analogue or digital form. The table shows some examples.

Transducer output actions

• open or close curtains	• show a number
• make a lift move	• sound a buzzer
• dim a light	• flash a warning
• give a meter reading	• tell traffic to stop
• play a tune	• stop the bus
• turn wheels	• push an object

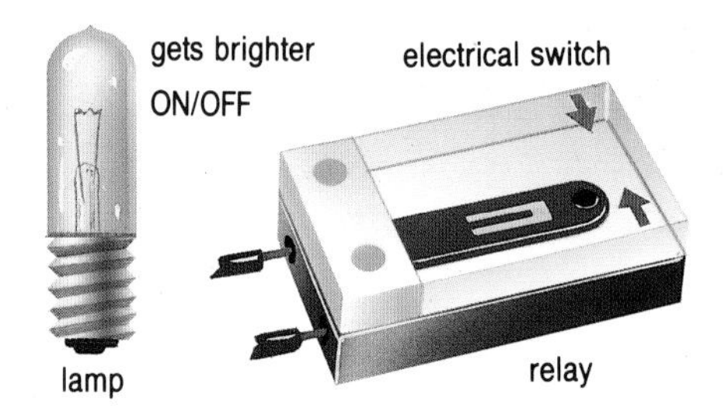

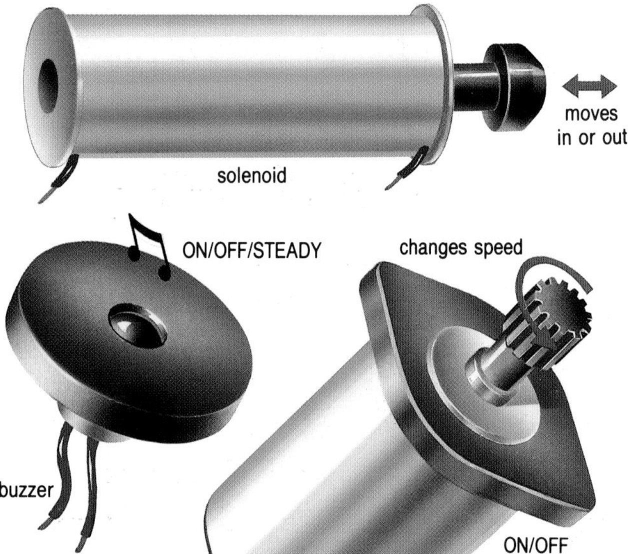

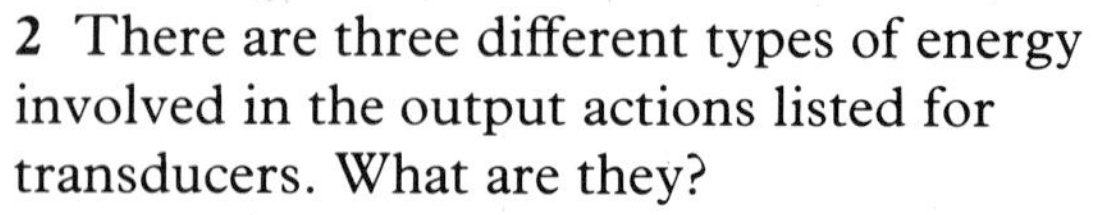

1 Which column of the table describes digital outputs? Give a reason for your answer.

2 There are three different types of energy involved in the output actions listed for transducers. What are they?

3 From the transducers shown select the most suitable for each of the output actions.

4 Which of the transducers were used as both digital and analogue output devices?

5 Name an output transducer not shown, which produces heat energy.

One light led to another

A light can be a very useful output device for drawing your attention to something or for warning you of danger. The Four Star Electronics Company decided to make an advertising badge which had four lights on it. Their design team collected some information about filament lamps, light emitting diodes and resistors. They asked a trainee technician to try out some circuits to find out the best way of arranging the components. A 6 volt d.c. power supply was to be used.

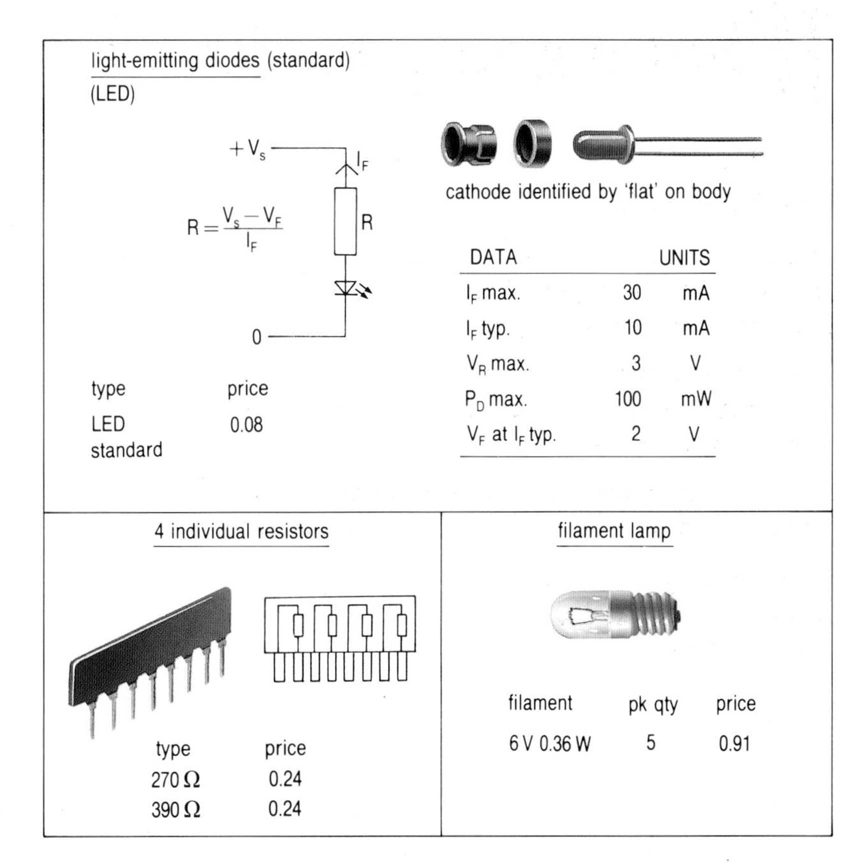

DATA		UNITS
I_F max.	30	mA
I_F typ.	10	mA
V_R max.	3	V
P_D max.	100	mW
V_F at I_F typ.	2	V

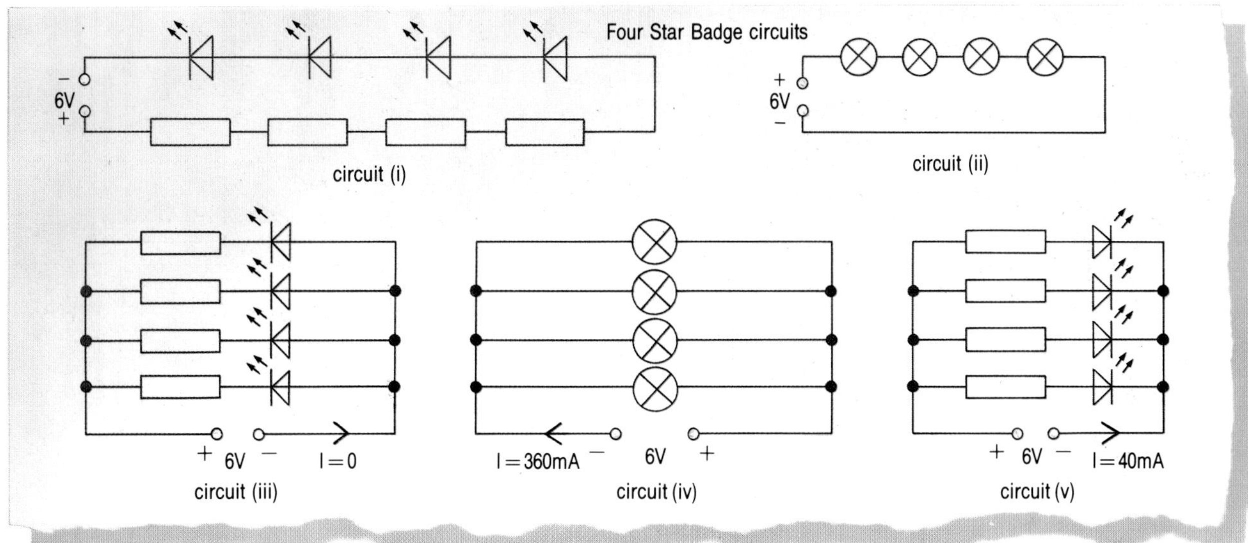

1 Draw the symbols for a filament lamp and a light emitting diode.

2 Find out the name of the material used to produce light in the light emitting diode.

3 Draw a simple circuit diagram which shows how a LED can be operated from a d.c. supply. Make sure that the positive and negative terminals are marked.

4 What two advantages do LED's have over filament lamps as light sources?

5 Which one of the technician's circuits did not work? Why not?

6 Why do you think circuits (i) and (ii) were not used?

7 Why did the technician choose circuit (v) instead of circuit (iv)?

8 Use the information given to calculate the value of resistor needed for a LED to operate with a current of 10 mA.

Putting on a display

Lamps, LCD's and LED's

Electronic measuring instruments need some way of displaying numbers. A **lamp** might seem to be the most obvious component to use but lamps use too much energy and are easily damaged.

A **liquid crystal display** (LCD) uses only a tiny amount of energy, as it does not give out any light – it only reflects it. In the dark therefore a digital wrist-watch needs a small filament lamp to illuminate the display. In addition, the LCD does not work well at low temperatures and can be damaged quite easily but it is very popular despite these problems.

For many purposes a **light emitting diode** (LED) has the best of both worlds – it can be seen easily and it has quite low power consumption.

Stopwatch with LCD.

Digital clock with LED display.

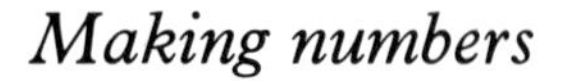

Making numbers

An arrangement of seven segments, forming a figure eight, can be used to make the numbers zero to nine. This is called a *seven segment display*. The same arrangement is used for both LCD and LED displays. In the LED display a single diode is used to illuminate each of the segments which, for convenience, are labelled 'a' to 'g'.

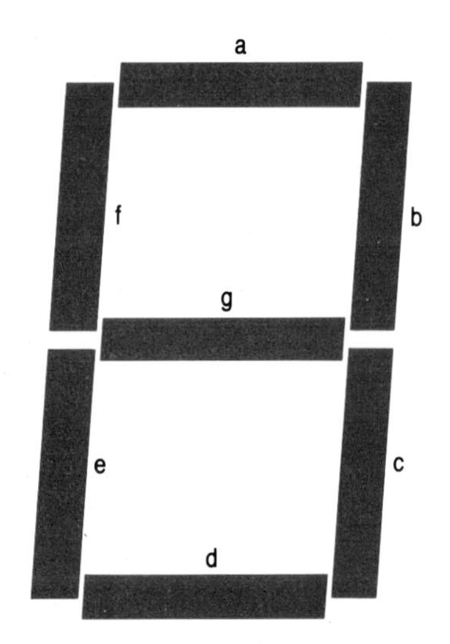

Seven segment display.

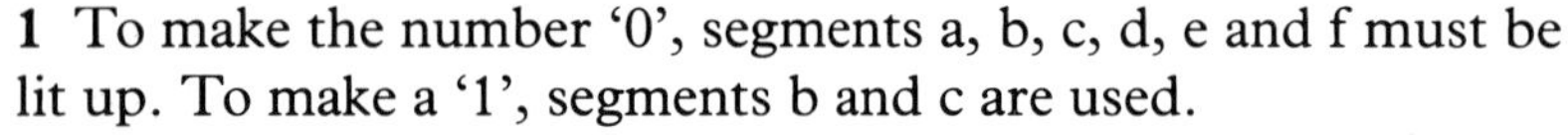

1 To make the number '0', segments a, b, c, d, e and f must be lit up. To make a '1', segments b and c are used.

Number pattern	Segment a	b	c	d	e	f	g
0	*	*	*	*	*	*	
1		*	*				
2							
3							
4							
5							
6							
7							

(* = lit up)

Copy and complete the table of number patterns and segments. Two have been done for you.

2 Find out which letters of the alphabet you can make with this set of segments. Can you make the letters of your own name (use capitals or lower case letters)?

An electronic code

The number on a seven segment display can be changed using seven separate switches, each linked through resistors to the LED's. One side of each switch is connected to the positive side of the supply, so that when the switch is closed the circuit is complete and the LED comes on. Each LED has its other end (cathode) connected to the negative side of the supply. When the switch is closed the input to the LED is said to be 'high' or at 'logic 1' and when the switch is open the input is 'low' or at 'logic 0'.

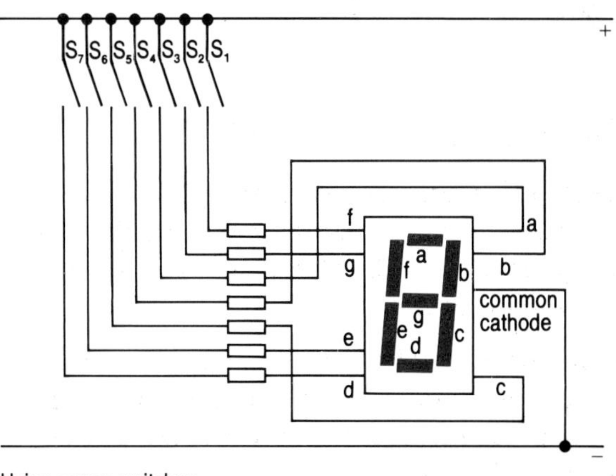

Using seven switches.

It takes quite a lot of effort to switch on each number pattern in this way. Fortunately the number of switches can be reduced from seven to four if a decoder chip is used. This chip uses a code which needs only four inputs to the seven segment display. The code is called the **Binary code**, and in it each of the decimal numbers (0, 1, 2, 3, 4, 5, 6, 7, 8, 9) is represented by a pattern of 0's and 1's. Each pattern has four binary digits or '**bits**'.

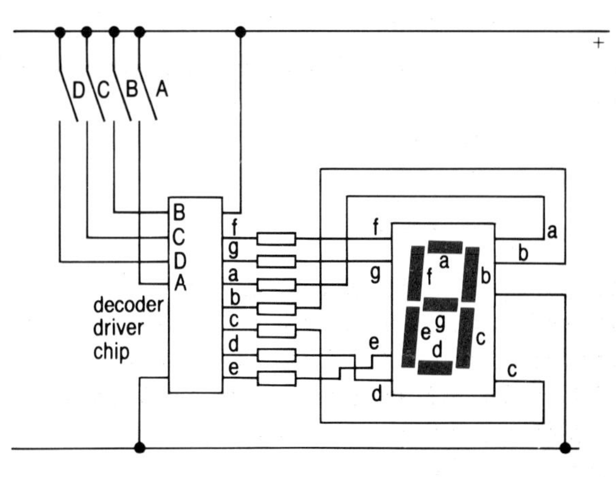

Switches with decoder driver chip.

An electronic decoder

To operate the decoder four switches A, B, C and D are used. When a switch is connected to the 'high' side of the supply, it produces a '1'; and when it is open (input 'low') it produces a '0'.

The 4 bit number formed by the four switches is then decoded by the decoder chip and the seven segment display shows the decimal number. For a number greater than 9, two or more 4-bit binary numbers are needed. The number 38, for example, is represented by 0011 1000. This needs two 7-segment displays to show it.

Decimal number	"bit value" 8	4	2	1	Check sum
0	0	0	0	0	= 0
1	0	0	0	1	= 1
2	0	0	1	0	= 2
3	0	0	1	1	= 2 + 1
4	0	1	0	0	= 4

When each digit of a decimal number is coded in 4-bit binary numbers like this, the coding is referred to as Binary Coded Decimal (BCD). This binary system allows easy conversion back to the decimal form for display purposes.

3 What is the binary code for decimal six?

4 What does BCD input 0111 display?

5 What is this 2-digit decimal number 0101 0111?

Sensing what's happening

Input transducers

An input transducer is a device which detects a tiny physical quantity such as sound or light and converts it into something which makes it easier to measure or control. Normally it is electricity which is produced. Some devices generate electrical energy directly from the input energy – no additional power is needed. As electronic circuits need only minute voltages and currents to operate them the devices described on this page are particularly suitable for use as input transducers.

Name of Device	Output voltage
Thermocouple	3 mV (0·003 volts)
Solar cell	500 mV (0·5 volts)
Microphone	5 mV (0·005 volts)

The **thermocouple** consists of a pair of wires of different material welded together to form a junction at the end. It converts heat to electrical energy, and it can operate from −200°C to 400°C. Thermocouples can be very small and sensitive.

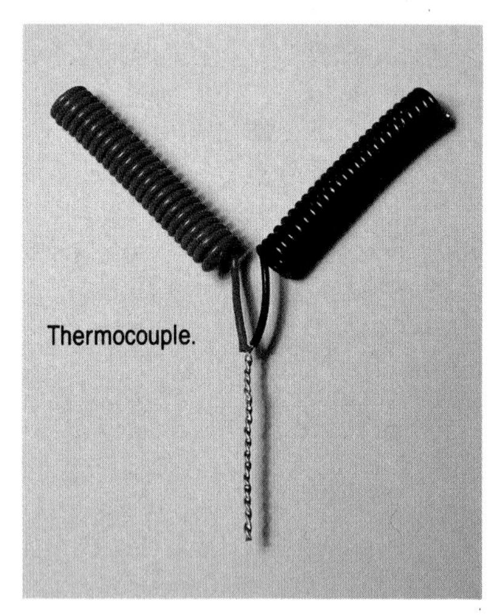
Thermocouple.

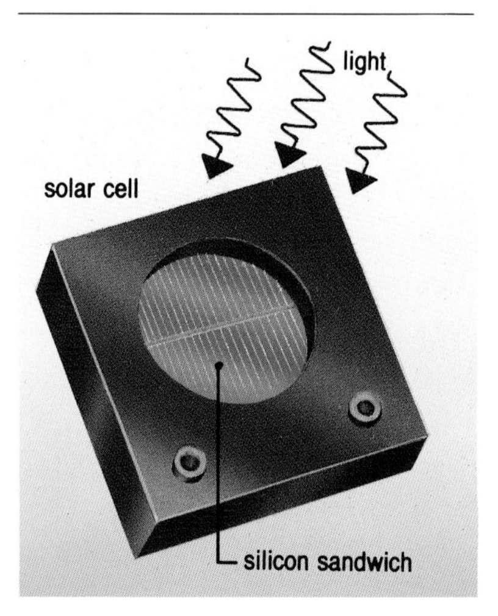

The **solar cell** is a device made from a very thin silicon sandwich which converts about 15% of the light energy it receives into electrical energy. It is very sensitive and its output depends on the brightness and colour of the light it receives.

The **microphone** contains a magnet and a coil. When you speak into the microphone, the coil moves and generates a small electric current which depends on the loudness and the frequency of the sound.

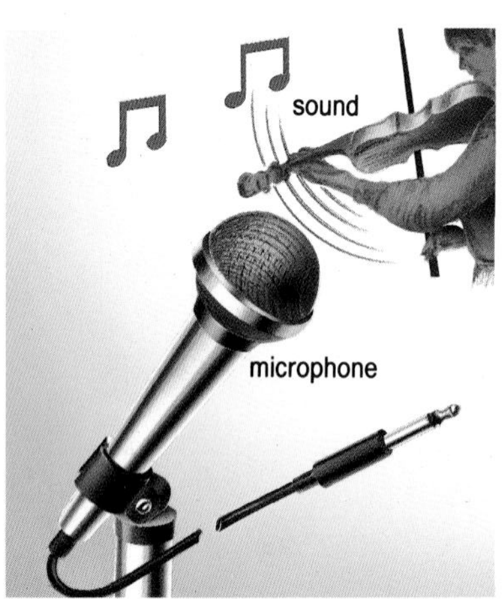

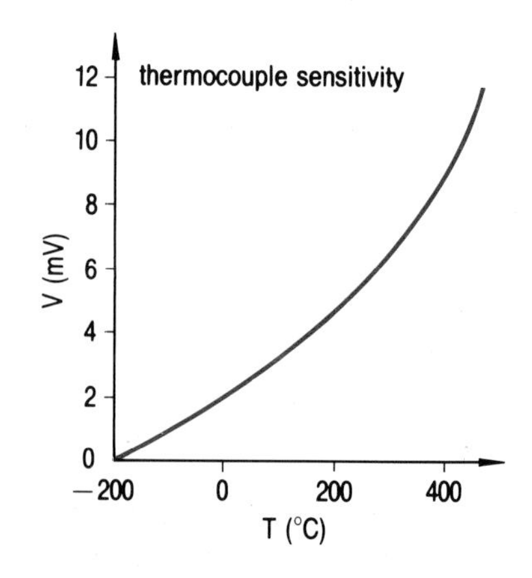

1 What energy changes take place in each of these devices?

2 State an advantage and disadvantage of each device.

3 Are the outputs analogue or digital?

4 Name two thermocouple applications.

5 The graph shows how a thermocouple voltage changes with temperature from −200°C to 400°C. Describe how the sensitivity changes.

What's needed?

When deciding which input transducer to use in a circuit the first question which must be answered is: 'what form of energy must it detect?' Then the following points should be considered.

Size Will it detect enough input energy?
Will it fit into the available space?
Cost Will it be too costly?
Could a cheaper one do just as well?
Speed Will it react quickly enough to detect the energy change?
Power Will it need a battery to help it work?
Will it generate its own voltage?

When choosing a transducer we need to accept that all these factors may not be ideal. We need to decide which matter most.

input device	size	cost	speed	power
thermocouple	••••	••	••••	••••
rod termistor	••	•••	••	••
bead thermistor	•••	••	••••	••
LDR	•••	•••	•	••
solar cell	•	•	••	••••
photo-transistor	••••	•••	•••	•••
photo-diode	••••	••	••••	•••
microphone	••	•	•••	•••
push switch	••	•••	•••	••••
touch switch	••	••••	•••	••••
micro-switch	••	••	••••	••••

•poor • •fair • • •good • • • •very good

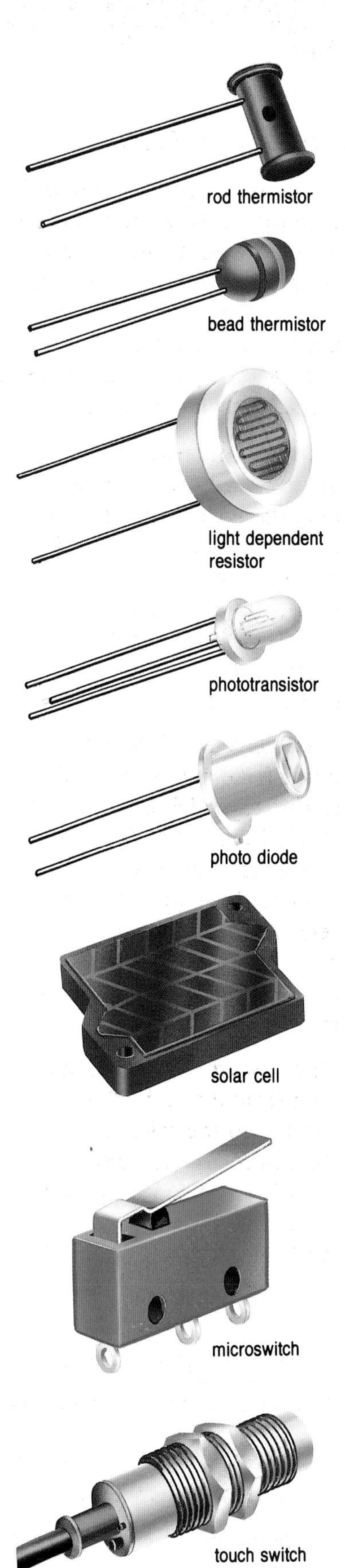

6 The input devices have been grouped according to the kind of energy they deal with. Name these forms of energy.

7 Explain why a light dependent resistor is chosen to be used for a hand held photographic light meter.

8 Explain why a photo-diode might be selected for use in a price bar-code reader.

9 Which sensor would you use for a doctor's electronic thermometer ? Explain why.

10 Which type of sensor would you choose for switching on a bedroom lamp? Explain why.

11 Which sensor would be best for a thermostat? Explain your choice.

Altering the input

Dividing the voltages

Light-dependent resistors and thermistors are examples of **input** devices which work because their **resistance** changes. To 'process' information, however, it is often a change of **voltage** that is needed at the input. We can get a change of voltage from a change of resistance by using a **voltage divider** and a battery.

A voltage divider is made from two resistors (R_1 and R_2) in series. R_1 could be fixed and R_2 changeable, or it could be the other way round.

When two resistors are connected in series with a voltage supply, the sum of the voltages across each resistor is equal to the supply voltage. The size of the individual resistances determine how the voltage is divided up. The larger the resistance the greater the voltage across it.

Instead of having two separate resistors, it is possible to have a voltage divider resistor with a sliding contact. This can be adjusted to give the required ratio of resistances (and hence voltages).

If one of the resistors is replaced by a switch we can obtain a 'digital' voltage. With the switch open the voltage at point X is +5 volts since the voltage across the resistor R_1 is 0 volts (current I = 0). When the switch is closed the voltage at point X is 0 volts since there is no voltage across the switch (resistance R_2 = 0). In practice, the voltage at X is said to be 'high' when the lower section of the divider has a much greater resistance than the upper section.

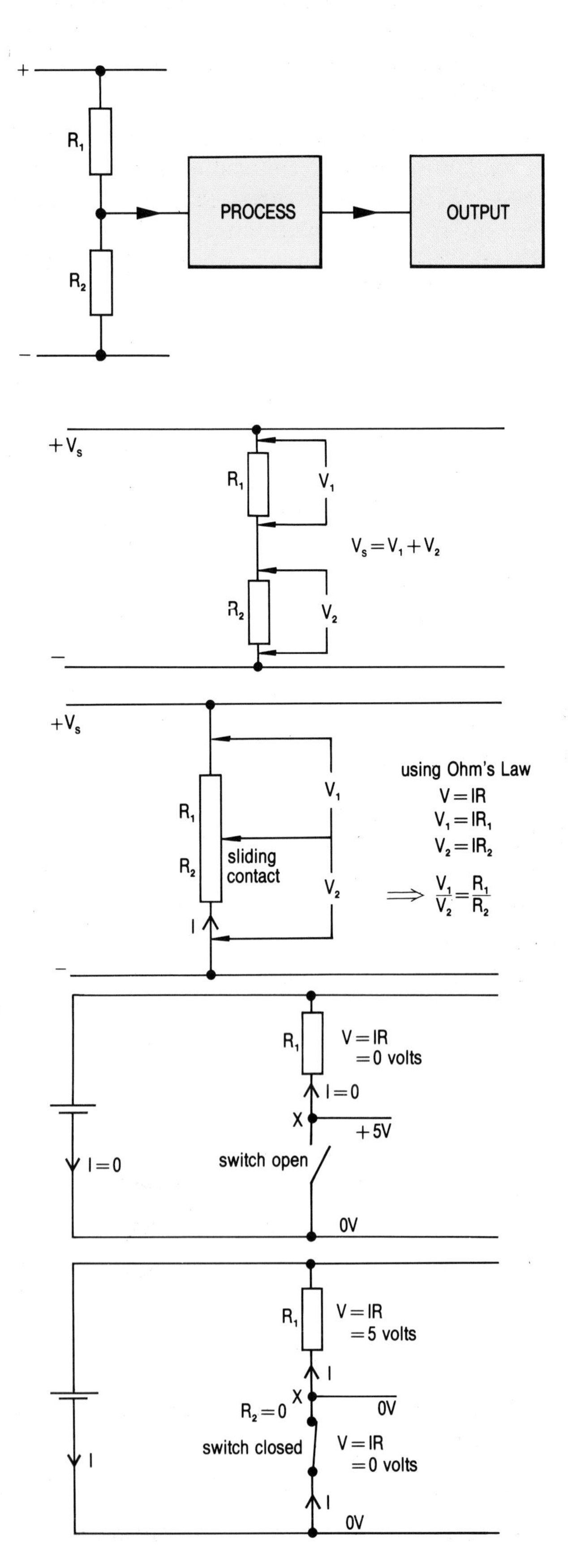

Light dependent resistor

The resistance of a light dependent resistor depends on the level of lighting. When the light level is 1000 lux its resistance is 130 ohms but when the light drops to 50 lux its resistance is 2.4 kilohms

Len designed a simple light meter which consisted of a light dependent resistor and a fixed resistor as a voltage divider. He connected a voltmeter across the fixed resistor to give a light reading. Details are given on this page.

Thermistors

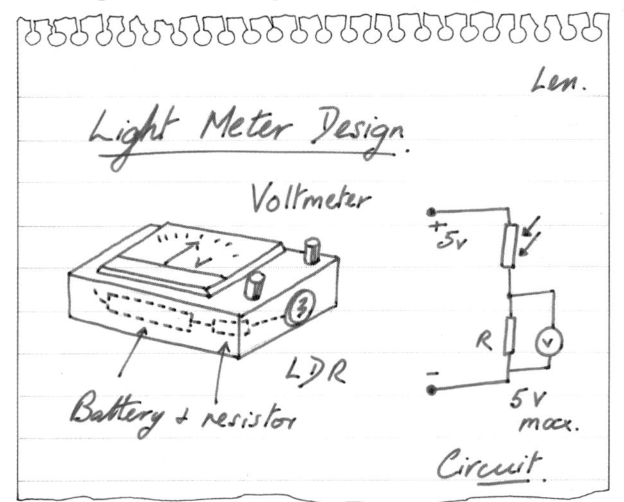

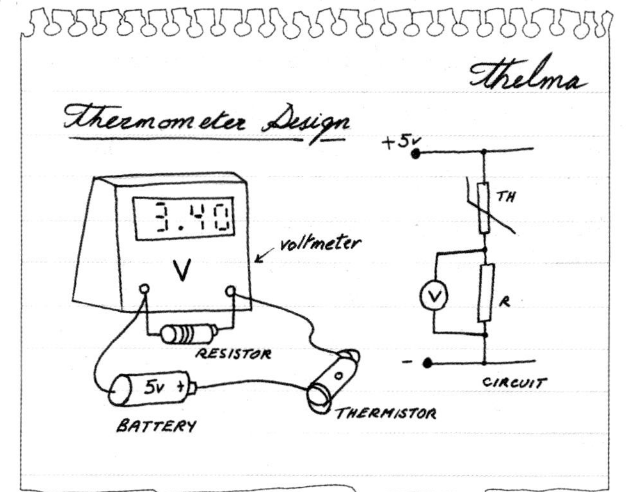

Thelma's thermistor thermometer

The resistance of a thermistor depends on the temperature. If the temperature is high then its resistance is low and when the temperature is low its resistance is high (eg. 380 ohms at 25°C; 28 ohms at 200°C).

Thelma has designed a simple thermometer using a voltage divider arrangement with a 470 ohm resistor and a rod type thermistor. She connected a voltmeter across the 470 ohm resistor to give a temperature reading.

1 Give one reason why voltage dividers are used.

2 Why is the voltmeter connected across R and not the LDR?

3 A resistor and LED form a voltage divider. When a current of 10 mA flows through the LED it has a voltage of 2 volts across it. What value of resistor is needed in this voltage divider if the supply is 5 volts?

4 Why is the voltmeter connected across the resistor and not the thermistor?

5 What are the voltmeter readings at 25°C and 200°C?

6 Calculate the current in the thermistor at 200°C.

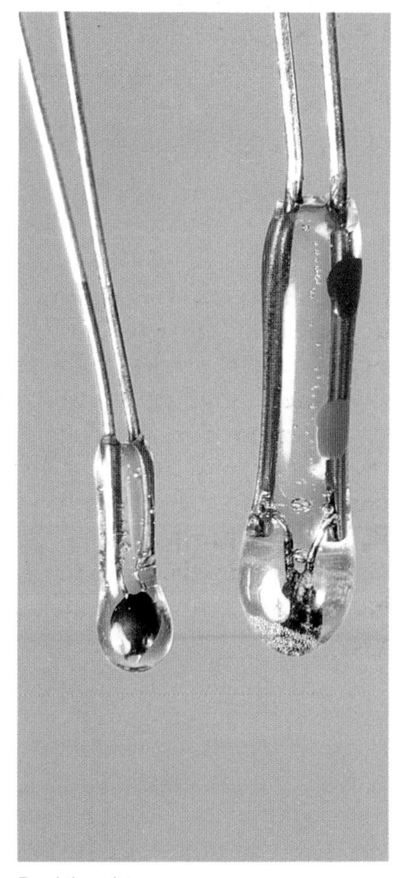
Bead thermistors

Delaying the action

Who needs to wait?

Energy saving scheme 'gets the bird'

In an attempt to save money by making sure that the street lamps only came on when it was dark enough, Crowborough Council agreed to a recommendation from the Engineers' Department to install 'light-sensor switches' on top of all the street lamps.

Local residents are not happy: 'The lights are coming on at odd times during the day' said Mrs Potts 'they just don't seem to be working properly with all this fancy "electronics" they've added – it's just a waste of money!' The Council are to receive an urgent report from the Engineers' Department.

CROWBOROUGH DISTRICT COUNCIL

Public Works Department

Problem	The lights are switching on too early and also at odd times of the day
Cause	Crows are landing on the light sensors which are mounted on top of the lamps. The circuit has been designed so that when the light falls below a certain level the lamp is switched on. The light sensor only needs to be shaded for a few seconds to activate the lamp circuit.
Solution:	Build a time delay into the lamp switching circuit. This will prevent the lamp from turning on unless the light level stays low for at least five minutes. The crows seldom stay that long on top of a lamp.

Signed: Phil.A.Ment (Lighting Engineer) P.A.Ment

The problem illustrated above requires a solution which involves no more than adding a few extra components to the existing electronic circuits.

Capacitors

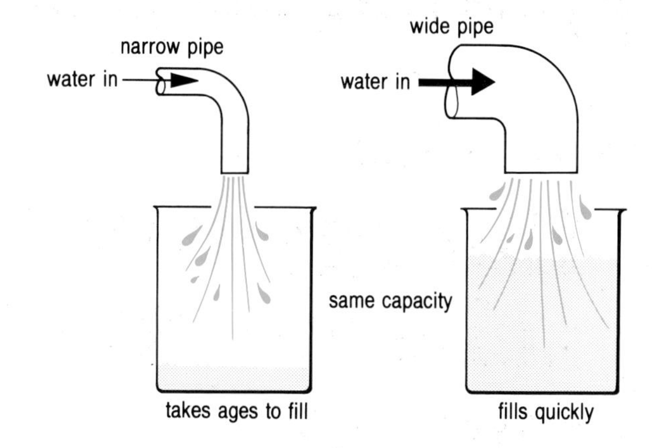

It is sometimes more convenient if an electronic circuit does not react immediately, but *delays* its action – using a component called a 'capacitor'. This is a kind of electrical 'storage tank' which can be 'filled' or charged with electricity. A **capacitor** gradually 'charges up' when it is connected to a battery or power supply. As it is being charged, the voltage across the capacitor gradually increases and eventually reaches the same value as the supply.

The length of time required to 'charge' it up to the supply voltage depends on the size of the capacitor, measured in farads (F), and the size of the resistor placed in series with it.

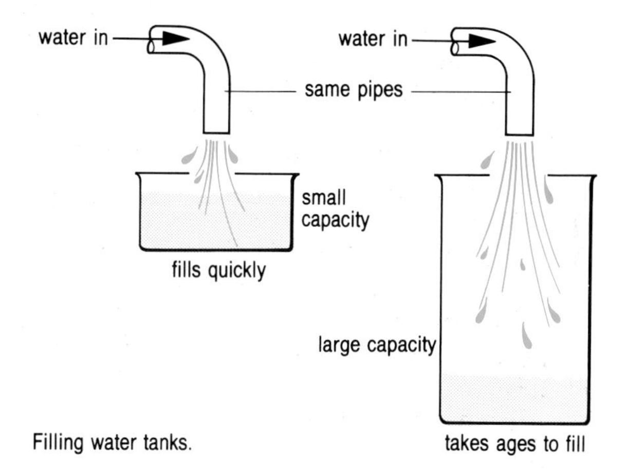

Filling water tanks.

As we have seen it sometimes helps to compare electrical systems or components with water supply systems.

Compare the capacitor behaviour described, with the time taken to fill a tank of water. The time depends on the capacity of the tank (how much it can hold), and also on the bore of the water pipes used to fill it. The greater the tank's capacity and the greater the resistance in the pipes (i.e. the smaller the bore) the longer it takes to fill the tank.

Similarly, the bigger the value of the capacitor and the larger the value of the resistor, the *longer* it takes to charge the capacitor to the supply voltage. A capacitor can be combined with an input transducer to delay the processing of the output actions. This is particularly useful for circuits such as burglar alarms. It can also be used in a circuit which must only be switched on for a certain time.

The circuit is arranged so that the capacitor charges up to the voltage level needed to operate the 'process' part of the system. Choosing suitable capacitor and resistor values gives the correct time delay.

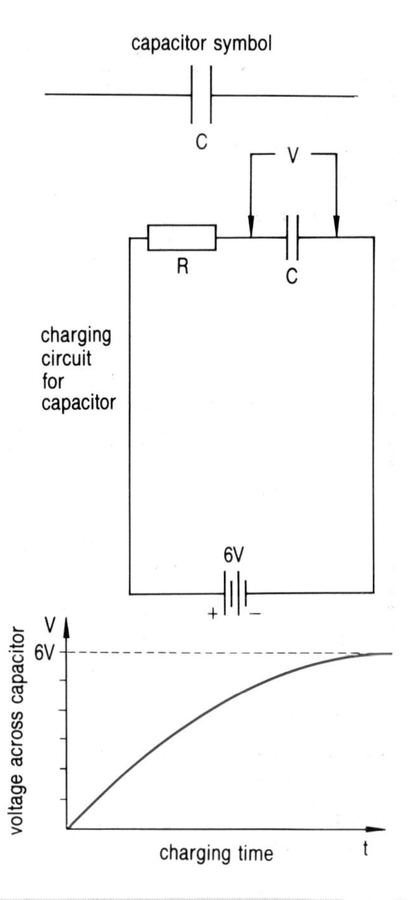

1 Draw the symbol for a capacitor.

2 What combination of components will provide a longish time delay?

3 In what way might the Lighting Engineer's suggestion help?

4 In practice most street lights are controlled by electric clocks. Explain whether you think the suggested solution is a good one.

5 Try to find out what is meant by the 'time constant' for a capacitor – resistor circuit.

Transistor switching

About transistors

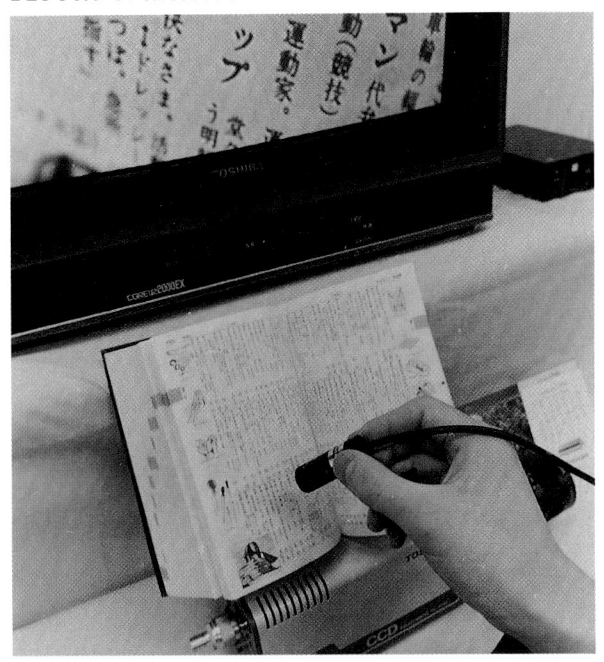

The camera used to scan this Japanese dictionary is only 45 mm long and 16.5 mm in diameter.

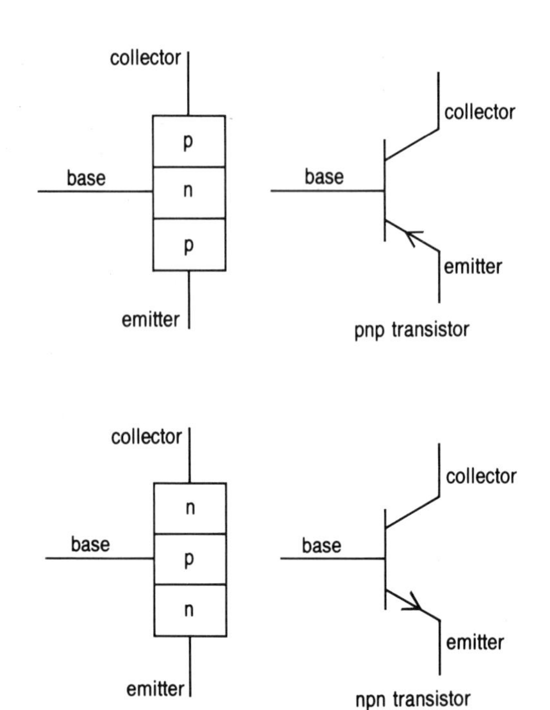

Transistors – semiconductor sandwiches.

Transistors were invented about 40 years ago and since that time they have become smaller and smaller. They often form a tiny part of a much more complicated 'chip', or 'integrated circuit', and allow very small systems to be developed. A transistor is constructed from a specially treated (or 'doped') **semiconductor** – either germanium or silicon. A semiconductor can be made so that negative charges move through it (n-type). Or it can be made so that positive charges move through it (p-type). Sandwiches of these materials form npn or pnp transistors. Each transistor has three terminals called the **base**, **collector** and **emitter.**

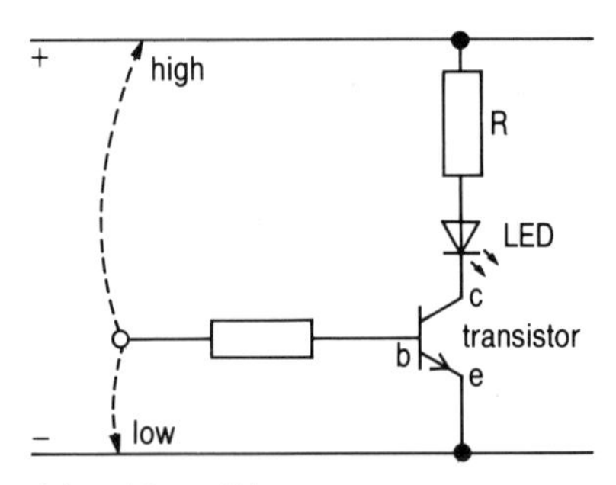

A transistor switch.

Making them switch

Transistors can be used as **electronic switches**. A simple way of seeing the switching action of a transistor is to connect a resistor and LED from its collector to the positive side of the supply, and its emitter to the negative side of the supply. When this is done the LED is not lit. If a resistor attached to the base of the transistor is connected to the negative side of the supply the LED remains 'off'. If instead it is connected to the positive side of the supply then the LED is turned 'on'. So, by making the base voltage positive the transistor conducts current and the LED is 'on'. In this circuit the LED is used to indicate whether the transistor is switched 'on' or 'off'.

Transistor switch/indicator board.

A little more detail ...

If a voltage divider is connected into the transistor circuit we can find out more about the switching action. By moving the contact X towards the positive side of the supply, the voltage applied to the transistor base is gradually increased from 0 volts. At first the voltage across resistor R_L shows no change (almost zero) and then suddenly it increases to +5 volts and remains at that value. The transistor is acting as a voltage controlled switch.

The voltage V_L across R_L changes sharply as the base-emitter voltage (V_{be}) goes through a very small range of values. A small voltage change at the base of the transistor results in a large voltage change across R_L as the voltage swings from 'low' to 'high'. This is also indicated by the LED as the transistor switches from OFF to ON. If the input voltage is 'high' then the output voltage (across R_L) is also high.

The voltage divider can have an input device in it – this is the 'input'. The transistor is the 'process' block of the system.

The resistor R_L can be replaced by a device such as an LED – this is the output.

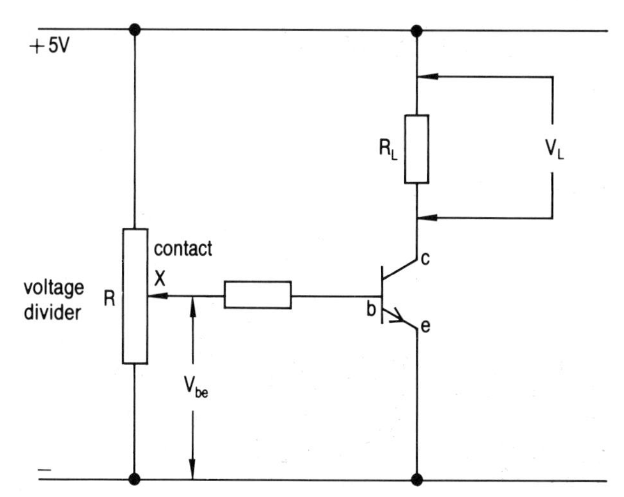

Transistor as voltage-controlled switch.

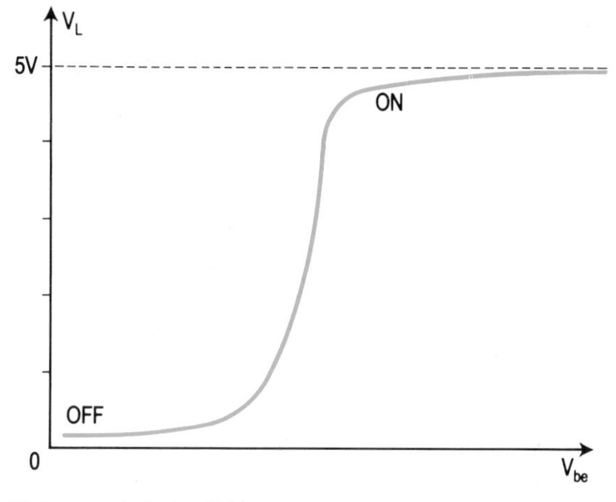

Voltage controlled switching.

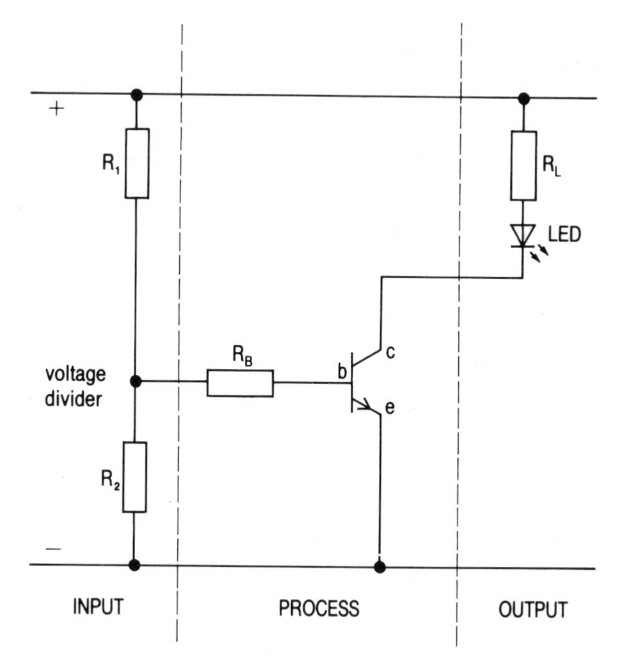

Transistor as a 'process' block.

1 Draw the symbol for a transistor and label each terminal.

2 Should the transistor base be 'high' or 'low' to switch the transistor on?

3 When the transistor is 'on', is the voltage between the collector and emitter high or low?

4 Why do you think that the transistor switch/indicator board illustrated has a 1 kΩ resistor connected to the base?

5 Draw a graph to show how the collector-emitter voltage varies as the transistor switches on.

Using transistor switches

Alarming circuits

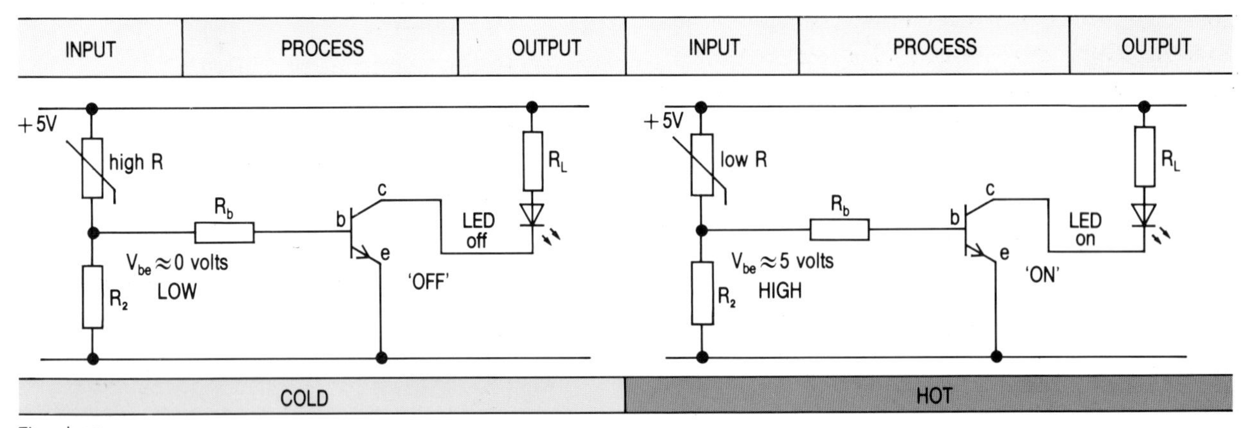

Fire alarm.

A fire alarm system needs an **input**, a **process** and an **output**. If the fire is to be detected quickly then a bead thermistor seems to be the best input device to use since it reacts quickest to a change in temperature. The transistor is the process section and in this case, for simplicity, we will use an LED as the output device. The thermistor is placed in series with a resistor as a voltage divider input to a transistor. When the thermistor is cold it has a high resistance and the voltage is divided so that there is a very small voltage applied to the base of the transistor and it is 'off'. If the temperature rises the resistance of the thermistor falls to a very low value and the voltage is divided with a much greater voltage applied to the base so that the transistor conducts and the LED is on.

A frost alarm for a car would probably use a rod thermistor as it is less easily damaged and cheaper. This time the thermistor would be placed in the lower arm of the voltage divider. When it is hot the thermistor has a low resistance and the base voltage is low – so the transistor is 'off'. If the thermistor cools enough its resistance rises until the voltage across it is enough to switch on the transistor. The LED indicates frost.

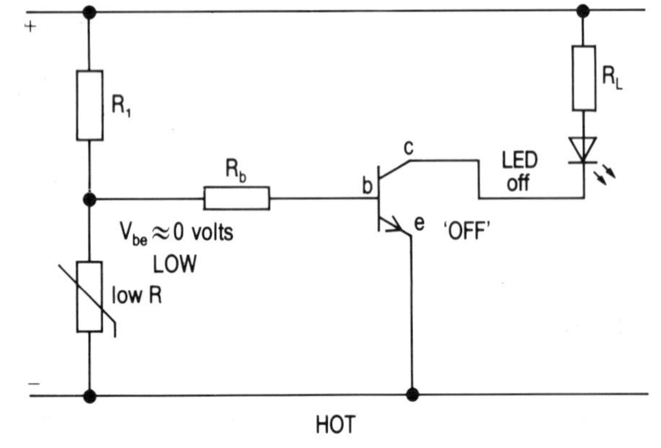

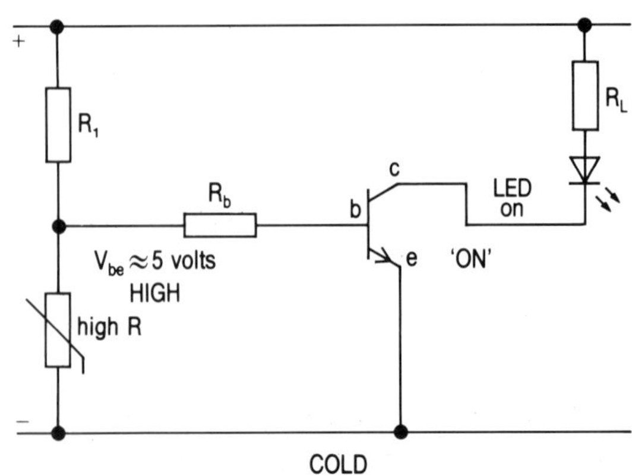

Frost alarm.

Making adjustments

An alarm circuit must respond to the correct conditions. A fire alarm should not go off because the weather is warm, or wait until the building has burnt down! We can set the circuit so that the desired rise in temperature switches the transistor on. This 'sensitivity' adjustment is made by altering the value of the resistor in the voltage divider.

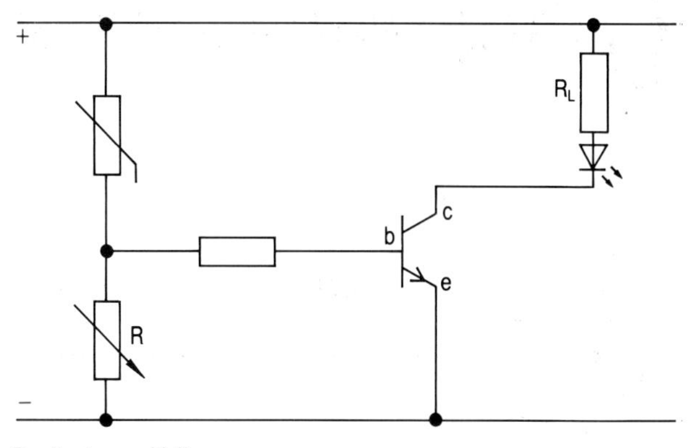

R adjusts sensitivity.

A similar adjustment is carried out when a particular level of lighting is needed for a LDR in a burglar alarm system.

When the variable resistor R in the burglar alarm circuit is adjusted until the LED is just 'off' there is not enough voltage at the base to switch on the transistor. If a burglar shades the LDR this increases its resistance and so the voltage applied to the base of the transistor becomes high enough to make it conduct. The presence of the burglar is indicated by the LED.

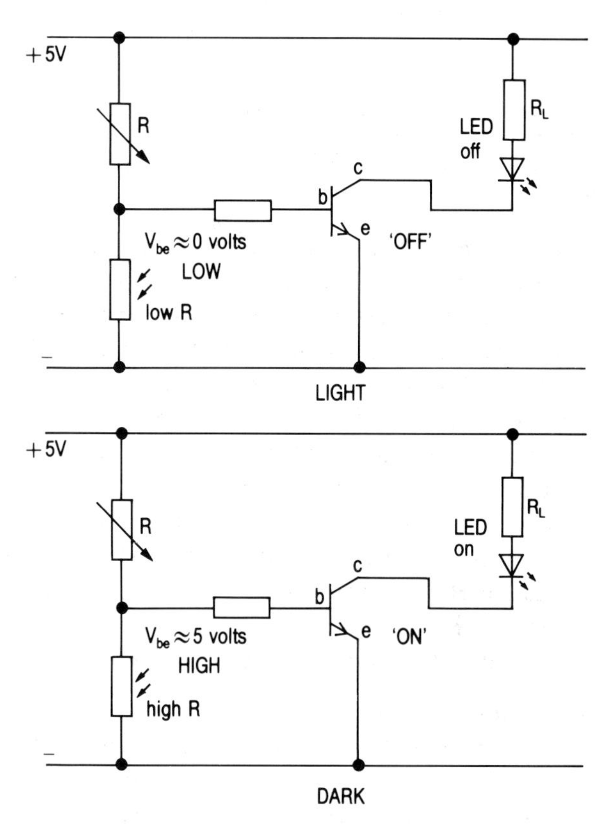

Burglar alarm circuit.

The switching action of a transistor can be delayed using a capacitor and a resistor in the voltage divider. As the capacitor charges up, the voltage across it increases. Before the power switch S_1 is closed, in the circuit shown, the voltage across the capacitor is zero and so the transistor is 'off'. When the power supply is connected the capacitor slowly charges up until the voltage across its terminals is high enough to switch the transistor. The LED then comes 'on'.

The length of time delay is decided by the values of the capacitor and resistor. By putting a variable resistor in the voltage divider we can alter the time delay.

›› *Make up as many circuits as you can from those mentioned on these two pages.*

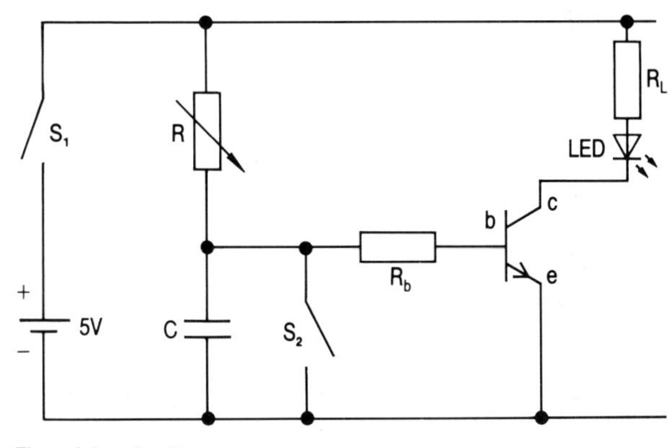

Time delay circuit.

1 Draw a circuit for a device which would waken you at daybreak.

2 What do you think the other switch (S_2) in the time delay circuit could be used for?

3 Explain how a circuit might be arranged to dim your headlights automatically when another car approaches.

Logically speaking

Switch controlled circuits

Dear Mr Duff,

I am writing about my new car, which seems to have a strange fault.

When I drive up to my garage at night and open the car door, a bleeper starts to sound. I cannot understand why. Is this normal?

Yours sincerely,
Maurice I. Tal.

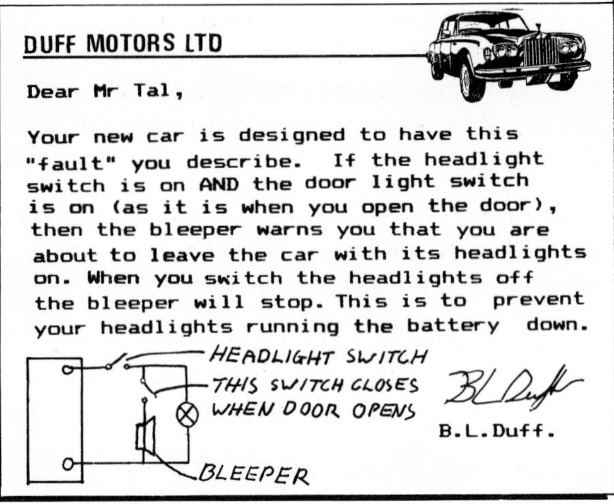
DUFF MOTORS LTD

Dear Mr Tal,

Your new car is designed to have this "fault" you describe. If the headlight switch is on AND the door light switch is on (as it is when you open the door), then the bleeper warns you that you are about to leave the car with its headlights on. When you switch the headlights off the bleeper will stop. This is to prevent your headlights running the battery down.

B.L.Duff.

The switching arrangement described above is called an AND circuit.

A circuit which can be operated by *either* of two switches is called an OR circuit. An example of this is the courtesy light inside a car. It will come on if the switch in the right-hand door OR the switch in the left-hand door is operated.

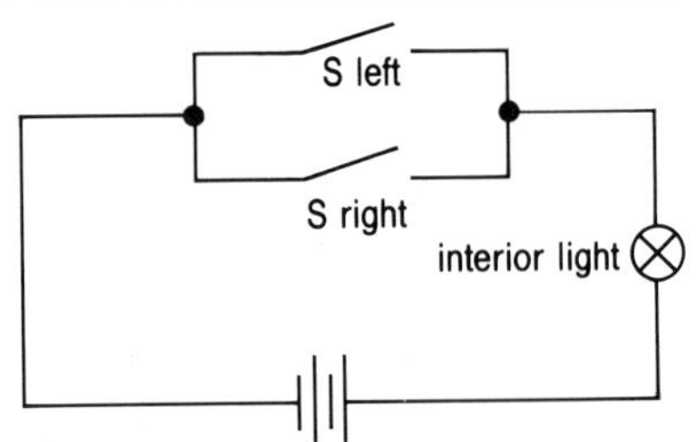

Using more transistor switches

Electronic systems often have to operate from a number of inputs too. Transistors can operate as switches in the process section. They can be combined to provide electronic AND and OR circuits. These are called **logic gates** because they control the passage of electrical signals from one part of the system to another in a logical way.

By using two transistor switches, T_A and T_B, in series to form the upper half of a voltage divider we can make an AND circuit. When their bases are positive ('high') they conduct. Point X becomes 'high', the transistor T_Z is switched on and the LED lights.

If T_A AND T_B are 'on' T_Z is 'on'. This circuit is called an AND gate.

Transistors T_A and T_B connected in parallel can make an OR circuit. When either T_A OR T_B is 'on' the output transistor T_Z is 'on'. This is called an OR gate.

A single transistor, T_A, in the lower half of a voltage divider makes another type of logic gate called a NOT gate. The output is always the reverse of the input. This gate is also called an **inverter**. When T_A is 'on' it conducts so that point X is 'low'. This is connected to the base of T_Z and so the output of this 'output indicator' is 'low' and the LED is 'off'. When T_A is 'off' point X is 'high' and the output LED is on.

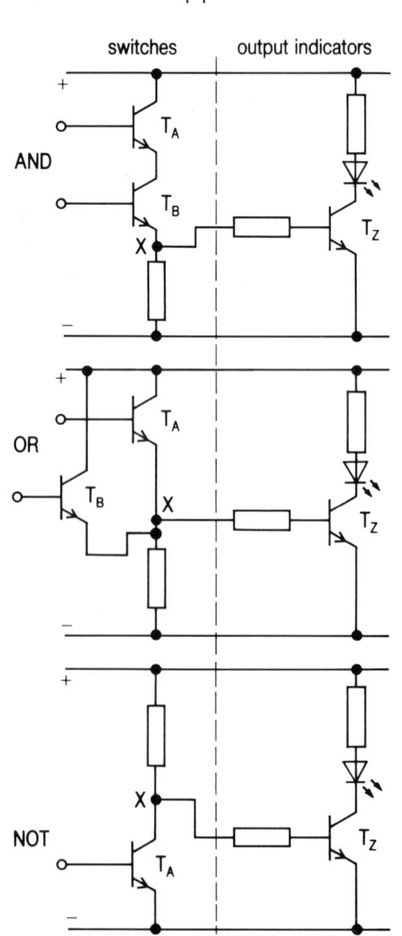

Nothing but the truth

The internationally accepted American Standard symbols for these gates are shown. No power supply is shown with these symbols when they are given in circuit diagrams. The 'A' and 'B' refer to the inputs of transistors T_A and T_B. 'Z' refers to the output indicator transistor T_Z.

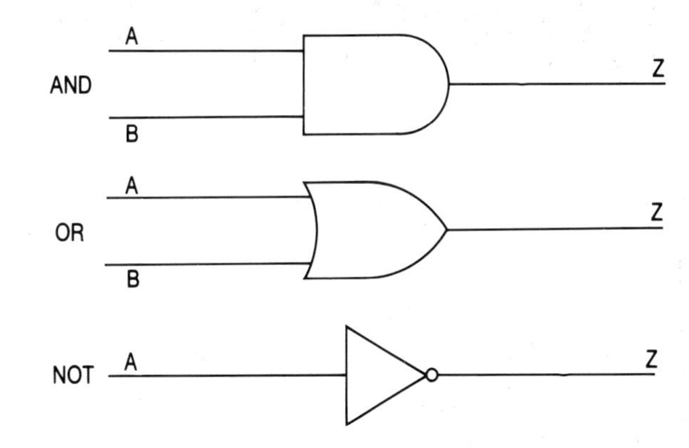

A table is often used to show what happens at the output of a logic gate for all the possible inputs.

A '1' indicates that a transistor is 'on' and a '0' that it is 'off'. (Also called 'true' and 'false'.) The table is called a **truth table.**

INPUTS		OUTPUT
A	B	Z
0	0	0
0	1	0
1	0	0
1	1	1

AND

INPUTS		OUTPUT
A	B	Z
0	0	0
0	1	1
1	0	1
1	1	1

OR

INPUT	OUTPUT
A	Z
0	1
1	0

NOT

Electronic circuits can be designed by considering the 'logic' of the system – what outputs are needed and which inputs are available. The truth table can then show which logic gates to use.

Mr Tal's car system is shown on the right.

This is immediately recognised as the truth table for an AND gate. The process section of this system could therefore be carried out using an AND gate.

Head-lamp	Door	Bleeper	Head-lamp	Door	Bleeper
OFF	SHUT	OFF	0	0	0
OFF	OPEN	OFF	0	1	0
ON	SHUT	OFF	1	0	0
ON	OPEN	ON	1	1	1

Mr Tal's car system.

The AND, OR and NOT logic gates described are available commercially in multiples of four or six gates in **integrated circuit** packages (i.c.'s) or 'chips' as they are commonly called. Only one of the AND gates in the chip would be used for this application.

1 What logic gate represents the switching circuit for an automatic electric kettle?

2 Write out a truth table for the interior light switches on two car doors? What gate is needed?

3 Draw and label the symbols for the three logic gates.

4 Write out the truth tables for the AND, OR and NOT gates.

5 Draw the truth table for the system shown here on the right.

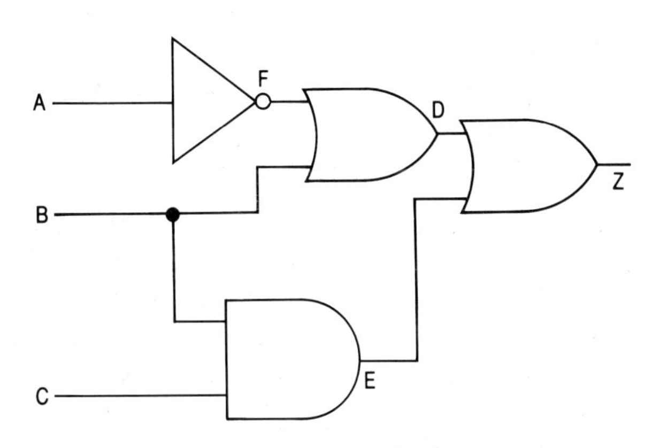

Using logic

Chips with everything

A small chip can contain a large number of transistors which are joined together inside the chip to form a number of logic gates Because it is so small a chip is a very convenient way of providing the logic switching needed in many applications. The photograph on the right shows just one chip at high magnification.

Washing machine

In an automatic washing machine the use of logic gates makes the system cheaper and more reliable.

Action : the machine agitates at the correct temperature.

Inputs : the machine must be switched on – use switch.
the temperature must be correct – use thermistor.

Process : both inputs are needed at the same time.

Output : the motor starts when the water is at the correct temperature.

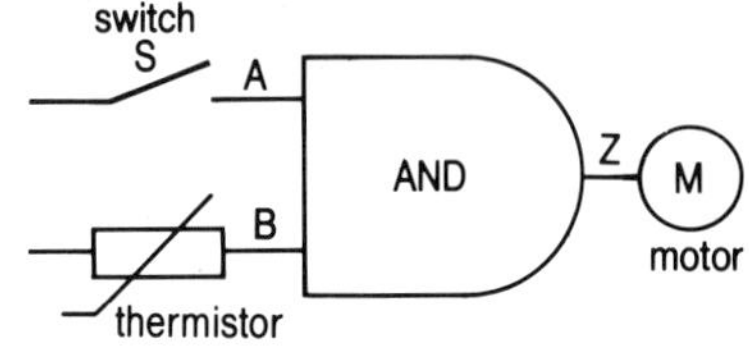

Washing machine logic.

INPUTS		OUTPUT
SWITCH	THERMISTOR	MOTOR
ON (1)	COLD (0)	OFF (0)
OFF (0)	COLD (0)	OFF (0)
ON (1)	HOT (1)	ON (1)
OFF (0)	HOT (1)	OFF (0)
A	B	Z

Automatic parking light

An automatic parking light can also make use of logic gates in chips.

Action : a red parking light switches on when the car is parked at night.

Inputs : light level is to be detected – use LDR (high resistance in darkness). The system can be turned off when the car is in the garage – use a switch.

Process : both inputs are needed at the same time.

Output : parking light.

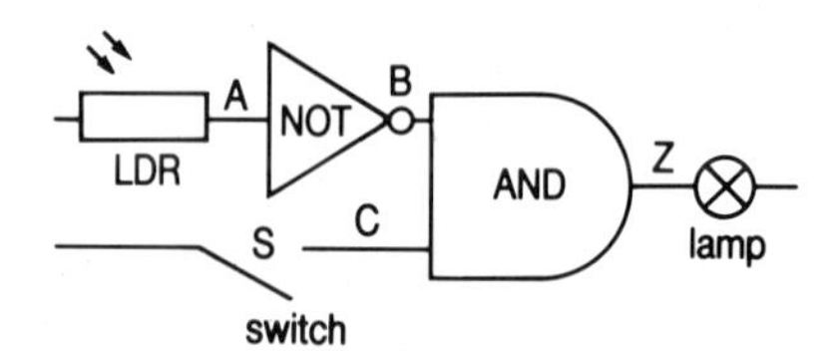

Parking light logic. Two inputs to give '1' output.

	INPUTS			OUTPUT
	LDR		SWITCH	PARKING LIGHT
DAY	LIGHT (1)	(0)	OFF (0)	OFF (0)
	LIGHT (1)	(0)	ON (1)	OFF (0)
NIGHT	DARK (0)	(1)	OFF (0)	OFF (0)
	DARK (0)	(1)	ON (1)	ON (1)
	A	B	C	D

Optional extras

The 'warm, sunny day' detector shown on this page shows two additional features.

The first is called an 'enable' switch (S_1). When closed it allows (enables) the circuit to operate. When the switch is open the circuit cannot be operated. The second is called a 'test' switch (S_2). By closing it the circuit is checked to see if the output operates correctly.

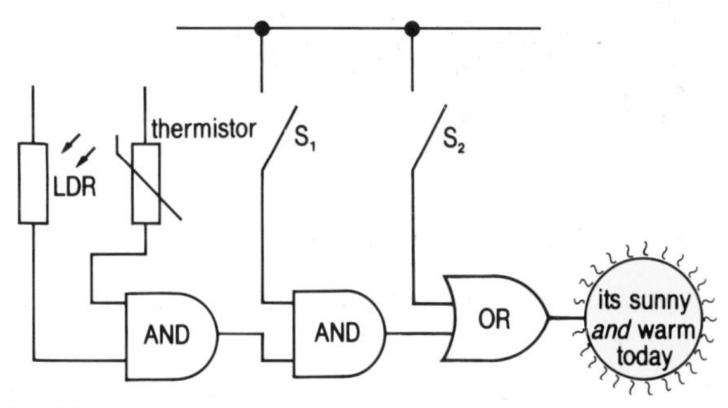

Circuit to detect a warm sunny day.

The future – no circuit boards!

Electronic components are usually mounted on printed circuit boards (PCB's) by inserting them in drilled holes and soldering them to the printed circuit.

There has recently been a move towards replacing printed circuit boards altogether and mounting the components directly on the inside surface of an appliance. Special small surface mounted devices (SMD's) have been developed which are often less than 1 square millimetre in area! They can be mounted on any surface that is firm enough and can cope with temperatures up to 250°C. This means that a radio or telephone could be practically empty – with circuits printed on the plastic casing!

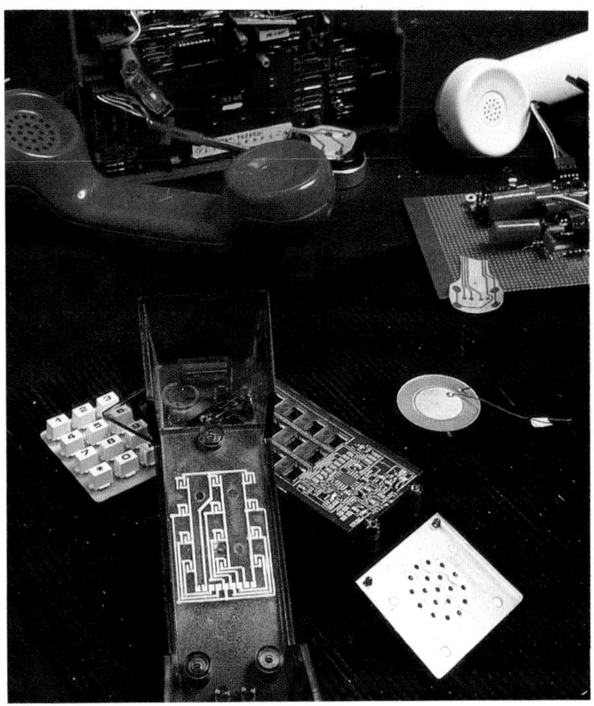

Phone with circuits printed on its case.

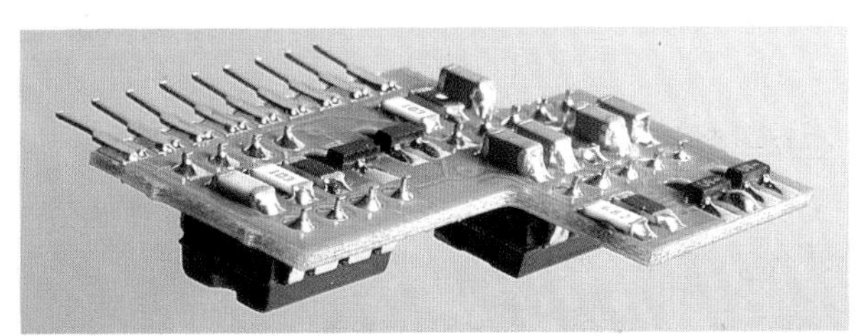

Surface mounted devices.

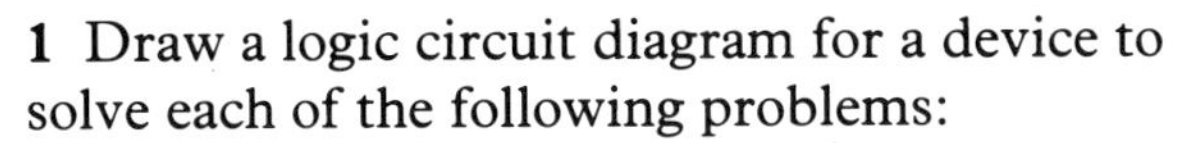

1 Draw a logic circuit diagram for a device to solve each of the following problems:
a) a buzzer must sound when an oven reaches the correct temperature;
b) a door buzzer has to operate only at night;
c) a frost warning must sound in a car only when the ignition switch is on;
d) a baby alarm has to sound when the baby cries;
e) the fan on a storage heater has to switch off when the temperature of the blocks is low;
f) an audible alarm is to sound only when a car door is opened and the headlights are on.

2 Suggest uses for each of the systems on the right.

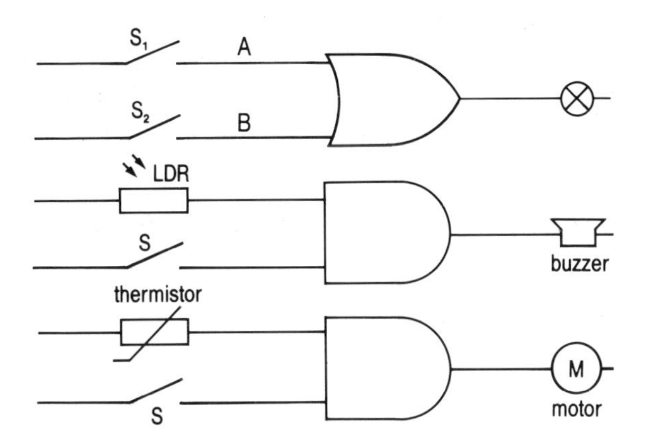

›› *Build some of the circuits from questions 1 and 2.*

Clocks without ticks

Clock pulses

Things must happen at the correct times in any electronic system. Imagine what might happen in a super-micro-electronic washing machine if this were not so! To make sure that each event does take place at the right time, electronic systems are controlled by a series of voltage pulses. The time between one pulse and the next is always the same, so they are called 'clock' pulses. The clock pulse voltage goes 'high' – 'low' – 'high' – 'low' (or on/off/on/off . . .) at a constant frequency.

The clock pulses can be fed to several gates at the same instant so that the processes don't get out of step. The gate shown here must wait for a clock pulse to arrive before it allows the other pulse through.

Pulse generator.

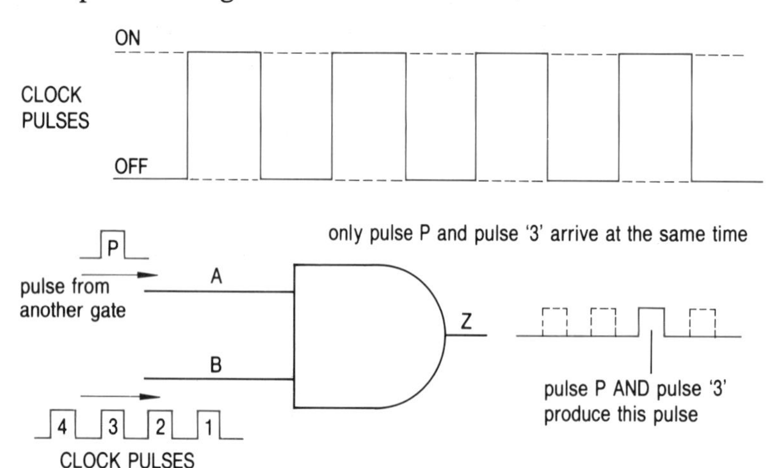

Pulser unit.

How they work

The time delay between clock pulses is produced in an oscillator (or pulser) circuit which contains a capacitor and resistor. The frequency of the pulses can be altered by changing the capacitor value or adjusting the resistance value using a variable resistor.

A simple oscillator can be built from a NOT gate (or inverter), a capacitor and a resistor. Starting with an uncharged capacitor, the NOT gate input is low and so its output is high. Current then flows through the resistor to charge the capacitor until the input becomes high and the output then goes low. At this stage the capacitor discharges through resistor R and the input goes low again, making the output high again. This action repeats continually, giving 'clock' pulses.

A pulser unit may also have a push button to allow single pulses to be sent out. This is useful if you want to observe the working of a circuit step by step.

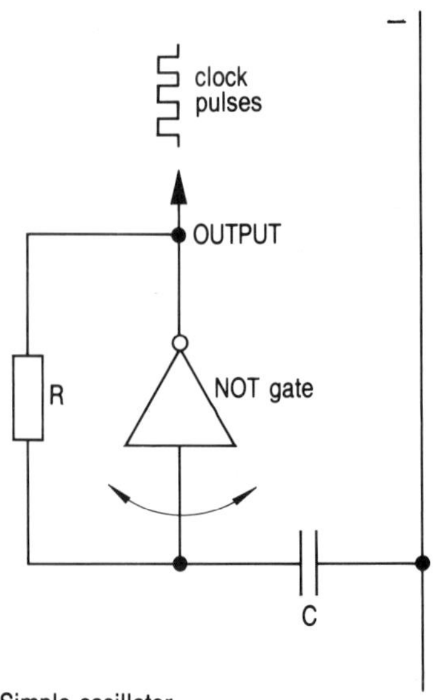

Simple oscillator.

Counting clock pulses

Before pulses can be used for timing they need something which can count them. Clock pulses can be counted by a special 'counter' chip (TTL 7493).

Each time a pulse is fed into the chip it sends out a set of four 1's or 0's, which make up a four binary digit number pattern (4-bit number for short). The four bit number is displayed by a set of four LED's connected to the chip's outputs A, B, C and D.

The A, B, C and D outputs are shown in the table. Note that the 'A' output changes every pulse; the 'B' output changes every two pulses; the 'C' output changes every four pulses and the 'D' output changes every eight pulses. After 15 changes the display reads 1 1 1 1, which is 15 in binary code (8 + 4 + 2+ 1). The next pulse returns the display to 0000, where it can begin again. The counter counts in 15's.

If output pulses from B and D are fed to the two connections marked $R_{O(1)}$ and $R_{O(2)}$ on this chip the display resets to zero when the count reaches 10. You will see from the table of output pulse patterns that this is the first occasion when both the B and D LED's are lit. At this stage the two 1's travel to an AND gate in the chip so that it sends out a 'reset' pulse. The counter chip is thus set to count in tens.

Taking things a step further, if the capacitor or resistor in the pulse generator is adjusted to a suitable value the circuit can be calibrated. If the time between one pulse and the next is adjusted to 1 second for example, the circuit above would be capable of timing up to 10 seconds. A simple adjustment could of course reduce this to 1 second or 0.1 second for greater accuracy with short timings.

1 What does the 'reset' pulse do to the count?

2 Describe how the reset works?

3 What is the 4-bit binary code for 7?

4 What is the 5-bit binary code for 16?

5 How could you make the counter count in dozens?

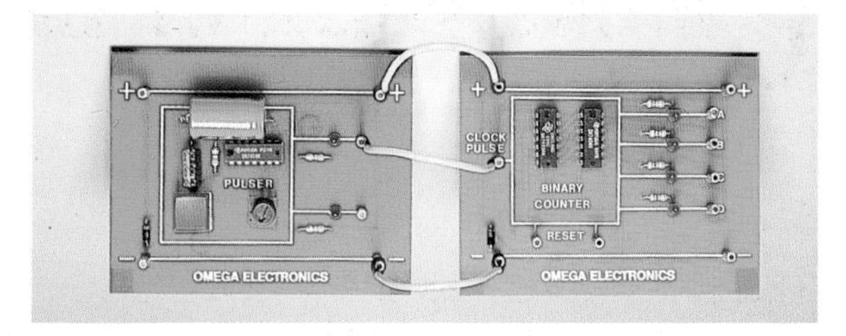

Pulser and binary counter.

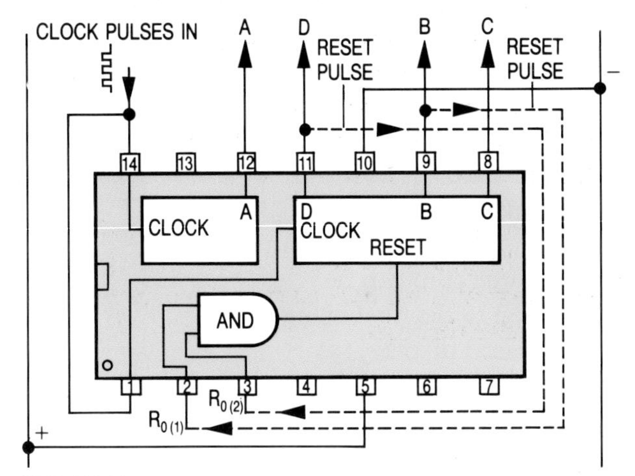

7493 4-bit binary counter.

input	output LED's			
pulse	D	C	B	A
0	0	0	0	0
1	0	0	0	1
2	0	0	1	0
3	0	0	1	1
4	0	1	0	0
5	0	1	0	1
6	0	1	1	0
7	0	1	1	1
8	1	0	0	0
9	1	0	0	1
10	1	0	1	0
11	1	0	1	1
12	1	1	0	0
13	1	1	0	1
14	1	1	1	0
15	1	1	1	1

›› Use a pulser unit as a reaction timer.

Counting

Going decimal

The 4-bit binary counter outputs, A, B, C and D from the counter chip can be converted to decimal using a decoder and display, (as discussed on page 101). Using the pulser and binary counter connected to the decoder-driver display, the numbers 0 to 9 can be displayed automatically, changing at the same rate as the pulser.

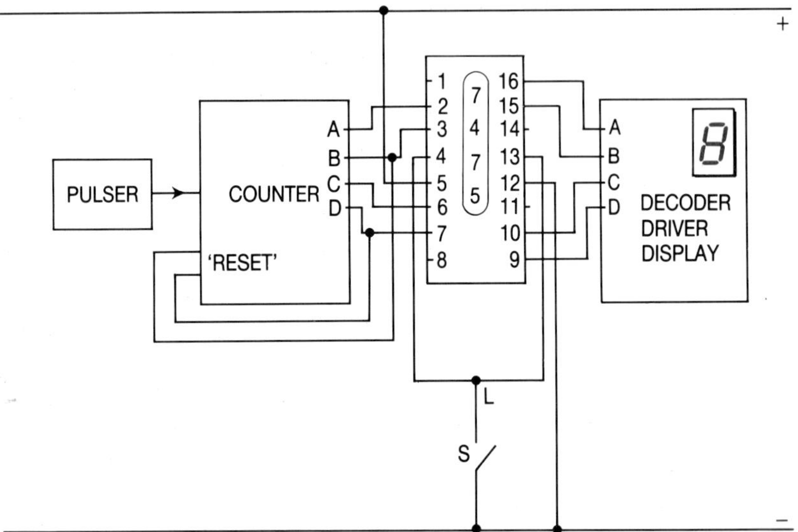

Counting for a minute

If the pulser is adjusted to give one second between pulses then the counting can be used for timing instead. A second 4-bit binary counter complete with decoder/driver and display is needed so that 60 seconds can be timed. The extra counter has its reset fixed at '6' by connecting its B and C outputs to its reset pins.

The second counter receives a pulse when the 'D' output changes from 'high' to 'low' – this happens when the 'reset' occurs.

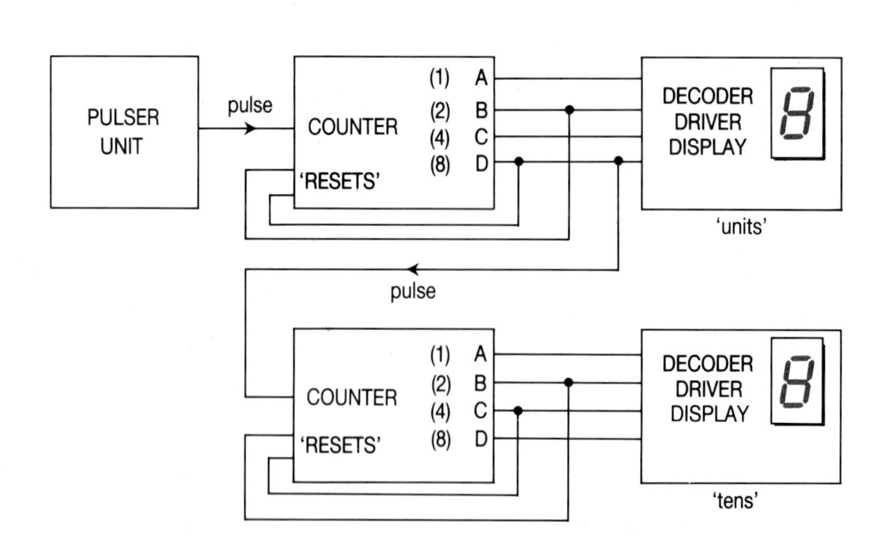

1 Why is the counter adjusted to a '6' reset?

2 How many counter/display circuits would you need for a minutes and seconds timer?

›› *Design and draw the circuit arrangement for a 12-hour display which also shows minutes and seconds.*

Using a pulser unit

The regular pulses of the pulser unit can be used to provide a repeating output for any alarm system. The Gas Board engineers often have to dig up the ground to investigate gas leaks. They leave warning lights which flash at night.

When darkness falls the light sensor output changes to '0', which is converted to a '1' by the NOT gate. This signal remains the same while it is dark and so pulses get through the AND gate and flash the light.

Circuits you can count on

How does a snacks machine know if you've put enough money in the slot? – The answer is that it counts it! The circuit diagram shows you the principles of such a system. The coins inserted in the slot are detected by a light sensor. Each time the light beam is interrupted by a falling coin the light sensor output signal changes from a '1' to a '0' and back to a '1'. If the counter input changes from a '1' to a '0' the number on the display increases by one. The counter display increases by one every time a coin has been inserted and so can show the total number inserted.

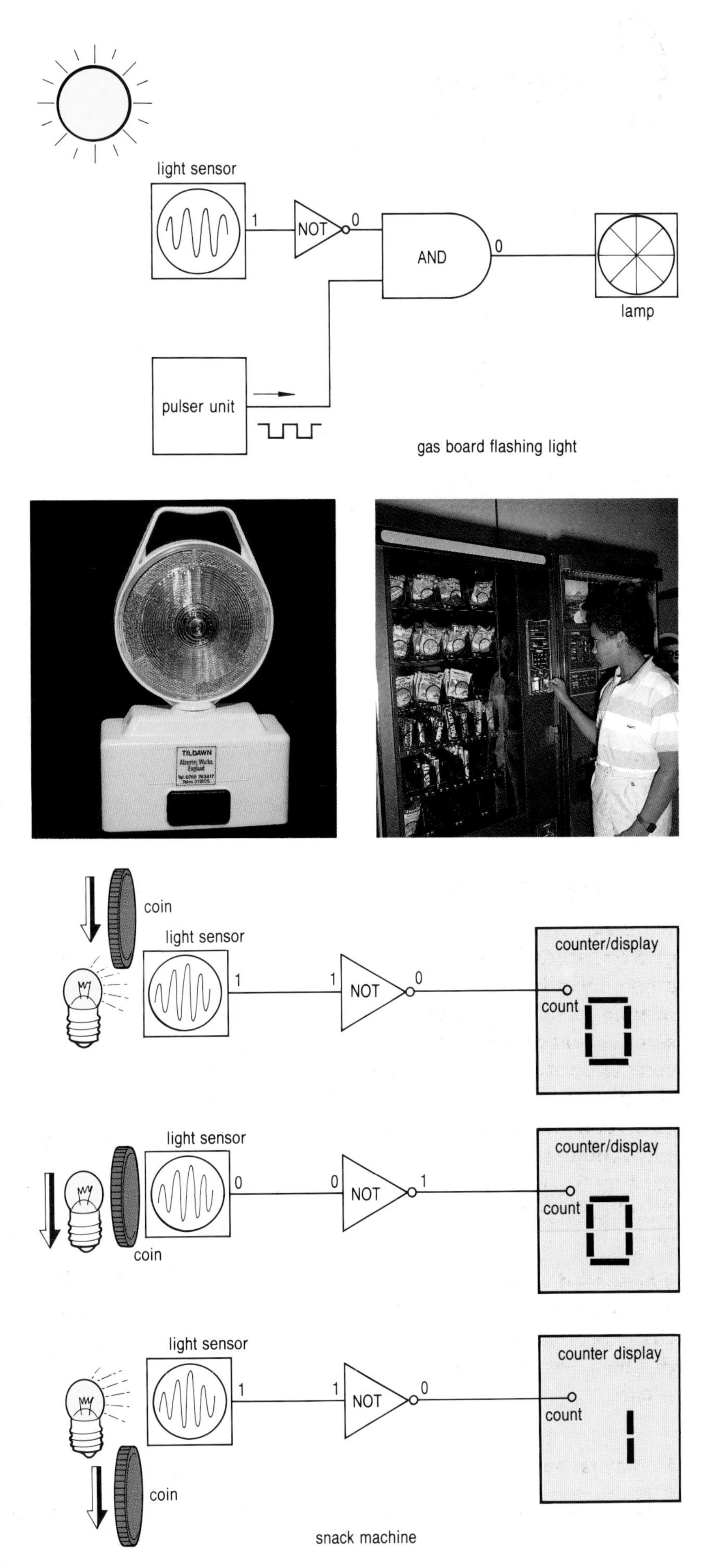

What's the gain?

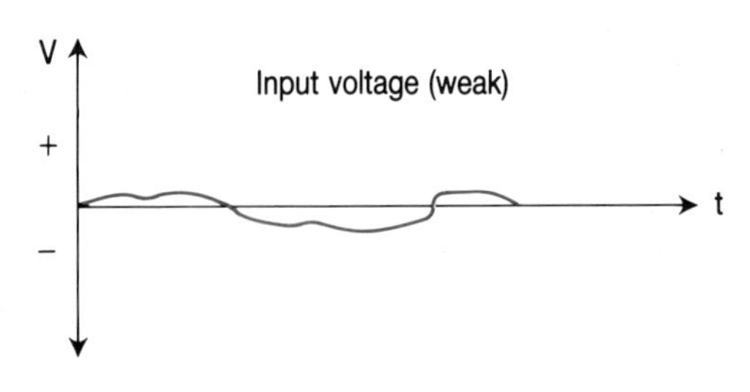

Weak signals

Some electronic sensors produce very weak electrical signals. A thermocouple, for example, might generate only a few millionths of a volt and yet it can be used as part of an electronic thermometer which needs a much larger voltage to operate its output display.

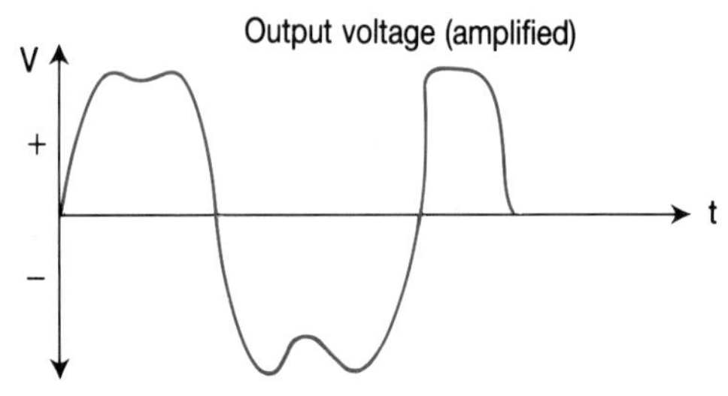

Weak electrical signals can be increased in strength using an electrical amplifier.

An amplifier which increases the size of the input voltage is said to have a 'gain'.

Amplifier chips

A relatively small 'chip' can increase the size of an input voltage by up to 100 000 times i.e. it can have a **voltage gain** as high as 100 000.

An 8-pin package containing an amplifier with about 35 components built into one small silicon chip a few millimetres square is quite common.

By adding relatively few extra components the package can become part of a low power audio amplifier, a brainwave amplifier, a radio receiver, a low temperature alarm, a diode thermometer, an optical communications system, a sound torch . . . the list is rather long! After almost twenty years of applications and improvements chips like these are still used in many systems because of their simplicity.

The symbol for an amplifier is shown below.

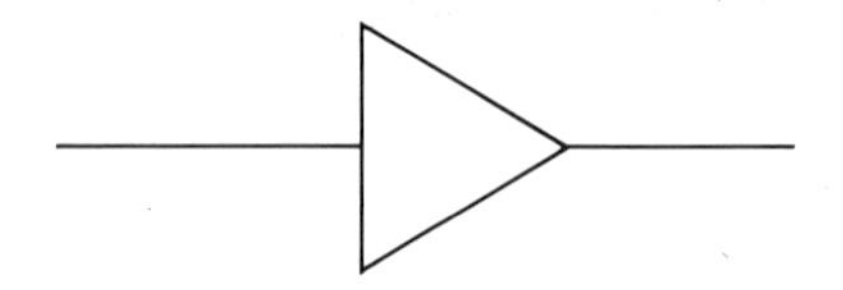

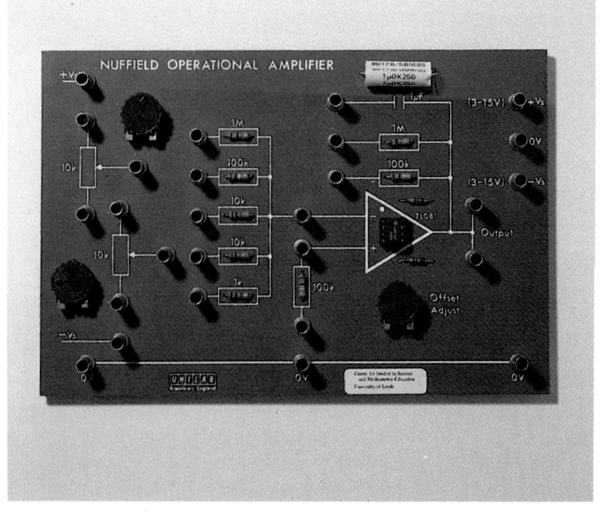

Using amplifiers

An obvious application of an amplifier which is able to provide a very large gain is in a hearing aid. Ken used standard electronics kit boards to show that the human voice can be successfully amplified using an amplifier chip. His report is given.

Ken's Investigation report

Title: Could amplifier chips be used in hearing aids?

Background: I discovered that the inside of a National Health Service hearing aid contains a few miniature transistors, resistors and capacitors. It also has a volume control, a loudspeaker, a microphone and battery (my grandfather's hearing aid). Could a 'chip' be used instead and so help to make a smaller hearing aid?

Experiment: Using the equipment shown, I found that the small voltage generated by a microphone could be amplified to make a loud sound in a set of earphones.

The amplified sound of my voice did not sound quite normal so I investigated how the amplifier reacted to various frequencies of sound as supplied by a signal generator.

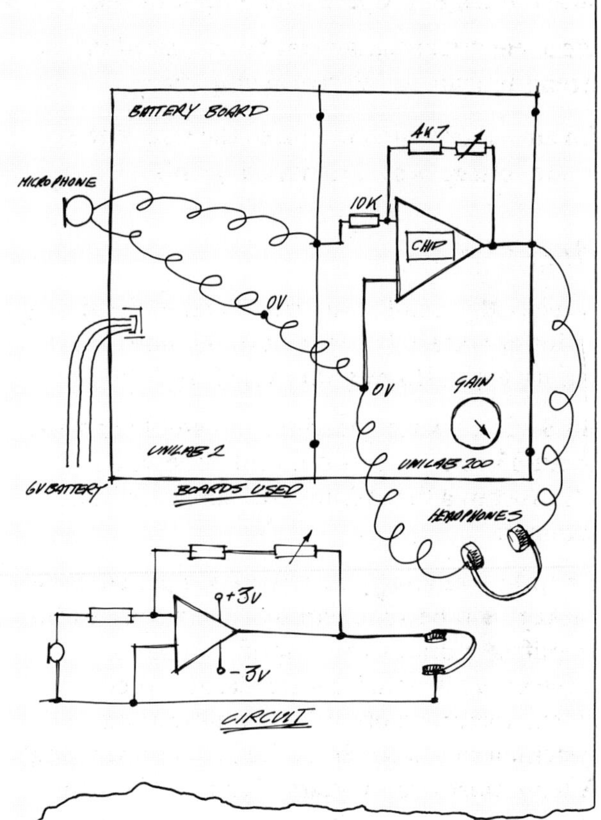

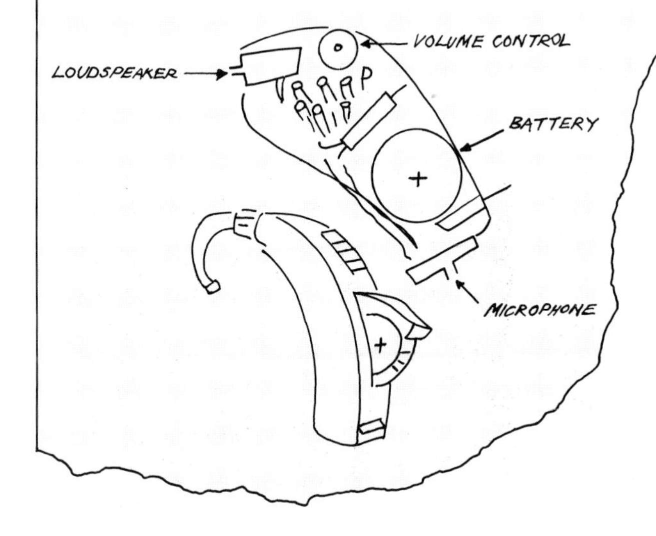

1 Do you think that 'chips' should be used in hearing aids? Give a few reasons for your answer.

2 Copy the following word search and find in it nine uses of amplifiers

E C Y J G V B Z X H
V E A R Y R O R E V
S D C G A E E A M Y
T J Q R D Y R O H G
E X H I A I C H O P
R S V L N R A H I X
E L P G E L G H D U
O D A T A U D H A D
C I N R L I H P R H
D I M G G C E S X H

TRANSPORT

5

The quest for speed

Keeping track of speed

Putting your foot down

Forces in action

Mass and movement

Getting to grips with friction

Making cars safer

The Earth's gravitational field

Energy transformations

More for your money

On your bike

Lifting the load

Stopping in time

Making work easier

Paying for performance

The quest for speed

High speed living

Speed affects all our lives. Car journeys are controlled by signs which tell us when to slow down, speed up, or when to keep our speed below a certain value. Trains carry passengers from city to city at high speeds and timetables tell when these trains arrive and depart. In order to ensure that timetables are reliable trains must travel at given speeds.

An Olympic runner can run one hundred metres (100 m) in 10 seconds. The runner is travelling at an average speed of 10 m/s. Humans can reach speeds of up to 12 m/s – but we can't keep it up for long.

In the 'quest for speed' a range of machines have been developed to travel at high speeds on land, sea and air.

In 1983, the world land-speed record was captured by Thrust 2 which travelled at an average speed of 1019 km/hr (283 m/s). The rules for such a record attempt are very strict. The car must be timed over a measured kilometre or mile, first in one direction and then in the other. Both journeys must be completed within one hour.

Form of transport	Typical speed (m/s)	World record (m/s)
Man walking	2	~4.5
Man running	7	~12
Bicycle	15	27
Car	25	283
Train	40	143
Speedboat	30	97
Hovercraft	20	47
Aeroplane	250	980

In 1889 this first Daimler reached a speed of over 20 km/hour.

In 1983 Richard Noble captured the world land speed record in Thrust 2 (1019 km/hr).

Every split-second counts

For thousands of years humans have tried to run faster and faster. Races are usually run over fixed distances so that only the times need be compared, not the speeds.

A complex system of electronic timing is used to measure the times of today's athletes. When the starter's pistol is fired, the hammer falls on two metal contacts, completing a circuit that starts the clock. On the finishing line there is a light-beam. When this beam is cut by the first runner to cross the finishing line, the clock stops. Electronic timing systems are used for all types of sporting events including skiing, athletics, and car racing.

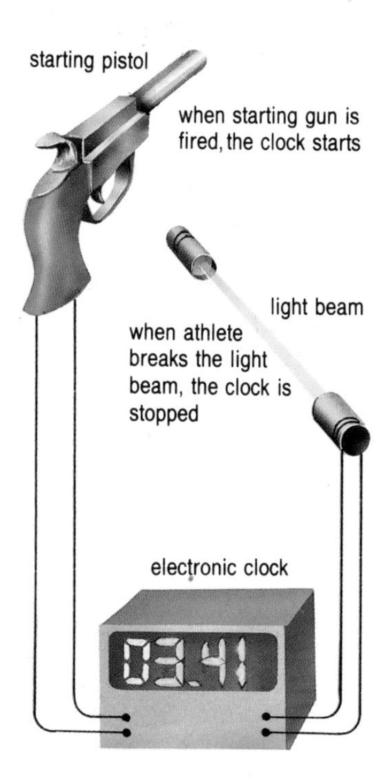

1 Use this table of data to produce a bar chart that shows how animals' speeds compare with man's running speed.

Animal	Recorded speed (m/s)
Horse	25
Cheetah	40
Swift	60
Antelope	35

2 This table shows part of the timetable for the Glasgow to Aberdeen Inter-City Express.

Station	Distance (km)	Depart Time
Glasgow	0	09.25
Perth	100	10.25
Dundee	142	10.49
Aberdeen	250	12.25

Use this data to find:

a) the average speed for the whole journey;
b) the stages of the journey with the highest and lowest average speeds.

3 The world record times for the 100 m sprint are 10.23 s for women and 9.83 s for men. Find the average speed in each case.

4 Suggest a reason why Thrust 2 had to be timed twice, first in one direction and then in the other.

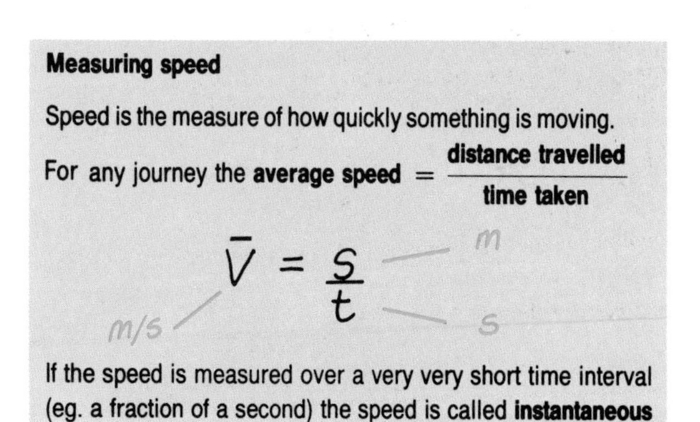

›› *Use an electric clock, with 'make-to-stop': 'make-to-start' connections, to design a timing system which starts when two contacts are made and stops when a light beam is broken.*

Keeping track of speed

The downhill racer

Have you ever watched a downhill ski event on TV? The skier bursts out of the starting gate and shoots off down the hill. Opening the gate automatically starts an electric clock. The time for the run is then recorded and displayed on the TV screen. The fastest skier will be the one who completes the race in the shortest time.

In addition, two light beams are often placed a short distance apart on the steepest part of the slope. By knowing this distance and by measuring the time the skier takes to travel between the two beams, the skier's speed at that place can be calculated.

As the skier opens the start gate the clock begins

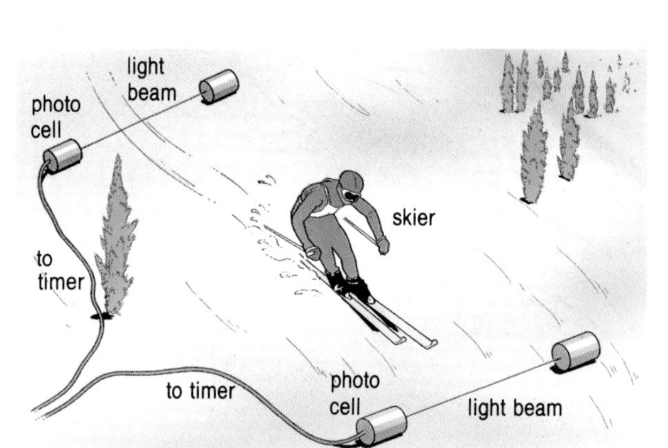

The skier's speed is measured by finding the time she takes to travel between the two light beams a known distance apart.

Watch your speed

Every car has a speedometer that tells the driver how fast she is going. A car speedometer records instantaneous speed and works as shown in the diagram.

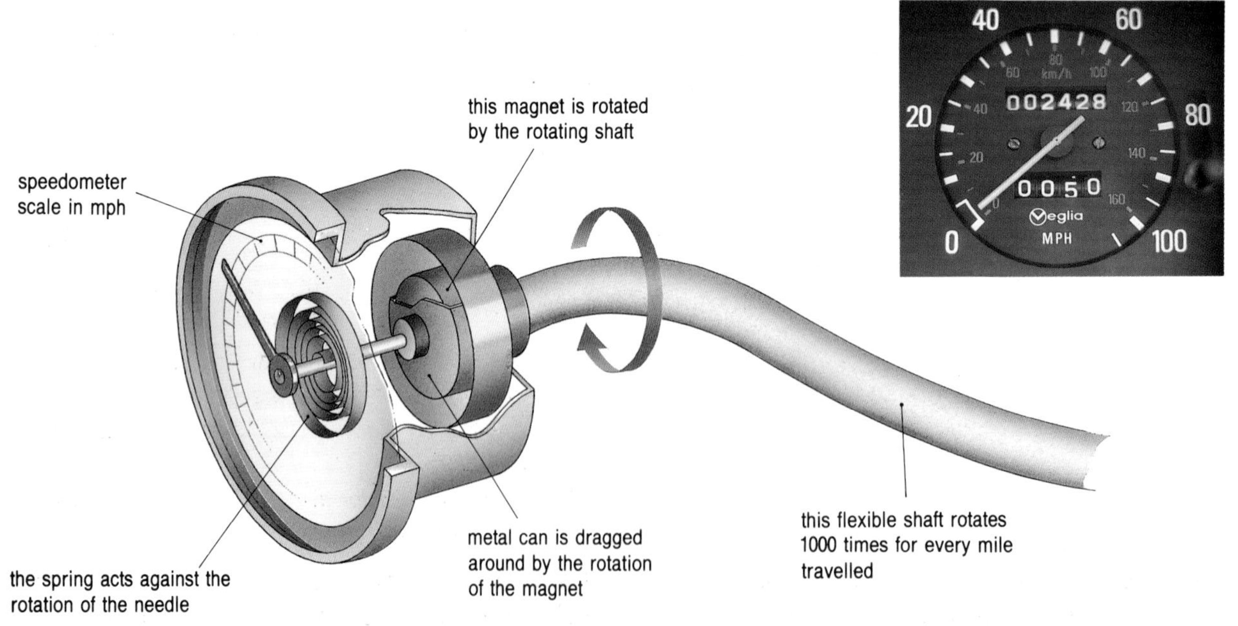

The faster the car goes the faster the shaft rotates and the more the pointer moves across the speedometer scale.

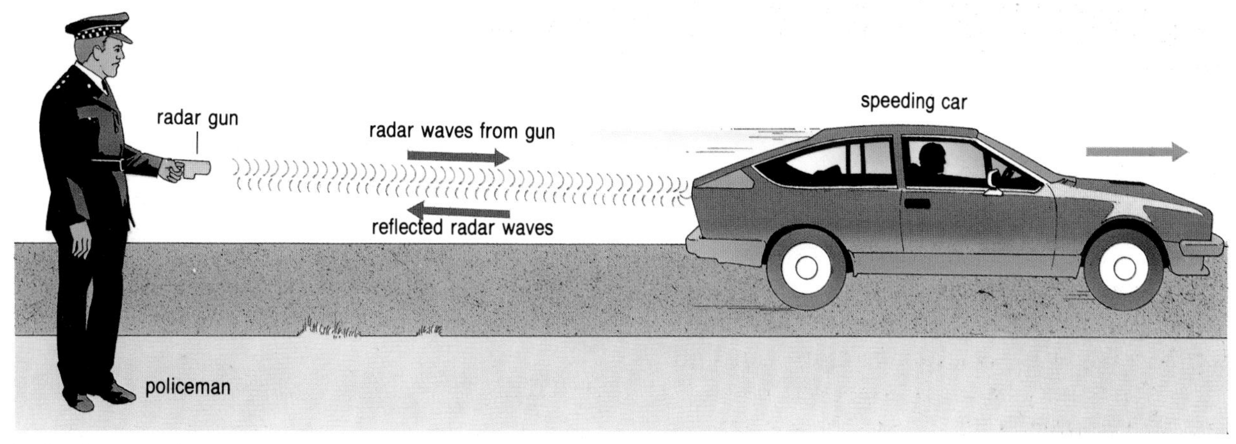

Finding the speed of a car.

Traffic cop

The traffic policeman uses a radar gun to find the speed of cars. The gun emits radar waves which are reflected off the target car. If the car is moving away from the gun, the reflected waves have a longer wavelength than those sent out. If the car is coming towards the gun, the reflected waves are shorter. The gun picks up the reflected waves, and from the difference in wavelengths of incidence and reflected waves, the speed of the car can be calculated and displayed. This method measures the instantaneous speed of the car.

Practical ways of measuring speed

Average speed is calculated by timing how long an object takes to travel a measured distance.

Photocell timer The timer measures the time, t, taken by the moving vehicle, of length d, to pass through the light beam.

$$\bar{v} = \frac{d}{t}$$

Ticker timer and tape The moving vehicle pulls a ticker tape through a timer making 50 vibrations per second.

The time for 5 spaces is 0.1s.

$$\bar{v} = \frac{d}{0.1s}$$

›› *A trolley runs down a slope. Devise one way of measuring experimentally: **a)** the average speed between two points, which are 1 metre apart; and **b)** the instantaneous speed at the bottom of the runway.*

1 At the end of a ski-race the three best times were displayed.

J Jones	2 min 31.15 s
B Brown	2 min 31.13 s
P Pearson	2 min 31.24 s

a) Which skier won the race?
b) Which skier had the highest average speed?

2 The graph shows how the speed of a car varies with time during a journey.

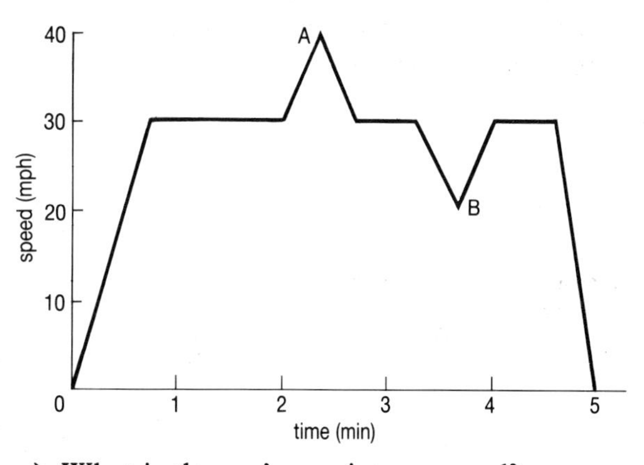

a) What is the car's maximum speed?
b) *Estimate* the car's average speed.
c) Explain why your answers in **a** and **b** are different.
d) Suggest an explanation for the shape of the graph at A and at B.

3 Use the speedometer scale on p.126 to draw a graph of speed in mph against speed in km/h. Hence devise a rule for changing mph into km/h.

Putting your foot down

Stepping on the accelerator

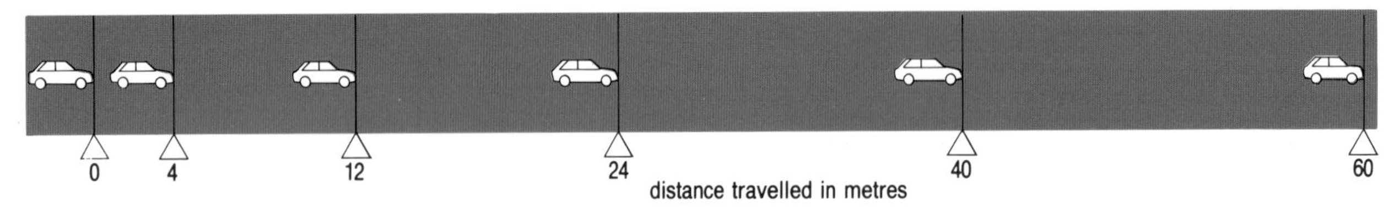

A car accelerating uniformly from rest.

Suppose we fix a camera to a tripod and then take a photograph of a car every two seconds. By combining all the photographs we would get a record of the car's movement. The diagram above shows that the car's speed is increasing by equal amounts every 2 seconds. The car is said to be travelling with uniform or constant acceleration.

A similar pattern of movement might be formed at the start of a 100 metres race.

This athlete is accelerating uniformly from rest.

Which car?

All new cars are provided with performance data so that buyers can compare them. Which of these cars can **a)** go fastest, **b)** accelerate quickest 0–60 mph, **c)** accelerate quickest 30–50 mph?

As well as being able to accelerate quickly from rest, a car must be able to accelerate from a steady speed. This allows it to overtake other cars quickly and safely. Performance data usually show how quickly a car can accelerate from 30–50 mph in fourth gear. Which car would you buy if the cost for each were the same?

max speed mph	94.9
0–60 mph sec	13.7
30–50 mph in 4th, sec	11.1
overall mpg	27.8
touring mpg	—

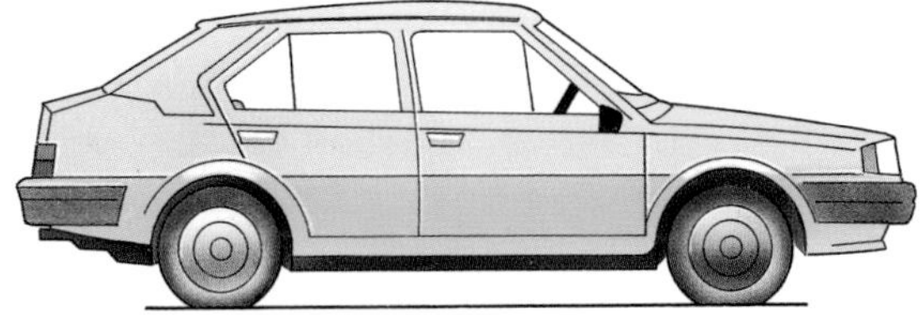

VOLVO 340 GL 5 DOOR

max speed mph	96.4
0–60 mph sec	12.0
30–50 mph in 4th, sec	12.3
overall mpg	31.2
touring mpg	38.4

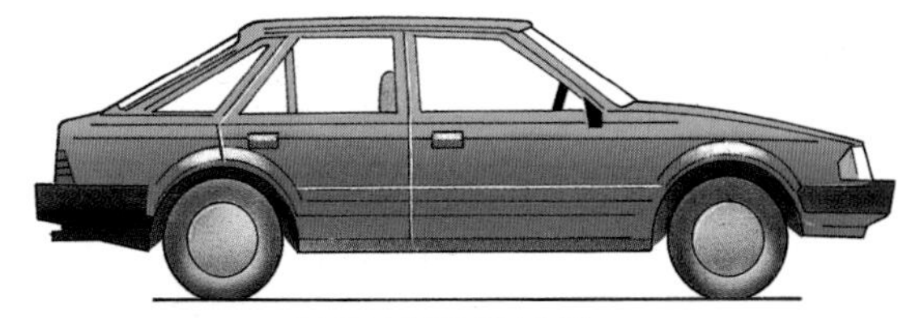

FORD ESCORT 1.3 GL

max speed mph	96.3
0–60 mph sec	12.3
30–50 mph in 4th, sec	11.3
overall mpg	34.4
touring mpg	39.1

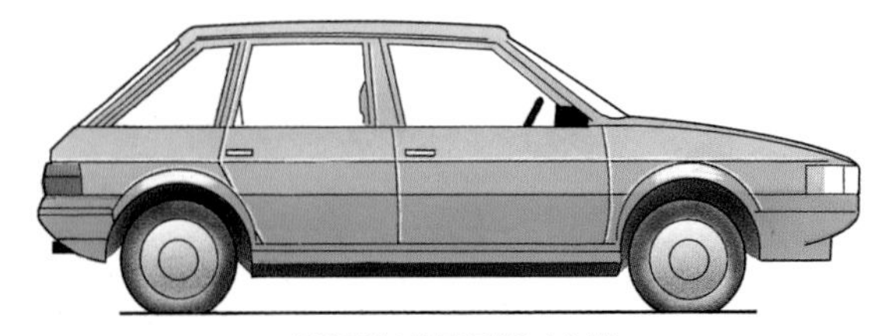

AUSTIN MAESTRO 1.3 HL

Stepping on the brakes

A car is made to accelerate by pressing on the accelerator; and made to decelerate by pressing on the brake pedal. The Highway Code gives information on how quickly a car can decelerate to rest, in an emergency and under ideal conditions. The faster the car is going when the brakes are applied, the longer the car takes to stop and the further it travels. We will return to the Highway Code in a later unit.

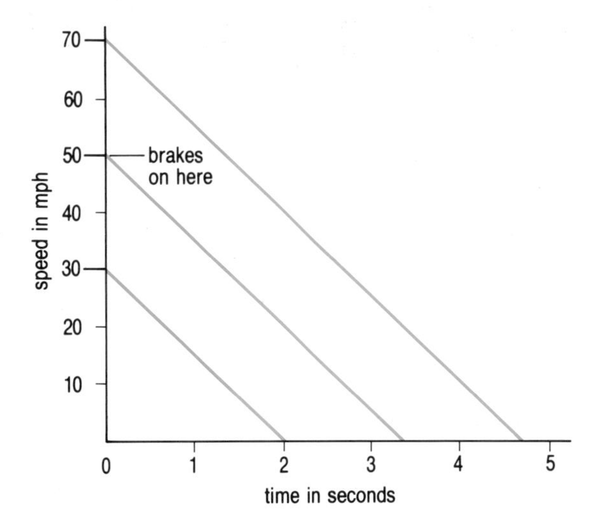

Speed-time graphs showing deceleration of cars going at different speeds.

›› *Use the above data to construct a record of a car's deceleration to rest from 30 mph, 50 mph and 70 mph. Use 1 second intervals.*

1 If a car accelerates uniformly from 0 to 10 m/s in 5 s what is its acceleration?

2 A train travelling at 30 m/s decelerates uniformly at 2 m/s². How long does it take to come to rest?

3 Use the data in the diagram of the accelerating car to:
a) calculate the average speed between photos;
b) draw a speed-time graph of the car's motion;
c) calculate the car's acceleration.

4 Use the data on p 128 to copy and complete this table.

Car	Acceleration 0–60 mph	Acceleration 30–50 mph
Maestro	m/s²	m/s²
Escort	m/s²	m/s²
Volvo	m/s²	m/s²

5 The Highway Code gives braking information 'under ideal conditions'. What do you think is meant by this? Describe conditions that would not be ideal and explain how the car's deceleration would be affected by these conditions.

Speed (mph)	Speed (m/s)	Time to stop (s)	Distance to stop (m)
30	13	2	13
50	22	3.3	36
70	31	4.7	73

Acceleration

If a moving object changes speed we say it is accelerating.

$$\text{acceleration} = \frac{\text{change in speed}}{\text{change in time}}$$

$$a = \frac{v - u}{t}$$

$$\textit{or}\ v = u + at$$

Speed-time graphs

This shows a vehicle moving at constant speed.

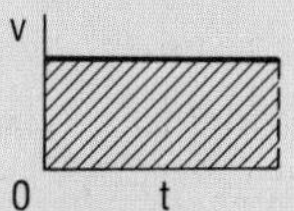

This shows a vehicle moving from rest with constant uniform acceleration.

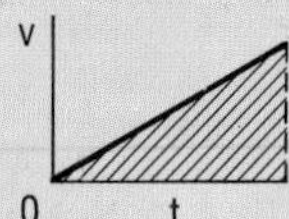

This shows a vehicle decelerating to rest with constant uniform deceleration.

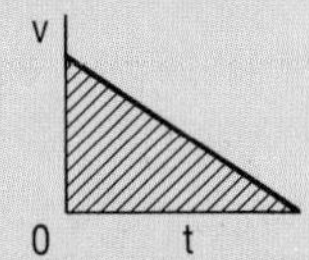

In each case the shaded area under the v–t graph is equal to the distance travelled.

Forces in action

Speeding up and slowing down

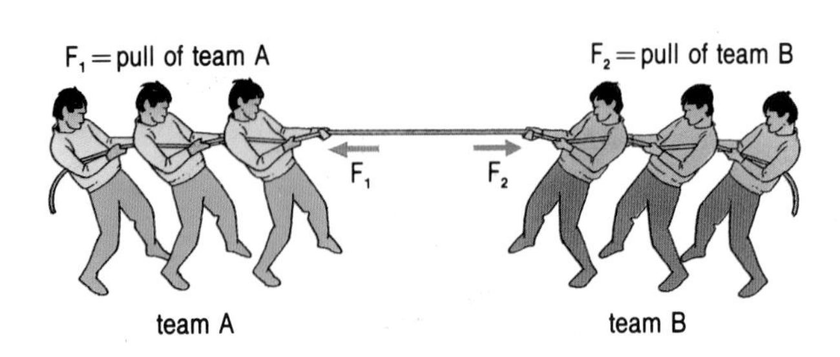

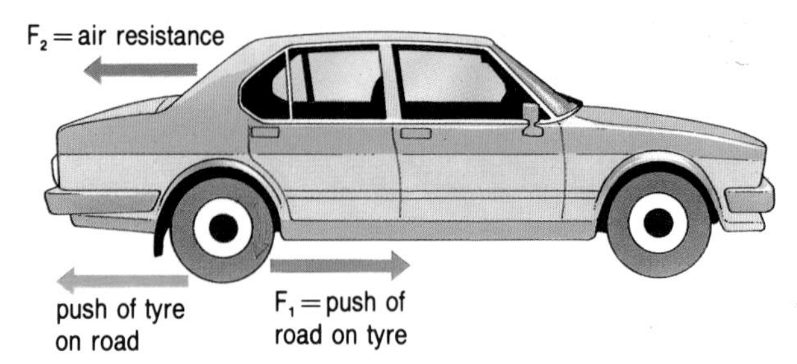

If, in a tug of war, both sides pull with equal force, nothing happens. But if one team exerts a larger force than the other both teams accelerate in the direction of this larger force.

In a speedboat the propellor pushes the water backwards. In return the water pushes the speedboat forwards (Newton's third law). As long as this forward force is greater than the backwards forces, the boat will accelerate forward.

Each diagram illustrates the basic principles of propulsion. If forward forces are greater than backward forces, the vehicle will accelerate forwards. If forward and backward forces are equal, or *balanced*, the vehicle will stay at rest or keep going at constant speed. If the backward forces are greater than the forward forces, the vehicle will slow down, i.e. decelerate.

The rear wheels of a car push backwards on the road. In return the road pushes the car forwards. As long as this forward force is greater than the backwards forces, the car will accelerate.

A jet engine ejects hot gases which are pushed backwards at high speed. The gases exert a forward force on the aircraft engine. Forward acceleration occurs if this forward force is greater than the backward forces.

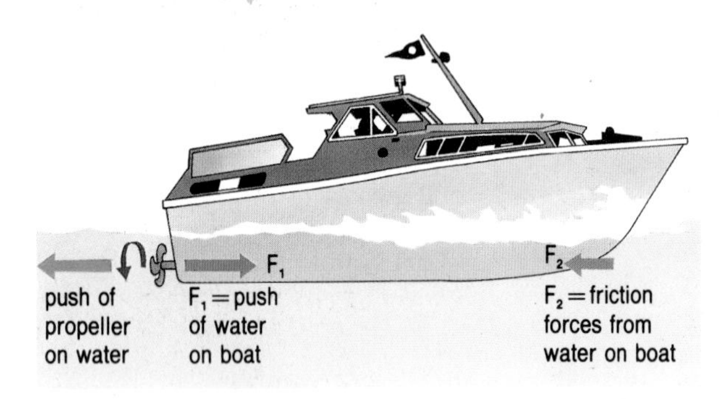

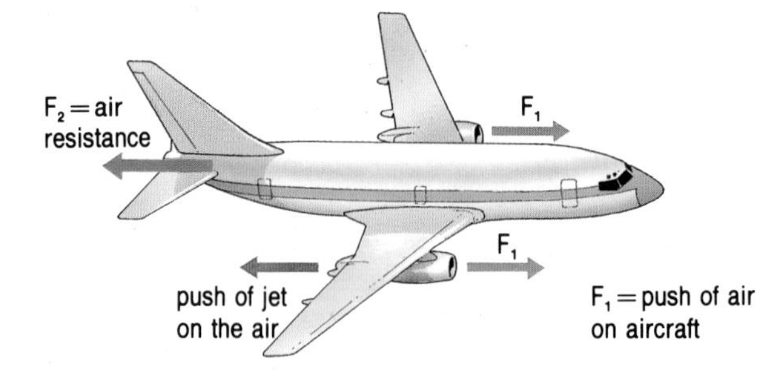

Changing direction

In order to turn a corner a cyclist or a motor cyclist leans into the turn. This makes the wheel of the bike push outwards on the road. The reaction force of the road pushing inwards on the wheels makes the bike turn in the required direction. Change of direction can occur without change of speed.

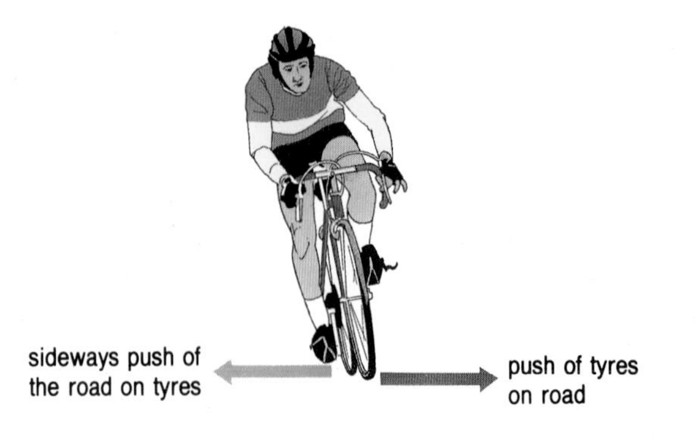

In aircraft turning is produced in much the same way. By moving the **ailerons**, one up and the other down, the aircraft is made to tilt. The lifting force on the wing now acts partly up and partly sideways. The upward force is needed to balance the weight of the aircraft (i.e. to keep it up), while the sideways force pushes the plane round in the required direction. In this example, both forces have changed the direction of the aircraft without changing its speed.

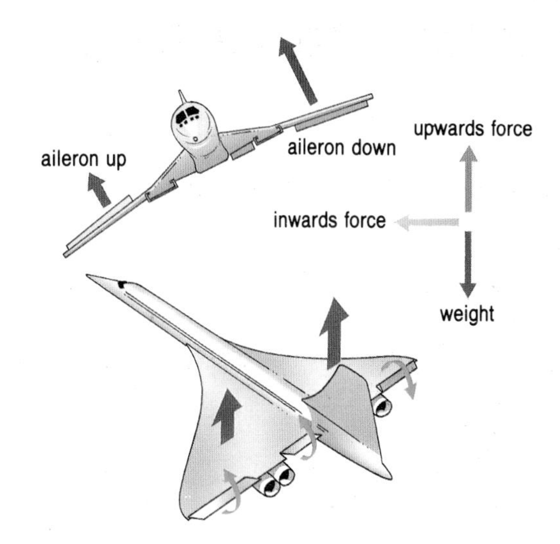

Force and shape

High-speed photography can show that during a squash match both racquet and ball change shape quite severely on impact – but only for a short time. When vehicles crash, the impact forces cause severe and permanent changes of shape. Cars are usually designed so that the passenger cabin will not buckle too much in a crash (p.136). All new car models are crashed under test conditions to see what effects the impact forces have on the car's shape. In the world of atomic energy, special containers, called Magnox flasks, have been designed to carry radioactive waste from nuclear power stations to processing plants. These are designed so that they are strong enough not to break open if they were involved in a rail crash.

1 Write down three effects that forces can produce.

2 Imagine a ball being dropped on to a hard surface then rebounding. For each stage of the ball's movement, the falling stage, the impact stage, and the rebounding stage, describe all the forces acting on the ball.
Describe the effect of each of these forces on the ball's speed, direction, and shape.

›› *As part of the testing programme for the Atomic Energy Authority's Magnox flasks, each flask is allowed to fall from a height of nine metres on to a hard surface. This test should ensure that it does not break open. Design and test a suitable container made of paper that will allow you to drop a raw egg from a height of one metre on to a hard floor without breaking.*

Forces

When forces act on something they can change:
a its shape;
b its speed;
c its direction.

Newton balance The Newton balance (or spring balance) is used to measure forces. The length by which the spring changes in a Newton balance is directly proportional to the size of the forces causing the change.

Weight The weight of a body is equal to the force of attraction caused by the Earth's gravitational field.

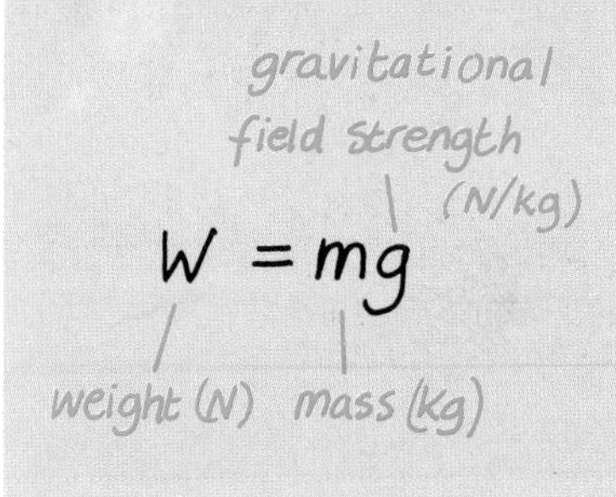

Friction Friction is a force which exists when two surfaces slide or try to slide over each other.

Newton's Third Law If A pushes B to the right, B pushes A to the left with the same size of force.

Mass and movement

Inertia

The largest man-made structure ever to move across the surface of the Earth was the Ninian Oil production platform. It needed eight large tugs to set it in motion and tow it at 2 mph from the yard where it was built to its final position in the North Sea. The journey of 516 miles took 12 days.

The Ninian central oil platform on tow. This structure has a mass of about 6×10^8 kg.

Objects that are difficult to start moving are also difficult to stop once they are moving. A loaded oil super-tanker, moving at 12 m/s, needs several kilometres before it can be brought to rest. It tends to keep on going. The property of matter that makes it stay at rest, or to keep on going (at constant speed in a straight line) is called **inertia**. A more common name for inertia is **mass**.

This moving tanker takes up to five minutes to stop.

The greater the inertia of an object the more difficult it is to accelerate or decelerate it. Inertia, or mass, is measured in kilograms.

The standard kilogram is a cylinder of platinum-iridium that is kept in Sevres in France. All other masses are compared against this standard mass.

Once moving, a large mass is difficult to stop.

Force overcomes inertia

Large aircraft normally need a long runway for landing. However, if a parachute is attached to the plane, a bigger braking force and therefore a bigger deceleration, is produced. This means a shorter landing distance is needed.

Suppose an unladen lorry can stop in 5 s from a speed of 30 mph. When it is loaded the lorry takes twice as long to stop. Its deceleration must be much less. This shows that the bigger the mass the smaller the deceleration – assuming of course that the braking force is the same each time.

If you have ever had the misfortune of having to push a car that won't start or that has broken down, you will know that it is very difficult to get it to begin to move. The greater the mass of the car, the less the acceleration. The bigger the push, the greater the acceleration.

These examples show Newton's Second Law in action. The acceleration of an object can be calculated using this equation.

$$a = \frac{F}{m}$$

a: metres per second per second (m/s^2)
F: newtons (N)
m: kilograms (kg)

By increasing the push and reducing the mass, a greater acceleration can be produced.

1 What is the acceleration of a mass of 5 kg when pulled by a 2N force?

2 What force is needed to accelerate a 20 kg mass at $0.2\,m/s^2$?

3 The supertanker shown above has a mass of 5×10^8 kg. If its deceleration is $0.8 \times 10^{-2}\,m/s^2$, what is the value of the decelerating force acting on it?

4 Railway signs, indicating control lights, are usually placed at least 1 km before the lights. On the roads, however, signs warning of traffic lights are placed only about 100 m before the lights. Explain this difference.

5 In a car crash, the occupants of the car appear to be flung forward as the car is suddenly decelerated.
a) Explain, in terms of physics principles, why this happens.
b) What would happen to the occupants in a stationary car if it were hit from behind? Explain your answer.

›› *Clamp a hacksaw blade to the desk. Attach a 50 g lump of plasticine to the end of the blade. Time 50 swings. Do the same for 75 g and 100 g lumps of plasticine.*
A *Describe why this experiment demonstrates the property of inertia.*
B *Take two lumps of plasticine of roughly the same size. Use the technique shown above to find which has the larger mass.*

Newton's laws of motion

Newton's First Law states that a body will stay at rest or keep on moving in a straight line at constant speed unless an unbalanced force acts on it.

Newton's Second Law states that the acceleration of a body is directly proportional to the force causing the acceleration, and inversely proportional to the mass of the body.

$$a = \frac{F}{m}$$

a: m/s^2; F: N; m: kg

Acceleration due to gravity

The weight of an object divided by its mass gives the acceleration due to gravity.

$$g = \frac{W}{m}$$

g: m/s^2; W: N; m: kg

Getting to grips with friction

Friction: friend or foe?

In the simple braking system of a bicycle, one surface (rubber) rubs against another (metal). The greater the force with which the two surfaces are pushed together the greater the friction.

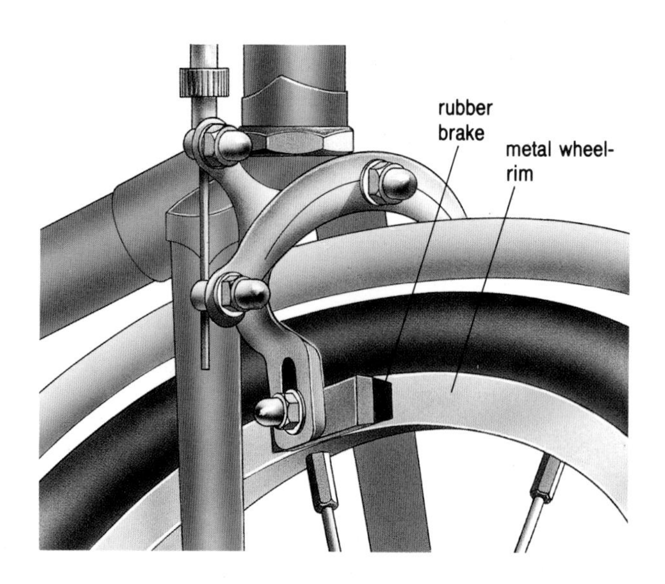

When a parachutist jumps from an aircraft, she will, at first, accelerate downwards. As her speed increases so will the air friction, until it exactly balances her weight. She will then be travelling at a steady speed – called the terminal velocity. When the parachute opens the air friction suddenly increases and the parachutist rapidly decelerates to a lower terminal velocity.

The aerofoil at the back of this racing car produces a large downward force on the rear wheels. This increases the friction between the wheels and the road enabling the car to turn corners at higher speeds without skidding.

Tyre and sports shoe manufacturers spend large sums of money trying to design products which will give maximum grip under all conditions. The patterns on car tyres are designed so that in wet conditions, water will be squeezed out. This improves the grip so much that it is now illegal to use tyres in which the grooves are less than 1 mm deep.

In most machines friction needs to be reduced between moving surfaces in order to improve performance, for example to reduce wear and tear and to prevent over-heating. Friction can be reduced by:

- lubricating;
- making surfaces smoother;
- changing shape (stream-lining);
- reducing the force of contact between the surfaces.

Any movement that depends on one surface pushing against another would be impossible without friction. Without friction it would be impossible to run, walk, cycle, or drive. Even if you could get a car going, it would be impossible to stop or to steer without friction.

A racing cyclist knows how to reduce friction by keeping a low profile.

No surface examined with a powerful microscope will ever appear perfectly smooth. So when two surfaces slide over one another some very tiny high points on both surfaces catch on to each other. This causes friction.

It is not only solids that have friction. A parachutist is slowed down by air resistance.

A boat also experiences opposite forces to its movements. Tiny 'whirlpools' (eddies) round the boat produce fluid friction or **drag**.

At high speeds the frictional forces on a boat can be very large. The hydrofoil overcomes this problem by using an underwater aerofoil to lift it out of the water at high speed. The smaller the area in contact with the water the less the drag.

Great care is taken to design a car so that the air flows smoothly over its surface. New cars are tested in wind tunnels to see if the air flow is smooth and does not produce eddies. This design process is called **streamlining.**

1 Make two lists showing ways in which friction can **a)** be helpful **b)** be a nuisance.

2 Draw a speed-time graph for a parachutist's fall.

3 Find out how an aerofoil wing works. Explain the difference between the ways it is used in an aircraft, in a racing car, and in a hydrofoil.

4 Explain why braking on wet surfaces is more difficult than on dry surfaces.

Making cars safer

'Gin a body hit a body'

A safe car is one that has a strong rigid passenger section which is not easily crushed. The front and rear sections that surround the passenger section are designed to crumple in a crash. In doing so these 'crumple zones' absorb much of the car's kinetic energy. Therefore as little energy as possible is transferred to the inside of the car, and so the passengers sustain as little injury as possible.

In this car the passenger section is made of strong steel struts. It also has steel tubes inside the doors to give good protection from a side-on collision.

The car has large rubber bumpers which absorb a lot of energy on impact, and the front and rear sections are designed to crumple on impact. As well as protecting passengers inside the car, some car designs protect the pedestrian who may be hit by a car. By having a 'shovel nosed' bonnet, any pedestrians hit by the car will be thrown on to the bonnet instead of being knocked down under its wheels.

Car being drop tested.

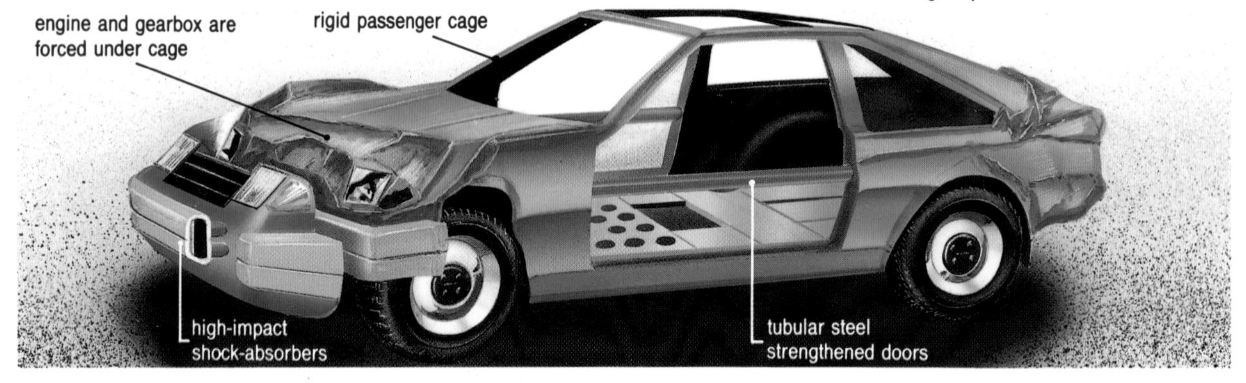

Design for safety

A designer has to think about three main problems when designing a car.

1 How to make a car look good.

2 How to make it perform well.

3 How to make it safe for driver and passengers.

If you were a car designer, what steps would you take to make a new car safe? The list shows some of the safety features that you may find in modern cars. Of the features listed here only front seat belts and good tyres are compulsory.

Modern features of car design
Effective braking system. Good tyres – minimum depth of tread. Seat belts – front and rear. Better design of passenger compartment. Shatterproof glass in windscreen. High impact shock absorbers in bumpers. Crumple zone in engine space. Shovel nose design for bonnet. Collapsible steering column. Non-rigid steering wheel. Padded dashboard. Head rests for driver and passengers.

Clunk Click!

If a car crashes while going at 40 mph, the passengers in the car will keep on going at 40 mph until a force acts to change that speed (Newton's First Law). This means that the passengers will crash into the parts of the car immediately in front of them. Seat belts help to prevent injuries in such a situation by providing decelerating forces which slow down the forward movement of the passengers. Even at relatively slow speeds, a driver or front passenger who is not wearing a seat belts may smash through the car windscreen.

Internal car design

Unfortunately even when wearing seat belts people in cars still get thrown violently forwards. A driver can suffer serious injury by being thrown hard against the steering wheel. To reduce injury, some steering columns are designed to collapse in an emergency. The driver then decelerates fairly slowly but safely as the steering column 'gives'.

Did you know that your head is one of the heaviest parts of your body? Because of its large mass, your head is more difficult to decelerate than the rest of your body. So in a crash the head appears to shoot forward and is then whipped back by the springlike action of the neck. Such whiplash effects can cause serious spinal injuries, but head rests can reduce this violent movement of the head.

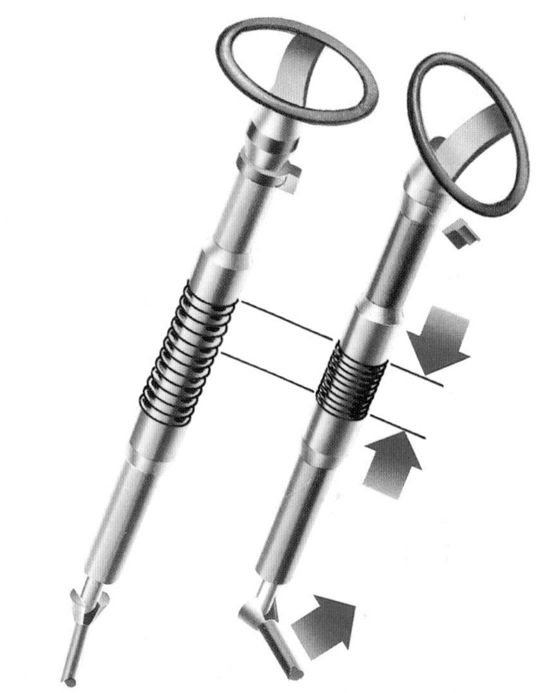

On impact this steering column collapses and the steering wheel tilts to avoid cutting into the driver's body.

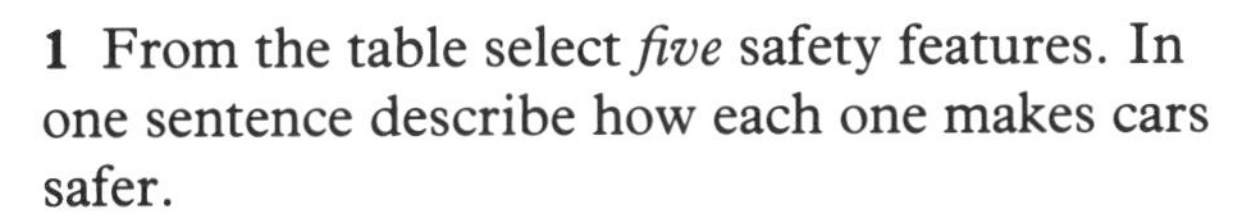

1 From the table select *five* safety features. In one sentence describe how each one makes cars safer.

2 Write a short paragraph describing the order of importance *you* would give to looks, performance, and safety, if you were designing a new car.

3 Make a list of any safety features you can see in the design of your bicycle.

4 A driver of mass 60 kg is moving at 20 m/s. He is stopped by a seat belt in 0.5 s. What force does the seat belt exert? (see p.133)

The Earth's gravitational field

Day in, day out

For thousands of millions of years gravity has made the Earth complete its elliptical journey around the Sun. Each orbit is 586 million miles long, takes $365\frac{1}{4}$ days, and involves the Earth moving at a speed of about 70 000 mph.

Twice every day the world's seas and oceans rise and fall by up to 10 m. This is how the tides are created. Gravity, then, is responsible for keeping the Earth in its orbit and for the ebb and flow in the tides.

What is gravity?

Every mass, whatever its size, exerts a force of attraction on every other mass. This force is called the **force of gravity**. Gravitational forces of attraction are very small and normally pass unnoticed. If, however, a mass is very large, like a sun, or planet or moon, the gravitational force can be very large and can exert a pull over very great distances.

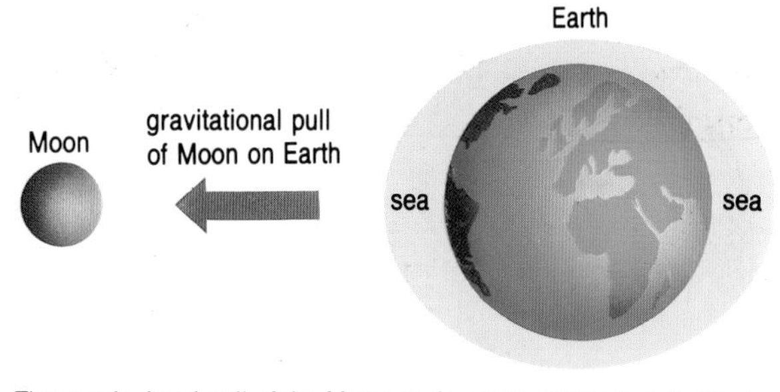

The gravitational pull of the Moon on the seas creates two 'bulges'. As the Earth spins these bulges appear to move around the Earth causing tides.

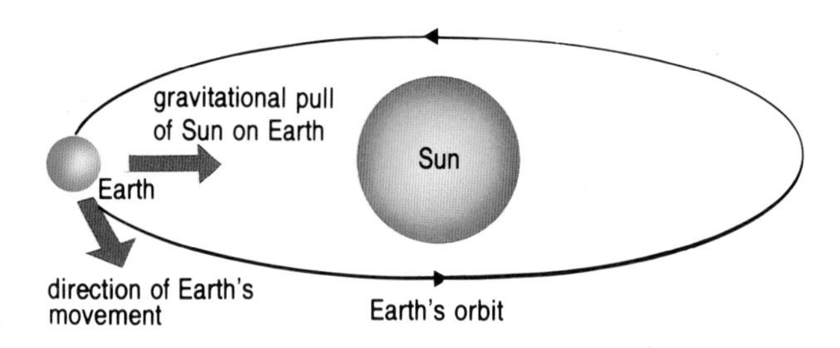

The gravitational pull of the Sun acts at right angles to the Earth's velocity. This results in the Earth following an elliptical orbit.

Driving the water cycle

One of the most important of nature's cycles is the water cycle. Without it the Earth would be a desert. The energy needed to drive this cycle comes from the Sun's heat, but gravity plays a key role in making it work. Each day about 10^{15} kg of water falls as rain throughout the world. An equal amount of water is evaporated each day by the Sun.

at high altitudes the water vapour condenses to form clouds

falling rain

rain water returns to the sea under the influence of gravity

the Sun evaporates water from the sea

A small part of the water cycle can be harnessed to provide electrical energy. Rain water that falls into a high loch, can be allowed to fall under the influence of gravity. The falling water is used to turn generators to produce electricity. This is the principle of hydro-electric power.

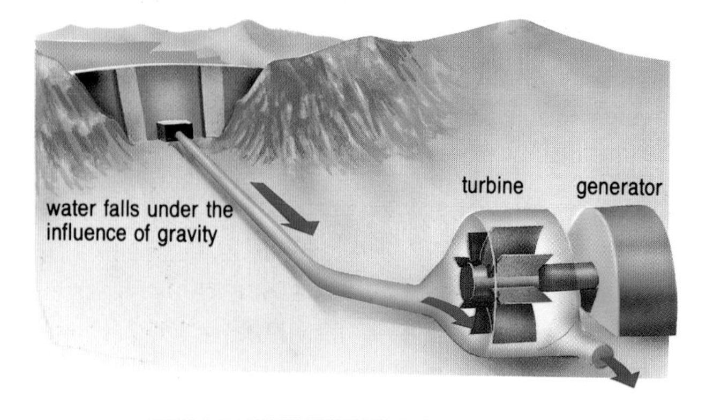

Gravitational field strength

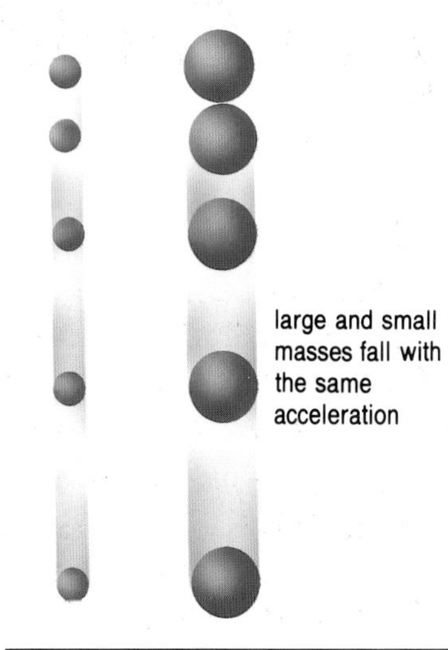

large and small masses fall with the same acceleration

Nearly 400 years ago Galileo discovered that all masses fall with the same acceleration. The downward force due to gravity acting on any mass is called its **weight**. Galileo's experiment showed that the ratio of weight to mass always gives the same answer. The quantity is called **gravitational field strength** and is measured in N/kg.

$$g = \frac{W}{m}$$

(g in N/kg, W in N, m in kg)

The value of the Earth's gravitational field strength is 9.8 N/kg. This means that the weight of 1 kg is 9.8 N.

The gravitational field strength of other planets in the solar system and on the Moon is different from that on Earth. Even on Earth, gravitational field strength varies a little from place to place. At the top of a hill, the gravitational field strength will be slightly less than 9.8 N/kg.

The acceleration due to gravity (g) is also equal numerically to the gravitational field strength. On Earth therefore $g = 9.8\,m/s^2$ (approximately $10\,m/s^2$).

Planet	Gravitational field strength (N/kg)
Earth	9.8
Mars	3.7
Venus	8.8
Moon	1.6

1 Find out what is meant by 'spring' and 'neep' tides. Explain how they arise in terms of the gravitational forces of the Sun and the Moon.

2 What is the weight of: **a)** a 5 kg mass **b)** a 20 kg mass, on a planet where the gravitational field strength is 6 N/kg?

3 On Planet X, the weight of an 8 kg mass was measured, using a newton balance, and found to be 96 N. What is the gravitational field strength on Planet X?

4 The Earth is slightly flatter at the poles than at the equator. Explain whether the force of gravity will be bigger at the poles or at the equator.

5 The experimental description on the right was found in Mary's lab book.
a) Use Mary's data to draw a graph of the length of the spring against its weight.
b) Can you suggest a more accurate way of describing the relationship between the spring length and the weight added?
c) Describe clearly how you would use the spring plus the graph to measure the force needed to pull a loaded trolley across the table at a constant speed.

Title : To find out how the length of a spring changes with different weights.

Spring
mass
ruler

What we did

We hung up a spring and measured its length. We then hung on 50g and measured its length again. We did this for masses of 100g, 150g and 200g and made a table of the results.

Results

Mass	Weight	Length
0	0	25 cm
50 g	0.5 N	30 cm
100 g	1.0 N	35 cm
150 g	1.5 N	41 cm
200 g	2.0 N	47 cm

We found that the bigger the force the longer the spring becomes.

Energy transformations

The bicycle race

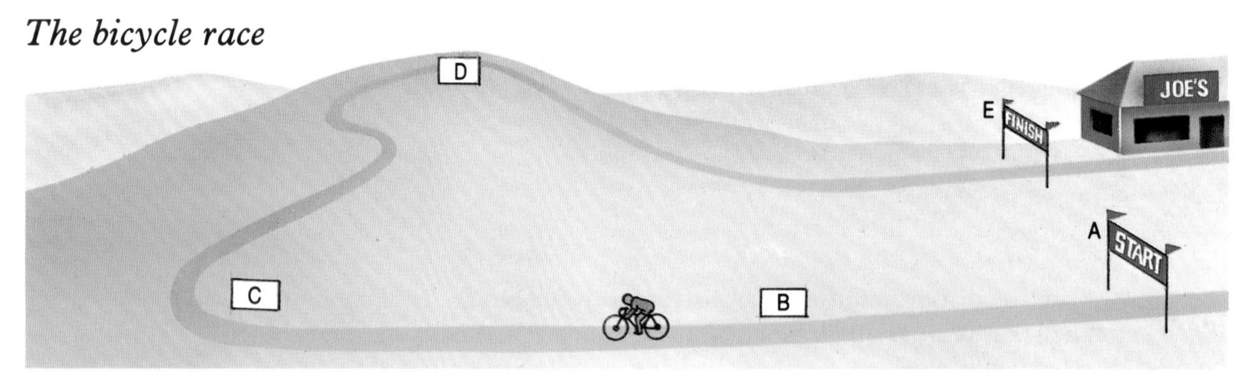

In a cycle race, Wasim started at A and accelerated to B. From B to the foot of the hill at C, the road was straight and flat so Wasim went at a fast but steady speed. Up the hill he went slowly but steadily. By the time he got to the top he was out of breath so he stopped briefly. Wasim then free-wheeled down the hill to the finish at E, where he braked and skidded to a halt. Wasim didn't win the race but consoled himself with a burger and a chocolate milk shake at Joe's cafe.

stage	type of movement	energy changes	
A→B	accelerating	chemical energy	→ kinetic energy → heat
B→C	constant speed	chemical energy	→ heat
C→D	constant speed	chemical energy	→ ______ → ______
D→E	acceleration then constant speed	potential energy	→ ______ → ______

1 Copy and complete the table for Wasim's journey.

2 This bar chart shows the energy Wasim supplied in stage AB. Copy and complete roughly the bar chart for the other stages.

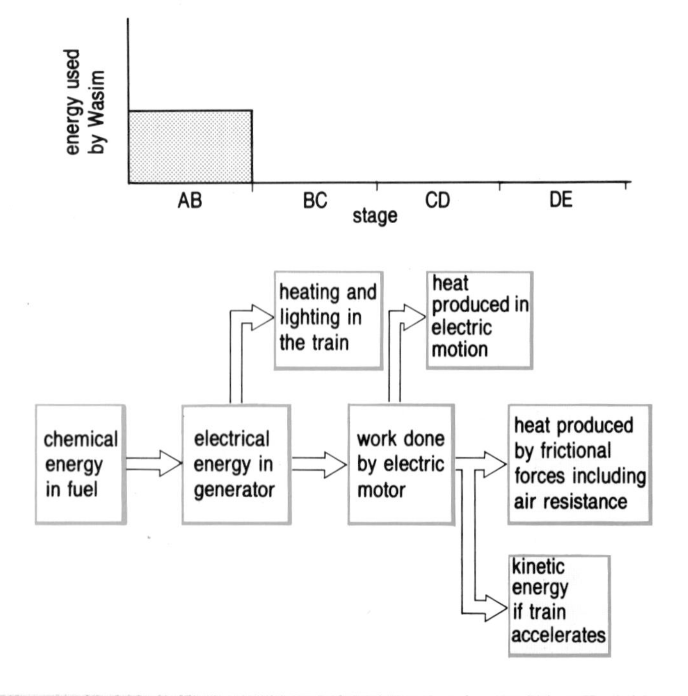

Electric trains

These are two common types of electric train.

a The diesel-engine train in which the chemical energy of the fuel is converted to various other forms of energy as shown in the block diagram.

b The electric train, in which the electrical energy used is picked up directly from an overhead cable system, or from a third rail laid between the other two.

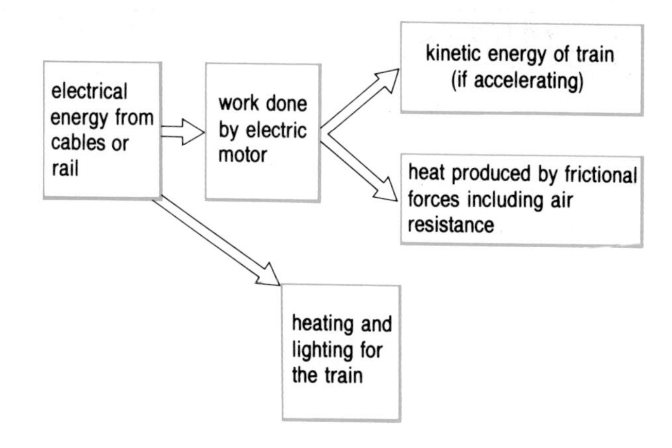

The diesel-electric train must carry all its fuel with it and so it needs to be filled up every few hundred miles. The electric train carries no fuel but can only run on lines that have been electrified, i.e. lines with an overhead electricity cable carrying 25 kV ac or a third rail carrying a few hundred volts dc.

3 Name one other type of train. Construct a block diagram showing the main energy conversions in this train.

4 Suggest why it would cost a lot of money to convert Britain's railway system to 'all-electric'.

5 Some people have suggested that all diesel trains should be replaced by electric trains because electric trains do not cause smoke pollution. What is your view? Give reasons for your answer.

Rocket engine

A rocket motor in its simplest form consists of two tanks. One tank contains fuel (e.g. liquid hydrogen) which is a source of chemical energy. The other tank contains liquid oxygen which is needed to burn the fuel to release its energy. The two liquids are vaporized then pumped into a reaction chamber where the mixture is ignited. The resulting explosion produces a great deal of heat which makes the gases from the burning fuel expand rapidly. These gases are then expelled at high speed from a nozzle at the rear of the rocket. Chemical energy is converted to kinetic energy.

The force of the gases being expelled from the rear of the rocket motor propels the rocket forward by an equal but opposite reaction force (Newton's Third Law).

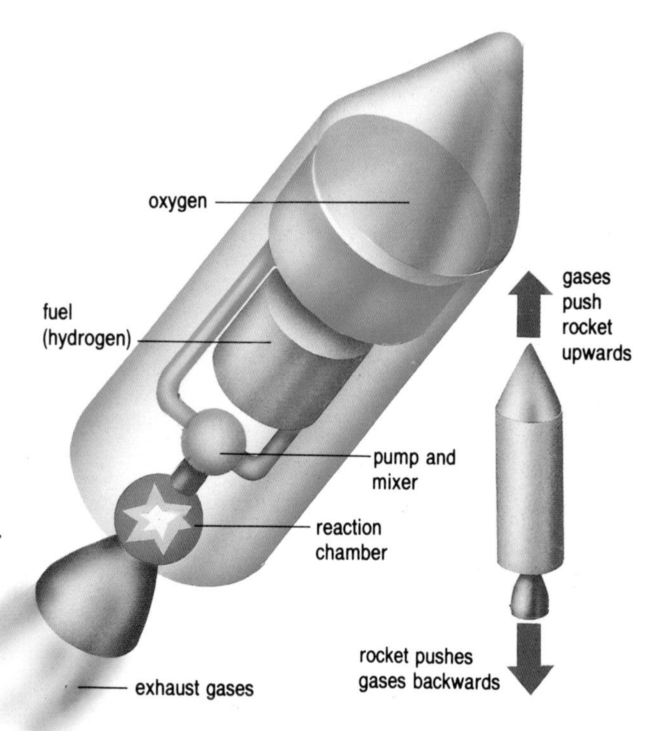

6 Suggest a reason why the fuel and the oxygen are pumped into the reaction chamber as vapours.

7 From the description of how the rocket works, draw a flow diagram showing the main energy changes that occur in the rocket motor.

8 Draw a diagram of a rocket about to take off. Show the forces acting in it. Explain in terms of these forces how take-off is achieved.

More for your money

Changing energy and losing energy

Most vehicles are powered by engines which use fuels such as petrol or diesel oil as their main source of energy. The engine uses this chemical energy to do work (force × distance).

This, in turn, produces:

- **kinetic** energy, when the vehicle is accelerating; and
- **heat**, because of the frictional forces acting against the moving vehicle.

Some of the original chemical energy is also changed into heat and sound by the engine.

In steam trains only about 7% of the fuel energy is changed into useful work. In modern diesel trains and in cars greater efficiencies can be achieved, but about two thirds of the original chemical energy of the fuel is still lost.

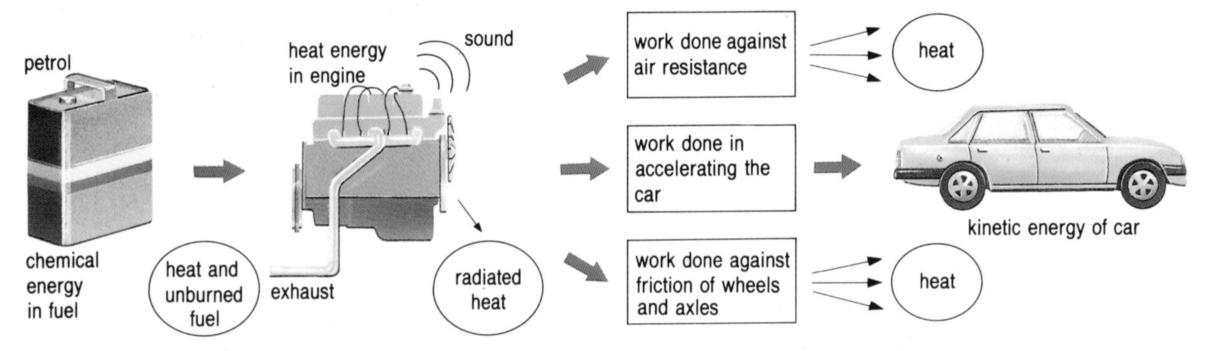

Energy changes in a car propulsion system.

Looking at fuel efficiency

Car manufacturers try to design cars which go further and faster and use as little fuel as possible. By law, every new car must be tested, and fuel data provided for the buyer. The table shows typical figures for these types of driving.

1 Urban or town driving.

2 Constant speed of 56 mph.

3 Motorway driving at 70 mph.

The examples opposite show that fuel consumption increases as the speed of the car increases. It also shows that driving in town, pushes up the fuel consumption as there are likely to be many stops and starts, i.e. decelerations and accelerations. The type of car and size of engine also affect these figures.

	Urban driving	Constant speed 56 mph	Motorway driving 70 mph
Car A	28 mpg	46 mpg	34 mpg
Car B	32 mpg	49 mpg	34 mpg
Car C	29 mpg	43 mpg	37 mpg

Better fuel consumption can be achieved in several ways.

1 Designing a more efficient engine.

2 Designing a car body that creates lower frictional forces.

3 Driving carefully at a constant speed.

By compressing the petrol vapour to a higher extent than before, modern car engines ensure that a higher proportion of the fuel is burned. This reduces waste.

It has also been discovered that by reducing the vibrational frequency of the pistons to less than 100 per second, better fuel consumption can be achieved.

Body shape is very important. Air must be able to flow smoothly past the car surface to keep the friction low. Even so, frictional forces increase greatly as the car's speed rises and this is why fuel consumption increases with speed.

Driving style also plays an important part, and the motto for the driver should be –

Drive gently and save fuel.

The most economical car in the world

Some family cars can now give the careful driver up to 120 kilometres per gallon (70 miles per gallon). Twenty years ago this would have seemed impossible. In a recent competition for car designers the winning entry achieved a fuel consumption of 2344 kilometres per gallon. However, in the USA a specially designed vehicle using a tiny diesel engine achieved the remarkable record of 5744 kilometres per gallon.

Driving hints for better fuel economy:
1 Keep tyres at the correct pressure.
2 Avoid fast starts.
3 Always accelerate smoothly.
4 Do not drive in low gear for longer than necessary.
5 Decelerate gently and avoid sudden braking.
6 Avoid frequently starting and stopping.
7 Remove roof racks when not in use.

1 Describe three ways in which energy is 'lost' in a car engine.

2 Give three reasons why town driving uses more fuel than open road driving.

3 Use the graph to answer these questions:

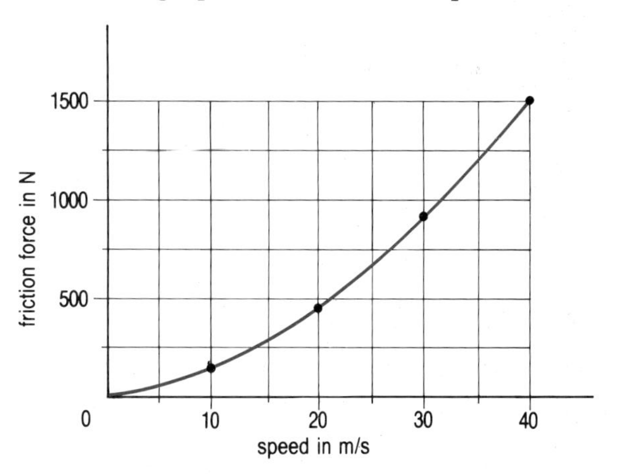

a) What is the frictional force when the car is travelling at 20 m/s and at 40 m/s?
b) Find the power of the engine when the car is travelling at 20 m/s and then when it is travelling at 40 m/s.
Use your answer to explain why more fuel is needed to travel at 40 m/s than at 20 m/s.

4 Official fuel consumption figures have been criticised by the Automobile Association because tests take place with *one* highly tuned car in ideal weather conditions. Can you suggest a better way to produce more typical fuel data?

On your bike

Energy transformations

The bicycle is the most popular form of transport in the world. Energy for propulsion is supplied by the cyclist so fuel bills are zero. The body acts as an engine in which chemical energy, stored in the fat and carbohydrates in the body's muscles, is used to drive the bicycle. Oxygen is needed to release this energy, just as it's needed to burn up other fuels.

Rotating the pedals with your feet makes the bicycle move forward. But the amount of forward movement depends on the gearing (p.151). The energy supplied by the cyclist's legs – the energy input – is used:

1 To do work in overcoming frictional forces. This produces heat.

2 To give the bike potential energy when going up a hill.

The bar chart shows the energy supplied by the cyclist each second. The darker blue areas show the power needed to overcome friction.

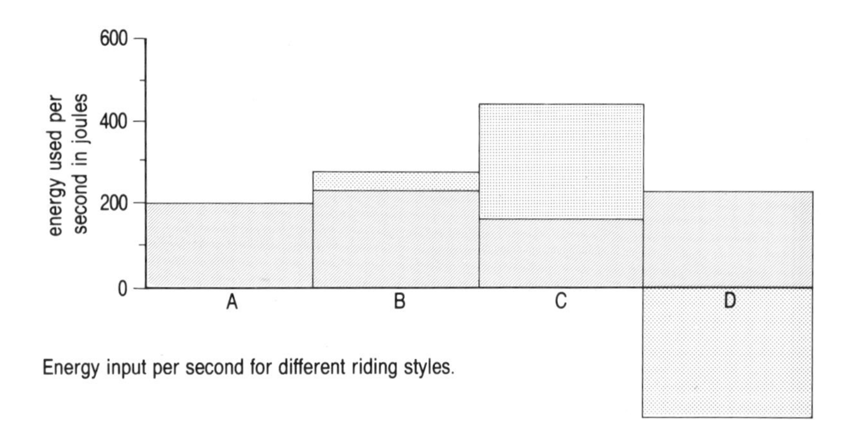

Energy input per second for different riding styles.

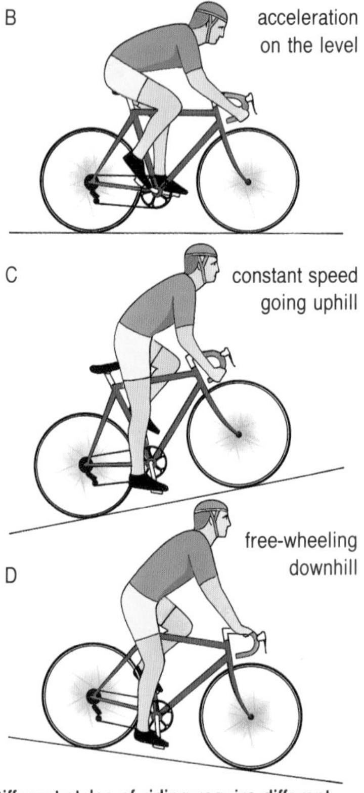

Different styles of riding require different energy inputs.

The body at work

The bicycle has been adapted to allow doctors to study the body as an energy converting machine (ergometer). In the bicycle ergometer one turn of the pedals make a point on the rim of the wheel move six metres. If it had been a normal bicycle it would therefore have gone forward a distance of six metres. A friction force can be created by applying a brake to the front wheel of the bicycle ergometer and this force (F) can be read from a scale. In tests the cyclist is meant to pedal at two revolutions per second so that the work done per second is:
$2 \times 6 \times F$ joules.

When working hard the body's muscles change chemical energy into other forms at a rate which is one hundred times greater than when the muscles are at rest. To work hard the muscles need more oxygen to burn up the body's fuel. This extra oxygen has to be supplied by the blood and so the body's pump – the heart – speeds up to increase the blood flow. The breathing rate also speeds up to supply more oxygen. The burning of fuel in the muscles produces waste products – in this case carbon dioxide and heat. These products are carried away from the muscles by the blood. The blood supplies oxygen to the muscles and removes carbon dioxide from them.

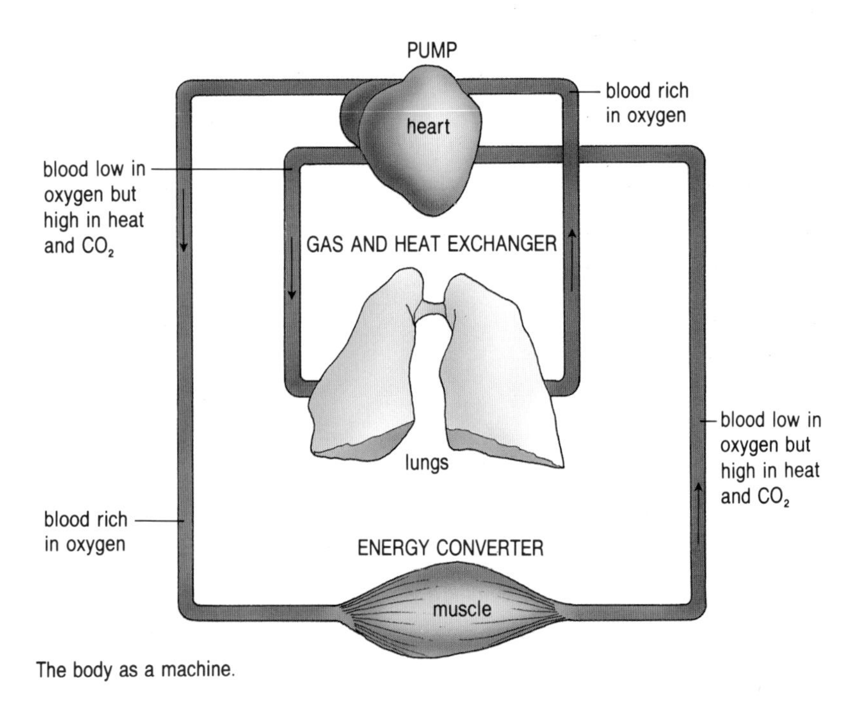

The body as a machine.

1 Produce a flow diagram showing how energy is changed for each type of motion shown in the diagram opposite.

2 In the ergometer described earlier a cyclist pedals at two revolutions per second, against a frictional force of 50 N. Calculate the rate at which the cyclist is doing work.

3 For these questions refer to the bar chart on p.144.
a) In movement A how much work is done per second? If the bike has a speed of 5 m/s what is the frictional force?
b) From the bar chart, write down the amount of work done per second against friction for movements A, B and C. Explain why these values are different.
c) What is the amount of work done per second in climbing the hill in C? If the mass of the cyclist and bike is 200 kg, what height is climbed per second?
d) In D, energy comes out of the system. Where does this energy come from? Work out how much kinetic energy is produced per second during movement C.

Work

Work is done whenever a force produces movement. The greater the force or the greater the distance moved, the greater the work done. Work is done when a force moves the point at which it acts through a distance.

$$E = Fd$$

E: newton metres (Nm) or joules (J); F: newtons (N); d: metres (m)

Power

Power is the rate at which work is done or energy is changed.

$$P = \frac{E}{t}$$

P: watts (W); E: joules (J); t: seconds (s)

Energy

Kinetic Energy

$$E_K = \frac{1}{2}mv^2$$

E_K: J; m: kg; v: m/s

Gravitational potential energy

$$E_P = mgh$$

E_P: J; m: kg; g: 10 N/kg; h: m

›› *Devise a way of finding the force of friction acting on a bicycle going at a speed of about 3 m/s.*

Lifting the load

Cable car family

There are many different ways of transporting people and materials from one level to another. To do this needs a great deal of energy for, as well as overcoming frictional forces that are found in all transport systems, the potential energy gained by the load must also be supplied.

In many systems this energy is provided by a stationary engine. This drives a continuous cable around pulley wheels at the top and bottom of a hill. Cabins or platforms to hold the passengers or materials are attached to this cable in a variety of ways to produce a wide range of transport systems.

In an aerial cable car a strong cable system, often made up of 2, 3, or 4 individual cables, is suspended from pylons. Cabins to carry the passengers ride along the cables in a set of grooved wheels. Another cable in the form of a continuous loop is driven by a stationary electric motor. This pulls the cabin up and down the hill.

In the San Francisco cable tram system, the cable pulling the trams is underground. Each tram is attached to the moving cable by a clamp which the tram driver can disengage in order to stop the train.

Escalators provide a moving stairway driven by a stationary engine. An ingenious system of flexible metal surfaces is attached to the cable so that on the upward journey they form a rigid stairway on which passengers can stand. On the downward part of the loop they fold into a flat belt.

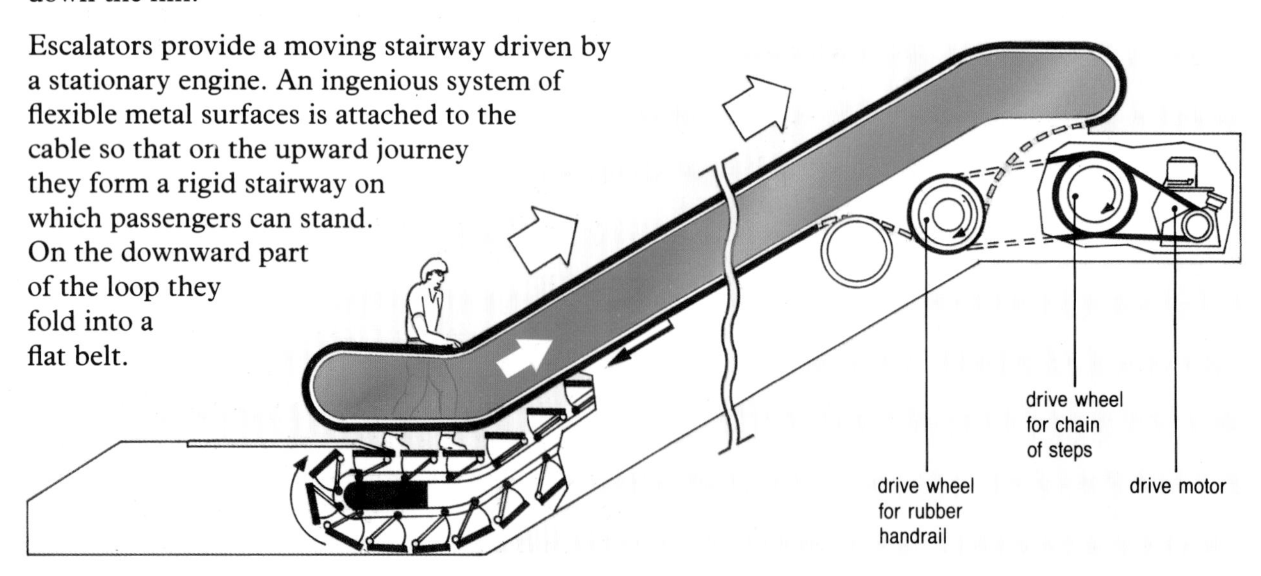

Gravity lends a helping hand

In many older houses, builders have made use of a simple energy saving technique in window design. To make windows easy to open and close, counter weights of exactly the same mass as the window are attached by cord to the window frame. These counter weights are hidden in cavities at the sides of the window. To open the window all the householder has to do is to provide sufficient energy to overcome friction. The potential energy needed to raise the window is provided by the potential energy lost by the counter weights.

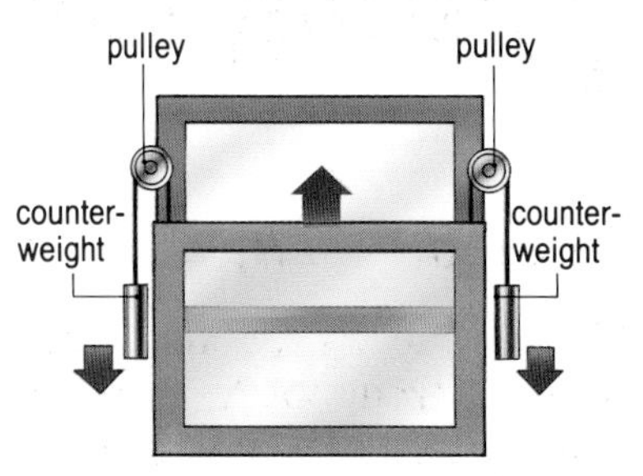

Potential energy gained by raising the lower window comes from the potential energy of the counter weights.

This principle has been used for many years in railcar design. Two railcars are attached to the same cable, one car goes up while the other goes down. The potential energy lost by the downward travelling railcar is used to provide some of the energy needed to raise the upward moving railcar.

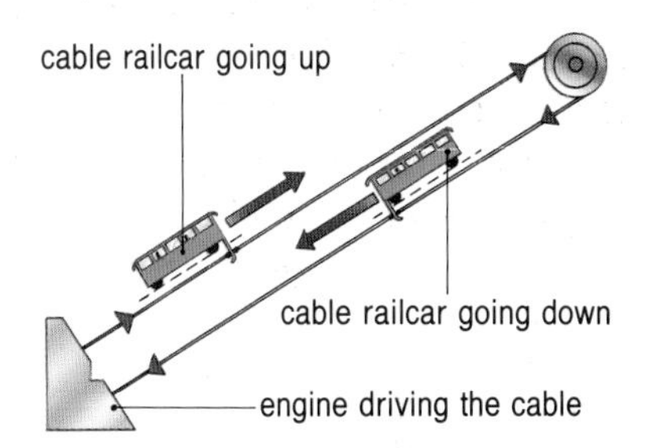

Energy lost by the descending railcar is transferred to the ascending one.

In one Portuguese system special water tanks are fitted in the railcars. When a car is at the top of the hill its tank is filled with water and when the car is at the bottom of the hill it is emptied. The extra mass gives the downwards-moving car additional potential energy and this is used to raise the other car.

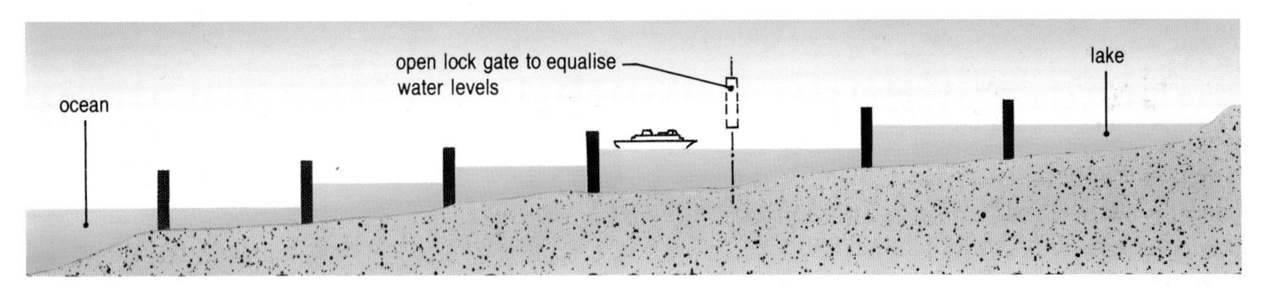

Another ingenious way of lifting large loads – in this case ships – up a hill is by using locks.

A canal is divided up into small sections with gates between each section. By opening each gate in turn, the water levels are equalised so that a ship can be raised or lowered a little at a time as it sails from one lock to the next.

The Panama canal allows ships to sail between the Atlantic and Pacific Oceans. Several series of locks are needed to lift the vessels up 50 m from one ocean to the other.

1 Suggest why cable systems for aerial cable cars are made up of several individual cables.

2 The longest aerial cable car system in the world is in South America. It can carry 45 passengers from an altitude of 1640 m to an altitude of 4764 m. It is 13 km long; the journey takes one hour and the maximum speed is 10 m/s.
a) What is the maximum change in height?
b) For a passenger of mass 50 kg what is the gain in potential energy?
c) What is this passenger's maximum kinetic energy?

3 A ski-lift carries a skier of mass 80 kg to the top of a ski-slope 500 m higher.
a) What is the skier's gain in potential energy?
b) Calculate the skier's maximum theoretical speed at the bottom of the hill.
c) Say why in reality it would be less than your calculated value.

Stopping in time

Speed is the killer

CARS CARNAGE TOPS CANCER

By CHRIS HOLME

CARS now kill more people in the prime of their lives in the North and North-east than cancer.

Road accidents are the biggest single cause of death in the 15 to 44 age group.

A total of 277 people in that group were killed on the roads in Highland and Grampian between 1981 and 1985 compared to 262 cancer deaths.

Speed is still the main killer – and the sophisticated engineering of modern cars makes it all too easy to allow the speedometer needle to creep up without you noticin

Car crashes are amongst the worst killers of the present day and high speed is the main cause. Many drivers go too fast and are unable to stop quickly enough in an emergency. However, a car going slowly can stop quickly after the brakes are put on. A car going at double the speed will take *more* than twice the time to stop and will travel *more* than three times as far, before stopping. The faster car is therefore more likely to hit something or someone before it comes to rest.

Thinking distance, braking distance and stopping distance

Imagine that you are driving along the road. Suddenly a child runs out in front of your car. You jam on the brakes ... Will you manage to stop in time? Let's have an action replay.

The child runs on the road. Your eyes record the scene. The information is passed to your brain, which then sends out a signal to your right leg to push hard on the brakes. All this takes at least $\frac{1}{2}$ second. This time is called your **thinking time** or **reaction time**. The distance the car travels in this time is called the **thinking distance**. The distance the car travels once the brakes are on is called the **braking distance.**

$$\text{stopping distance} = \begin{matrix}\text{thinking distance} \\ + \\ \text{braking distance}\end{matrix}$$

Shortest stopping distances

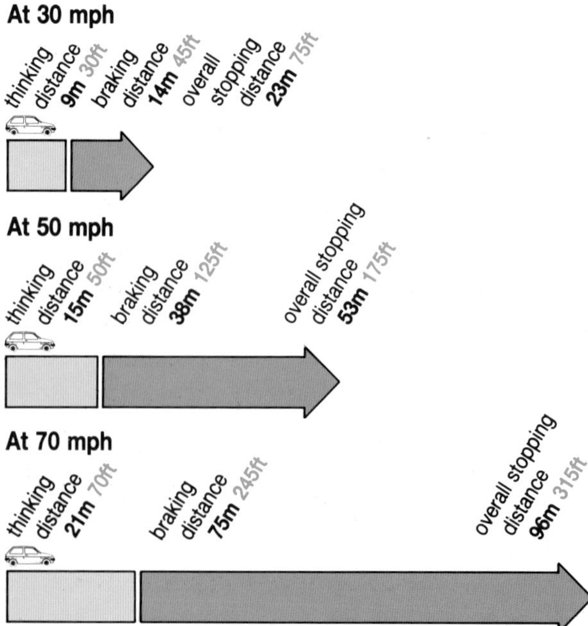

Stopping distance and the Highway Code

The Highway Code shows how a car's stopping distance changes as the speed increases. The thinking time is taken as $\frac{2}{3}$ second for all speeds. Braking time increases from about two seconds at 30 miles per hour to four seconds at 60 miles per hour.

Shortest stopping distances – in metres

mph	Thinking distance (m)	Braking distance (m)	Overall stopping distance (m)
20	6	6	12
30	9	14	23
40	12	24	32
50	15	38	53
60	18	54	72
70	21	75	96

On a dry road, a good car with good brakes and tyres and an alert driver, will stop in the distances shown. Remember these are shortest stopping distances. Stopping distances increase greatly with wet and slippery roads, poor brakes and tyres, and tired drivers.

Peter's Experiment

Here is an extract from Peter's physics note book.

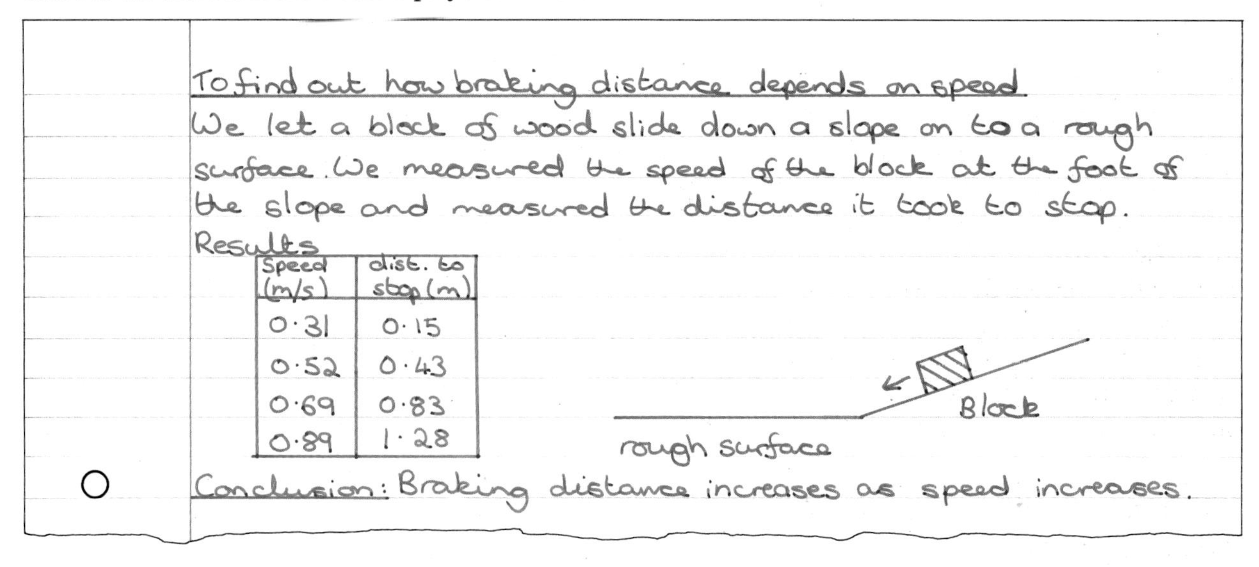

To find out how braking distance depends on speed.
We let a block of wood slide down a slope on to a rough surface. We measured the speed of the block at the foot of the slope and measured the distance it took to stop.
Results

Speed (m/s)	dist. to stop (m)
0·31	0·15
0·52	0·43
0·69	0·83
0·89	1·28

Conclusion: Braking distance increases as speed increases.

1 Hold a flat new £5 note just above your friend's open fingers. She has to catch it between her thumb and forefinger when you release it. Can it be done? Explain your answer.

2 a) Use the data in the table to make graphs of thinking distance against speed, and then for braking distance against speed for a car going at different speeds.
b) Show how you think these graphs would change if (i) the car had new tyres and, (ii) the driver had been drinking.

3 Given that kinetic energy E_k is $\frac{1}{2}mv^2$, calculate E_k for each of three cars which have the same mass of 1000 kg, but are travelling at three different speeds: **a)** 15 m/s; **b)** 30 m/s; **c)** 45 m/s.
Suggest why the stopping distance does not increase directly with speed.

›› *Describe one way of measuring the speed of the block at the foot of the slope. Explain how you could make the block reach the foot of the slope at different speeds.*

Explain how you would find the exact relationship between speed and braking distance using Peter's data. What is this relationship?

Making work easier

Using machines

The simplest machine of all is the lever; and it comes in many different forms. In the crowbar shown here a small force (the **effort**) is applied to the long end. This produces a large force at the short end. A heavy load can be lifted in this way, but in order to move the load a short distance, the effort must move a long distance.

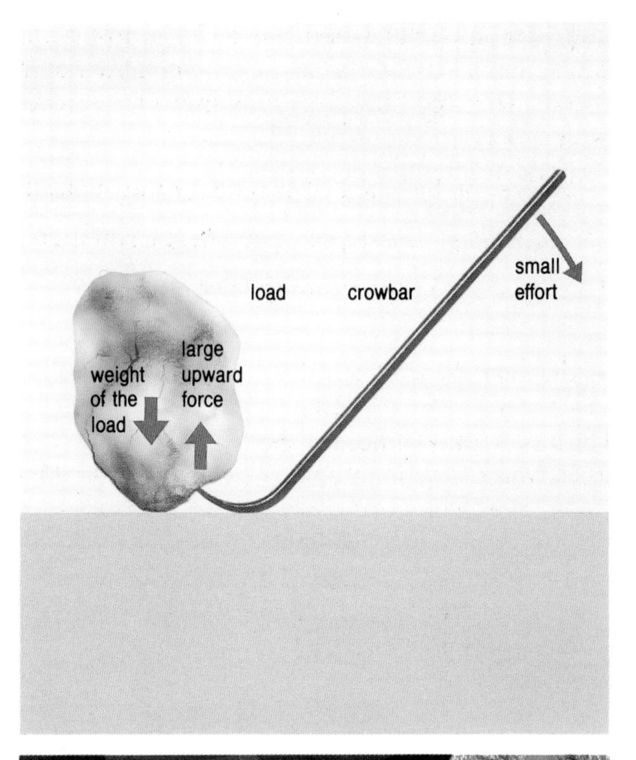

Since *work* = *force* × *distance*
a small force moving through a long distance can produce the same work output as a large force moving a short distance. Most machines are devices for changing work done from:

small force (f) × long distance (D) ,to

large force (F) × short distance (d).

If there is no loss of energy then:

$$P = \frac{F \times d}{t} = Fv$$

Although some men can lift a car single handed, not many of us can! To change a car wheel we need to use a jack. The effort needed to turn the handle of the jack is always less than the weight being lifted.

Another useful lifting machine is the block and tackle. In this example, the four supporting strings share the weight of the load. The operator need only pull with a force equal to one quarter of the weight of the load in order to lift it. In practice, the work input will always be greater than the work output, as the operator has to lift part of the machine as well as overcoming frictional forces at the pulleys.

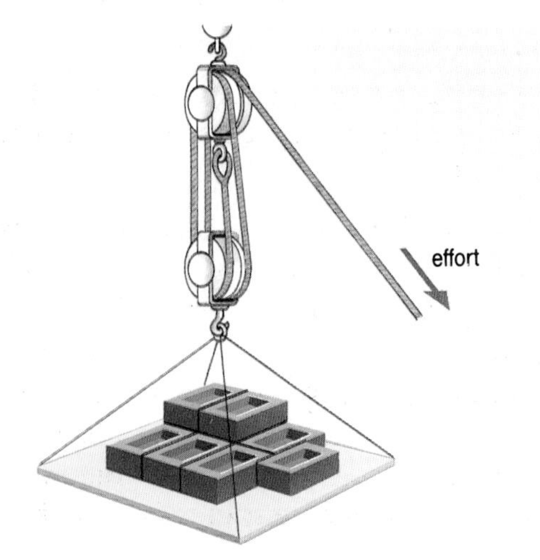

Up through the gears

Most modern bicycles have a range of gears that allow the rider to select one that matches riding conditions. A low gear is needed for hill climbing and for accelerating and a high gear is used for going at constant speed. The most common bicycle gear system is the 'derailleur' system shown on the right. As we have seen (p.140) the energy input to a bicycle can be used in three ways.

1 To give the bicycle kinetic energy during acceleration.

2 To give the bicycle potential energy when going up a hill.

3 To do work in overcoming frictional forces.

In low gear the cyclist will travel at a slower speed so that less work will need to be done against frictional forces. So, for the *same* work input, more energy will be available to produce kinetic energy, i.e. to accelerate the bicycle or to produce potential energy when climbing a hill.

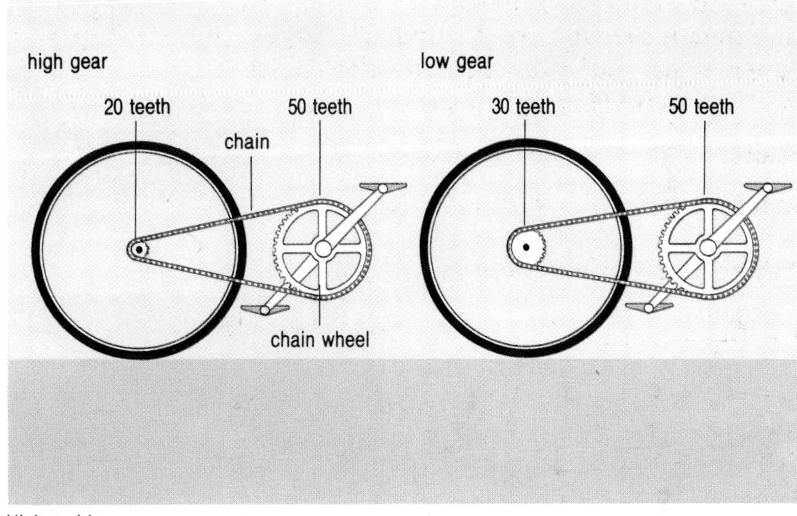

High and low gears.

›› *Use a bicycle with three, five or ten gears to find out the amount of forward movement produced by one turn of the pedals in each gear.*

1 Explain why a machine is a useful device for doing work even though it is less than 100% efficient.

2 In a machine the ratio load/effort is called mechanical advantage. Devise a method of finding the mechanical advantage of a car jack.

3 In a bicycle, the chain wheel has 60 teeth and the rear wheel sprocket 30. If the cyclist pedals at 10 revolutions minute, how many turns will the back wheel make per minute? Calculate the bicycle speed if the back wheel diameter is 80 cm.

High Gear

If a chain wheel has 50 teeth and the back wheel sprocket has 20 teeth, the rear wheel will make $2\frac{1}{2}$ (50/20) rotations for one rotation of the pedals.

Low Gear

If the chain wheel has 50 teeth and the back wheel sprocket has 30 teeth, one turn of the pedals will only produce $1\frac{2}{3}$ rotations of the back wheel.

work = force × distance
For a machine:
work in = effort × distance effort moves
work out = load × distance load moves in the same time
Work out is always less than work input.

Efficiency of a machine is $\frac{\text{work out}}{\text{work in}} \times 100\%$.

Paying for performance

Improved performance

The Rover Coupe Concept Vehicle (CCV) was first shown at the Turin Motor show in 1986. The car was designed with both performance and fuel economy in mind and the shape of the car was determined using a computer. The car is manufactured from a number of materials including steel, glass, plastic and ceramics. The choice of materials is crucial to fuel economy as more mass means more fuel used.

Powerful cars

A Ferrari is more powerful than a Fiat – that is, it can work harder and so do more work each second. In physics we define power as the work done every second.

$$\text{power} = \frac{\text{work done}}{\text{time taken}}$$

As the work done is calculated from the product force × distance, we have:

$$\begin{aligned}\text{power} &= \frac{\text{force} \times \text{distance}}{\text{time}} \\ &= \text{force} \times \frac{\text{distance}}{\text{time}} \\ &= \text{force} \times \text{speed}\end{aligned}$$

and so:

$$P = \frac{Fd}{t} = Fv$$

W N s m N m/s

Pence or performance

When choosing a car you decide between one which you can run economically and one which will perform well. The Oxford Miser and the Oxford Turbo represent these two cases. Here are their vital statistics.

- Miser

Engine capacity	=	1.2 litre
Mass	=	1 tonne (10^3 kg)
Power	=	45 kW (60 BHP approx)
Max speed	=	35 m/s (85 mph approx)
Acceleration	=	0–25 m/s (0–60 mph) in 15 seconds

- Turbo

Engine capacity	=	1.8 litre
Mass	=	1 tonne (10^3 kg)
Power	=	100 kW (130 BHP approx)
Max speed	=	50 m/s (120 mph approx)
Acceleration	=	0–25 m/s (0–60 mph) in 8 seconds

Facts and figures

From these statistics and Newton's Second Law, we can find out a great deal about each car. We will assume that the acceleration and the force producing it remains the same during the 15 seconds the Miser is speeding up.

1 Acceleration $= \dfrac{\Delta v}{\Delta t} = \dfrac{25}{15} = 1.67\,\text{m/s}^2$

2 Distance car travels during that time
$= \text{average speed} \times \text{time} = \bar{v} \times t$
$= 12.5 \times 15 = 187\,\text{m}$

3 Resultant (net or unbalanced) force causing acceleration $= ma = 1000 \times 1.67$
$= 1.67\,\text{kN}$

4 Work done during acceleration
$= F \times d = 1.67 \times 10^3 \times 187 = 313\,\text{kJ}$

5 Kinetic energy gained by car
$= \tfrac{1}{2}mv^2 = \tfrac{1}{2} \times 1000 \times 25^2 = 313\,\text{kJ}$

6 Power during acceleration
$= \dfrac{\text{work done}}{\text{time}} = \dfrac{313 \times 10^3}{15} = 21\,\text{kW}$

›› *Write an essay or draw a picture describing how you visualise transport in the year 2100.*

1 Why is the calculated power (21 kW) used to give the car kinetic energy much less than the rated power (45 kW) of the car engine? What happens to the extra energy supplied?

2 Why are the answers for work done and kinetic energy gained (4 and 5) the same?

3 Repeat these calculations for the Oxford Turbo.

4 Find the total force opposing the motions of **a)** the Oxford Miser and **b)** the Oxford Turbo when they are travelling at 'full power' at their maximum speeds.

5 The Turbo has more than double the power of the Miser yet the maximum speed is much less than double. Why is this?

6 We assume in the calculation that acceleration and force are constant during the acceleration time. What do you think happens in practice and how does this affect the situation?

ENERGY MATTERS

6

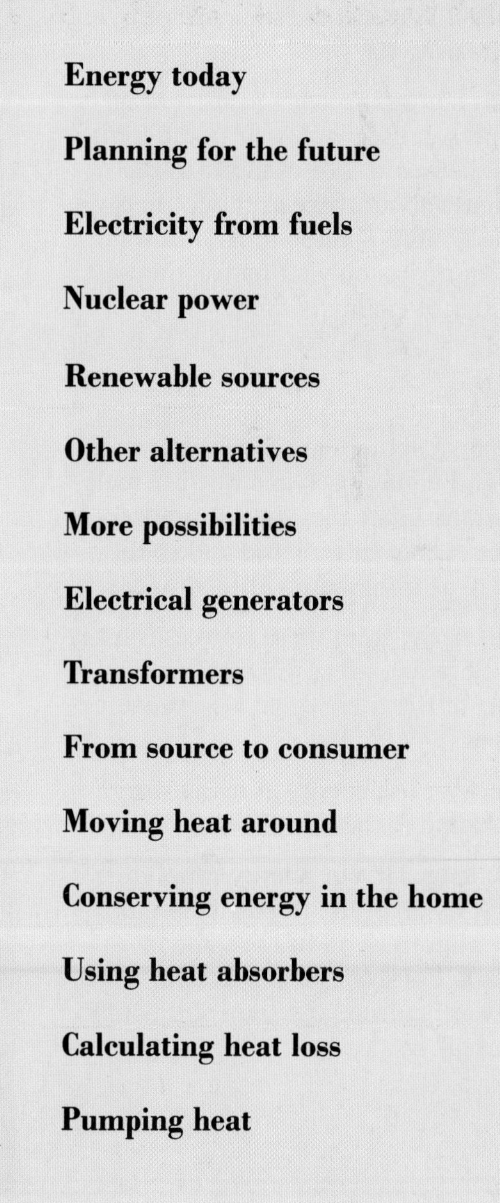

Energy today

Planning for the future

Electricity from fuels

Nuclear power

Renewable sources

Other alternatives

More possibilities

Electrical generators

Transformers

From source to consumer

Moving heat around

Conserving energy in the home

Using heat absorbers

Calculating heat loss

Pumping heat

Energy today

Energy slaves

To live an active life we each use energy at roughly the same rate as a 150 W light bulb. This energy comes from the food we eat.

But what about the energy that is used in our homes, our industries, and in running cars and buses? Taking all of these into account, energy is being used at a rate of about 6000 watts for every person in the United Kingdom. So each of us needs about 40 times the energy used by one human being. That is, we each use 40 'energy slaves' to give us the standard of living we've become used to.

The bar chart shows the average amount of energy one person uses each year in different countries of the world. For example in the USA people use about twice as much energy as people in the United Kingdom. And in some developing countries, only a tiny fraction of these quantities of energy is used.

In Britain we each have about 40 of these 'energy slaves' continuously working for us.

Using energy

Our main energy sources are fossil fuels such as coal, oil, and natural gas; and nuclear fuels such as uranium. Once they have been used these sources cannot be replaced and so they are referred to as **non-renewable sources.**

Some of the energy from these sources is used to generate electricity in power stations, and becomes part of the energy that we use in our everyday lives.

As consumers we use energy in many ways. You can see some of these in the table.

Any major changes in our energy supply will clearly be felt by all of us. In particular you can see how important oil has become in certain sectors. For example it's hard to imagine how agriculture and transport could cope without oil.

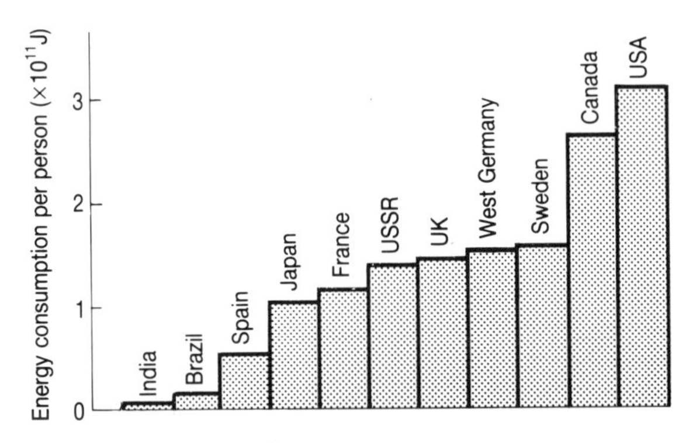

Industrial and domestic uses of energy.

Sector	Liquid fuel	Gas	Solid fuel	Electricity	Total	% by Sector
Industry	35 000	24 500	20 500	11 000	91 000	39.9
Transport	50 000	—	80	400	50 480	22.1
Road	38 500	—	—	—	38 500	
Rail	1500	—	70	400	1970	
Air	8000	—	—	—	8000	
Water	2000	—	10	—	2010	
Domestic	5500	24 500	16 000	11 500	57 500	25.2
Public Services	7000	3000	2000	2000	14 000	6.1
Miscellaneous	4500	3500	500	4000	12 500	5.5
Agriculture	2000	—	60	500	2560	1.2
Total	**104 000**	**55 500**	**39 140**	**29 400**	**228 040**	
% by fuels	**45.6**	**24.3**	**17.2**	**12.9**		**100.0**

Data based on official statistics – thousand tonnes of coal equivalent (10^3 tce).

Forecasting our needs

'Our main energy supplies will run out in fifty years unless . . .' (Worried conservationist, Bognor Regis)

Many energy forecasts are just like this. They are meant to change the way people behave in the hope that the prophecy will not be fulfilled!

Look at the graph shown here. It forecasts the demand for energy in Britain over the next few years. It also predicts that our supplies are going to decrease leaving an energy 'gap'.

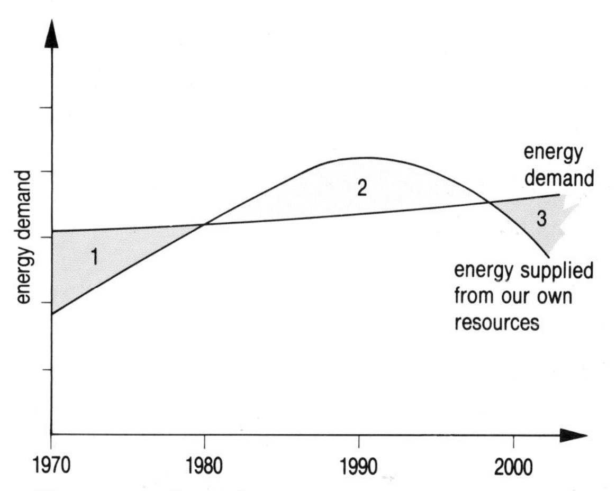

UK energy supply and demand. 1, an energy gap – filled by the import of oil. 2, an energy surplus – this allows fuel to be exported. 3, an energy gap – we will need to fill this gap by reducing our energy demand or by finding new energy sources.

1 Which two sectors use most energy?

2 Which source of energy is used most?

3 Which sector uses most oil?

4 What percentage of agriculture's energy comes from oil?

5 What do you think provides the energy surplus during the 1980s and 1990s?

6 What energy sources might fill the gap until alternatives are found?

'Low demand' for energy

GROWTH in UK energy demand during the rest of the century is expected to be below rates for Europe as a whole. This is in spite of Britain's position as the only major industrialised country totally self-sufficient in all forms of fuel supply.

'The Scotsman' 3 February 1987.

Planning for the future

Where will our energy come from?

It seems possible that by the year 2000 we will be using more energy. For this to be so the way we are using our energy sources at present will have to change.

The graph shows that the use of nuclear energy and coal is expected to increase worldwide. Oil will continue to be an essential fuel but coal will partly replace oil and gas for the generation of electricity. Coal may be also be used to produce synthetic oil and gas.

There is likely to be a growth in hydro-power particularly in South America, Canada and the USSR. Geothermal and solar energy are two examples of a wide range of *alternative sources* which are now being researched and developed. They are described in more detail on pages 164 - 169.

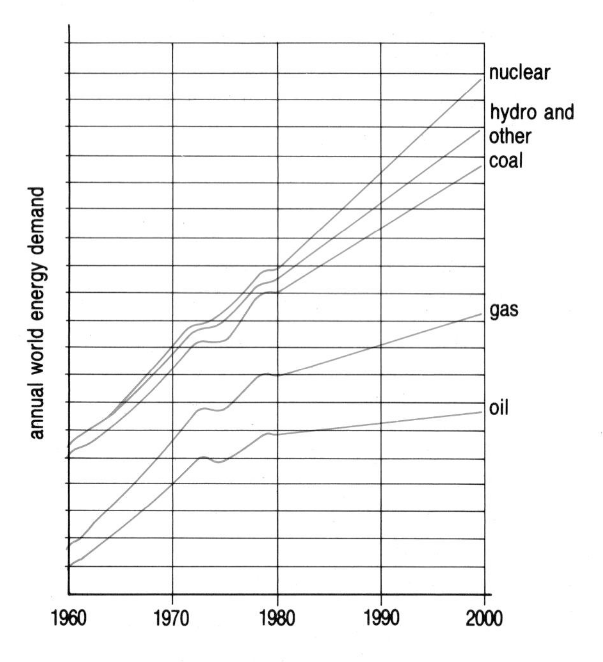

The graph shows how the major oil companies expect the world demand for energy will be met from different sources.

Save it!

Whatever decisions are made about our future energy sources it makes sense to use carefully the energy we have available today.

We can find ways of saving energy in our homes (page 178). Car manufacturers will continue to improve the design of engines to make better use of fuel. Chemical companies may develop economical alternatives to oil as a fuel.

Architects are designing better office blocks and factories which cut down on energy losses.

Chinese plan for conservation

The State Council has decided to stress both exploration and conservation of energy resources. In the short term, priority will be given to conservation. They plan that the amount of energy to be conserved annually in the five years of the Sixth Five Year Plan period will be the equivalent of 24 million tonnes of standard coal.

Factory saves £25,000 on energy

An energy economy drive at the York factory of Joseph Terry and Sons Ltd is yielding savings estimated at £25,000 this year.

Savings range from improvements in the efficiency of production machines to reductions in the power of lamps used for floodlighting.

The firm's main energy source is gas, costing £300,000 a year, and a target has been set to cut this by 10 percent.

Energy costs

The sources of energy which various countries choose to use will depend mainly on the natural resources that are available. However the sources of energy chosen will also be influenced by the attitudes taken to problems of pollution and safety. Some of these problems are familiar to us.

Coal stockpile.

Coal

The burning of coal releases many harmful chemicals. In Britain, for example, of the 4 million tonnes of sulphur dioxide poured into the atmosphere in 1983, 65% came from power stations. These chemicals return to the earth as acid rain which damages plants and wildlife. The chemistry of the soil is changed and trees are then more likely to be affected by disease and drought. Our lochs too can become lifeless as fish stop breeding.

Ways have been developed to clean up the smoke from coal burning power stations but this could add 10% to the price of electricity. We have a choice to make.

Acid damaged trees.

Oil

The use of oil also has its disadvantages. Just as with coal, the burning of oil produces fumes which pollute the atmosphere and the possibility of a major spillage from a tanker is always a threat to sea birds and to our coastlines.

Seabird killed by oil pollution.

Nuclear

Nuclear power would seem to offer a dependable supply of energy in sufficient quantity and at a reasonable price. However many people believe that the dangers of radiation are too great. The effects of an accident such as those at Three Mile Island in the USA and at Chernobyl in the USSR are feared. Nuclear fuel has to be reprocessed in order to use it again. This means that dangerous radioactive substances have to be transported across the country. Finally, the reprocessing also produces highly dangerous waste materials which have to be stored safely for hundreds of years.

Chernobyl disaster.

Electricity from fuels

Power stations

In most countries about 30% of the energy used is in the form of electricity. This proportion seems likely to increase. We depend upon electricity in our homes, in industry and in our transport systems. None of our essential services such as hospitals, broadcasting, water and sewage could operate without a regular supply of electrical energy.

Electricity is generated in large central power stations which burn oil or coal. Other power stations are driven by water flowing from reservoirs in the hills (hydro-electricity). In addition, some thirty countries have nuclear power stations.

A coal-fired power station.

Principles of operation

Most power stations, apart from hydro-electric, operate in similar ways. In each of them:

1 heat is produced to raise steam;

2 this steam is used to turn a turbine;

3 the turbine then drives a generator;

4 the steam is then condensed and;

5 it returns as water to the boiler or reactor.

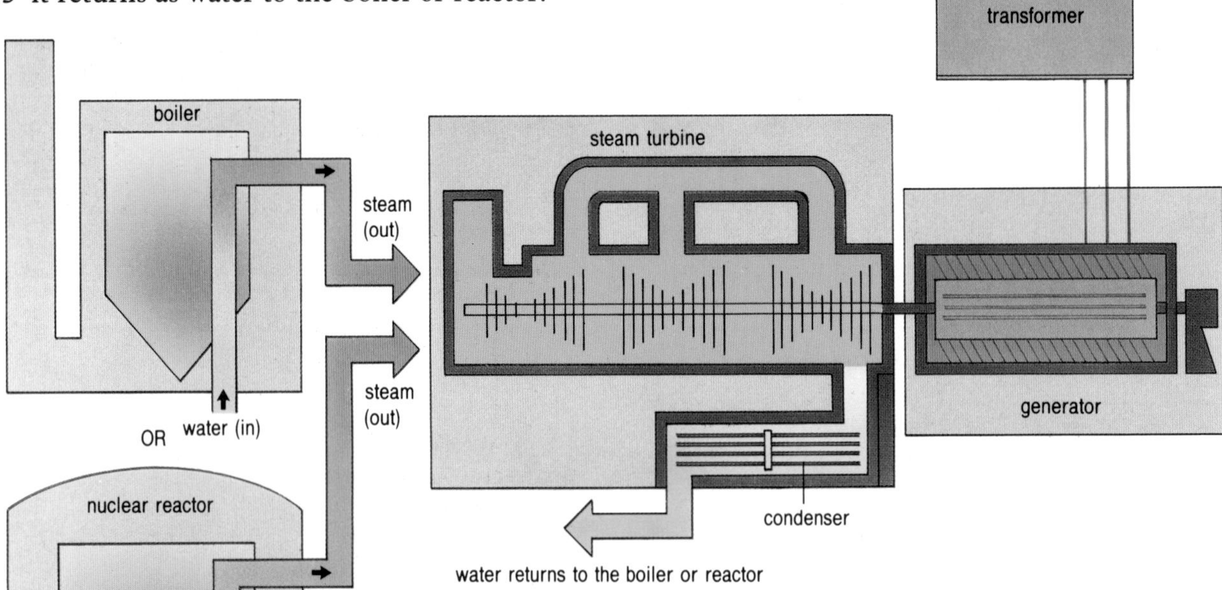

A power station is only about 30% efficient. This means that about 70% of the energy available from fuels used in electrical generation is 'lost'. It doesn't of course disappear but is changed into something of very little use – waste heat. This often merely heats water which passes through cooling towers or into rivers or the sea.

Combined Heat and Power (CHP)

A combined heat and power station generates electricity but it also uses the warm water which would normally go to waste.

As this water is not very hot it has only limited uses – for example in horticulture or fish farming. By using a special kind of steam turbine, however, the water comes out at a temperature high enough for it to be used to heat buildings. In making use of this waste heat a pipe network is used to circulate the water to factories, offices and homes. This type of distribution is known as *district heating*.

Nine cities in the UK have been surveyed in order to decide whether CHP should be introduced. We have fallen well behind many of our neighbours in Europe. For example, in Finland and Denmark more than 20% of their total heating requirement is provided from what was previously waste heat.

Country	Percentage of total heating requirements provided by 'waste heat'
Finland	40%
Denmark	25%
Sweden	20%
Germany	8%
UK	1%

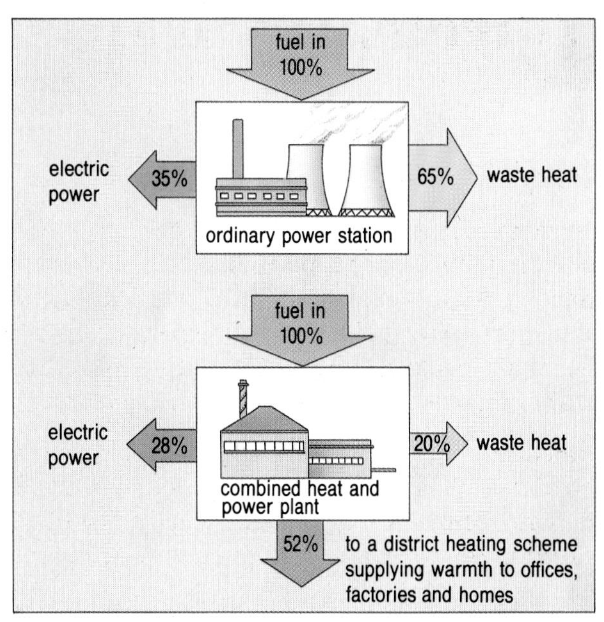

A combined heat and power (CHP) station generates electricity but also uses the warm water which normally goes to waste in an ordinary power station.

Go-ahead for first power station to recycle waste heat for industry

Britain's first purpose-built power station designed to sell the waste heat from electricity generation to power nearby industries has been approved by the Department of Energy. It will be sited in Hereford and built by the Midlands Electricity Board at a cost of £3.5m.

'The Times' 22 February 1978.

Units of energy and power

People working within different parts of the energy industry use different units of energy. This is very confusing!

Ideally we need to agree on one unit so that the different energy types can be compared. The table shows some of the units commonly used and their equivalent values in joules.

Units of energy and power

The basic unit of energy is the joule (J).
The basic unit of power is the watt (W) = 1J/s.

Multiples of the joule:

Kilojoule (KJ) $= 10^3$J
megajoule (MJ) $= 10^6$J
gigajoule (GJ) $= 10^9$J
the 'Q' unit $= 10^{21}$J

Energy source	Unit of energy	
Food	calorie	= 4.18J
Electricity	kilowatt hour (kWh)	$= 3.6 \times 10^6$J
Gas	therm	$= 1.06 \times 10^8$J
Coal	tonne of coal equivalent (tce)	$= 2.8 \times 10^{10}$J
Oil	tonne of oil equivalent (toe)	$= 4.76 \times 10^{10}$J

1 What is 1 tonne of coal equivalent in **a)** megajoules and **b)** gigajoules?

2 The power output of a modern nuclear power station is 1 200 000 000 W. What is this in megawatts?

3 Find the equivalent energy value of 1 tonne of coal (tce) in **a)** kWh **b)** tonnes of oil and **c)** therms.

Nuclear power

Fission

In power stations using coal, oil or gas the fuel is burned in the boiler to produce steam. In a nuclear powered station, however, it is a process called fission (meaning 'splitting') which produces the necessary heat in the reactor.

The nuclei of some atoms such as uranium-235 and plutonium-239 are very unstable. If a neutron hits the nucleus of one of these atoms it can be made to 'split up' and release energy. This splitting of an atomic nucleus is called **nuclear fission**. The fragments produced – bits of the original nucleus plus some neutrons – are called the **fission products**.

When the total mass of the original nucleus plus the bombarding neutron is compared with the total mass of the fission products it is found that some has been 'lost'. This change in mass explains why energy is released (E).

Albert Einstein was the first person to suggest that energy itself has mass (1905). This means that if an object gains energy its mass increases and if energy is lost its mass decreases. But these changes are usually very small indeed.

The fast reactor site at Dounreay in the north of Scotland. The waste heat produced is lost to the sea.

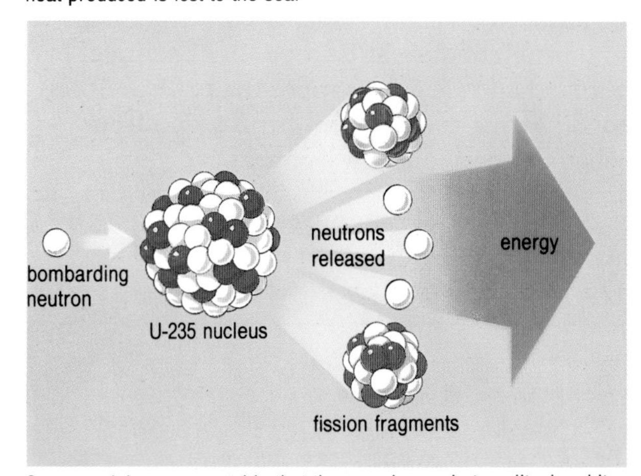

Some nuclei are so unstable that they can be made to split when hit by a neutron. This splitting of a nucleus is called 'fission'.

Chain reaction

The fission of only one atom of U-235 produces very little energy. However, some neutrons are released during fission and they can go on to cause fission in other U-235 nuclei. This leads to a **chain reaction**.

As a very large number of such chain reactions take place at the same time inside a nuclear reactor an enormous quantity of energy is released.

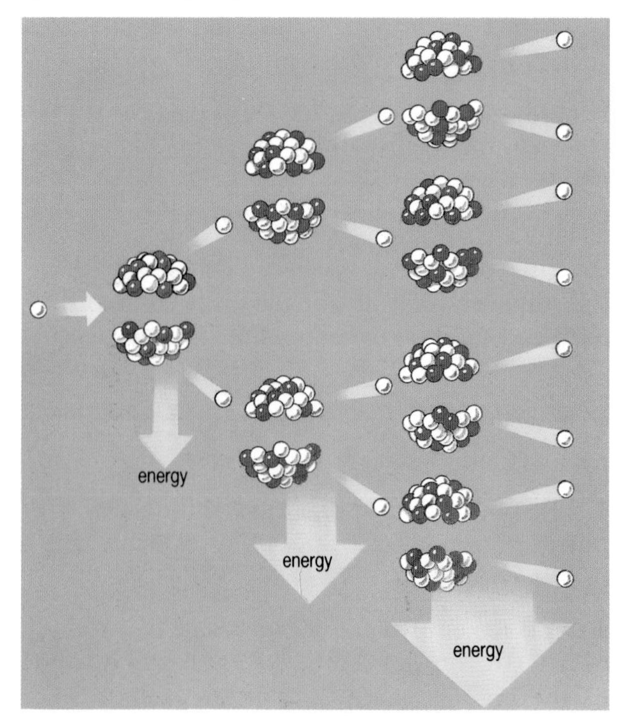

Nuclear reactors

A nuclear reactor replaces the boiler in a fossil-fuelled power station. The five main parts of a reactor are shown here.

1 Fuel: Uranium metal was first used as fuel but today uranium dioxide is more common. The dioxide powder is compressed to form solid pellets which are loaded into narrow tubes about 3.7 m long. These are called pins and are mounted side by side into cylinders to form the *fuel rods* for the reactor.

2 Moderator: It has been found that slow moving neutrons are more likely than fast neutrons to cause fission and keep the reactor going. As the fission process produces fast-moving neutrons they have to be slowed down. This is done by a material known as a moderator. The most common moderators used are carbon (graphite) or water.

5 Radiation shield: A very thick shield of steel and concrete is required to prevent any escape of neutrons or radioactive fragments.

control rods (3)
radiation shield (5)
coolant in
(4) hot coolant (out)
(1) fuel rods
(2) moderator

3 Control rods: To run the reactor safely we need to control the flow of neutrons. The control rods are made of a material, such as cadmium, which absorbs neutrons. By pushing the rods into the reactor neutrons are absorbed and the reaction slows down; by pulling the rods out the reaction speeds up.

4 Coolant: The heat produced by the fission reactions is removed by pumping a coolant such as gas or water past the hot fuel elements. The heated fluid is then piped from the core to a heat exchanger where it heats water to produce steam. The coolant then returns to the core of the reactor to be reheated. Nuclear reactors usually take their names from the kind of coolant they use: for example Advanced Gas-cooled Reactor (AGR) or Pressurised Water Reactor (PWR).

The core of the Dounreay reactor being loaded with fuel rods.

1 The fission of one U235 nucleus releases 3.20×10^{-11} J. If one gram of U-235 contains 2.56×10^{21} atoms how much energy could be released if all the atoms were split?

2 If 1 tonne of coal can produce 2.8×10^{10}J how much coal would be needed to produce the same amount of energy as 1 gram of U-235?

3 Imagine you are part of a group having to decide on the best site on which to build a new nuclear power station. List the factors that you would have to consider.

Renewable sources

Energy supplies

At present about 98% of our fuel comes from what are called **non-renewable** sources such as coal, oil, natural gas and uranium. These supplies are limited and so scientists and engineers are trying to develop other sources which are **renewable**.

Power from the tides

Around our coasts inlets and estuaries are being filled and then emptied each day as the tides rise and fall. This endless flow of energy is caused mainly by the gravitational pull of the Moon as it circles the Earth – but the position of the Sun and the rotation of the Earth also have an effect.

At parts of our coastline where the tides are particularly high we can harness this energy by building a barrage across the inlet. Gates in the barrage are open when the tide is rising, in order to fill the basin up to the high water level. The gates are then closed before the tide begins to fall.

Once the tide level outside the barrier has dropped sufficiently the water is allowed to flow out through large turbines which drive generators to produce electricity.

In this simple type the outgoing flow is used. However, the turbine blades can be reversed so that electricity can also be generated from the incoming tide.

More research on 'natural' energy

By David Fairhall, Energy Editor

For every pound the Department of Energy has spent on supporting research into 'renewable' sources of energy – such as solar, wind and wave power – the Government has put £60 into the nuclear power programmes. But the research budget for renewable sources has trebled since last year.

Fuel from crops and organic waste, the so-called biofuels, are rated by the department as 'a significant energy resource', which could produce the energy equivalent to 60 million tons of coal a year. Wave energy is described in the review as 'a substantial energy resource'. Solar water heating could contribute the annual equivalent in energy of between one and four million tons of coal by the end of the century, the department estimates.

'The Guardian' 9 May 1980.

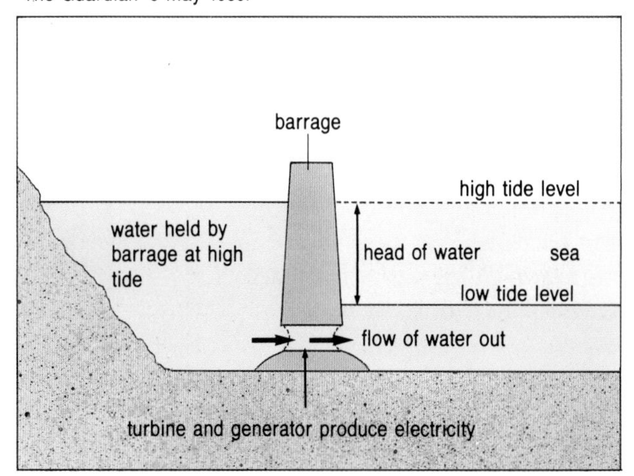

When the sea level outside the barrage has fallen with the tide, the trapped water is then released through the turbine and generator to produce electricity.

The 240 MW tidal power station on the Rance estuary in Brittany (France). Plans have been made for such a scheme in the Seven Estuary. If built this would generate the equivalent of 6% of the UK's annual electricity needs.

Hydro-power

The energy of flowing water can be stored by building a dam across a river at a suitable point. The stored water can then be released through turbines which drive generators and produce electricity.

Today hydroelectric stations supply nearly a quarter of the world's electricity. This will increase in the future particularly in Third World countries where there may be a need to combine hydroelectric power projects with water supply projects.

In our own country the need for electricity is not steady throughout the day or indeed throughout the year. The diagram shows how a *pumped storage scheme* can be used to store energy until it is needed.

Power from the waves

Energy from the sun heats the atmosphere and this produces winds which whip up waves in the sea. In the waters to the north-west of Britain we have some of the world's most energetic wave sites. If a way could be found of extracting this energy up to a fifth of our electricity needs could be met.

Any wave-power device needs to cope with waves of different heights, frequencies and directions. Some types use 'ducks' or 'rafts' which are moved up and down by the waves. The energy of the waves can then be converted to electrical energy.

Calculating power from a hydro-scheme

Power input = potential energy transformed per second
= mgh

m = mass of water falling per second (kg/s)
g = gravitational constant (10 N/kg)
h = head of water i.e. the height fallen

Power output (watts) = power input × efficiency of the turbine and generator

The efficiency of the turbine and generator in a modern hydroelectric station is about 90%.

Hydroelectric power station in Mexico.

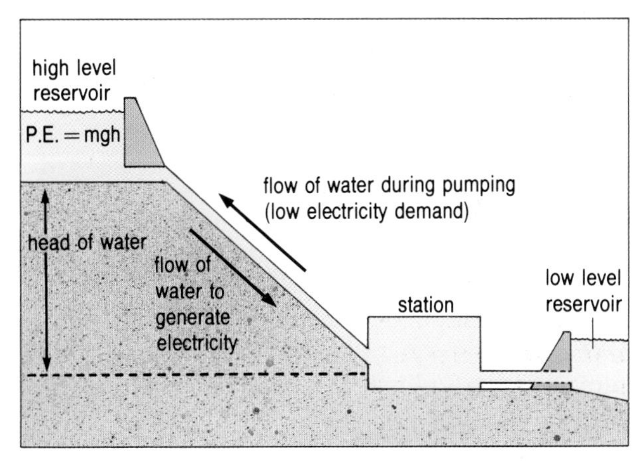

A hydro-pumped storage scheme. When electricity is being generated the mass of water falling per second = water current (m^3/s) × water density (10^3kg/m^3).

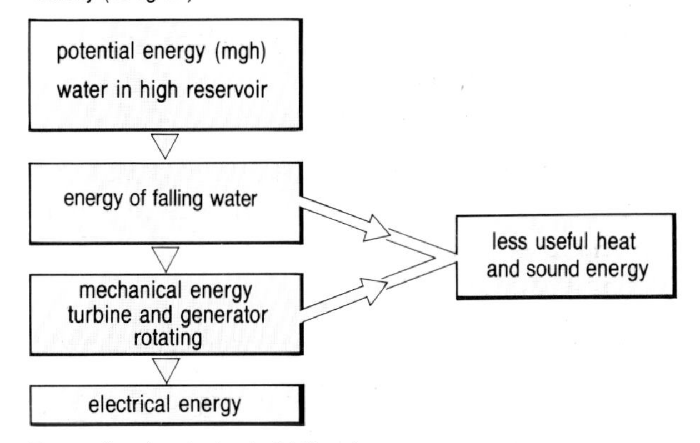

Energy flow in a hydroelectricity scheme.

1 What mass of water must flow each second (in kg/s) to generate 2.25 MW if the water is stored behind a dam 50 m high? Assume an efficiency of 90%, g = 10 N/kg.

2 A wave-power device is found to be able to extract 50% of the energy carried in waves. If this were set in waves having a power of 80 kW per metre of frontage what would be the power output along a 25 km site?

Other alternatives

Blowing in the wind

Windmills were grinding grain in seventh century Persia. Today the windmill is one of our most promising renewable energy sources. Wind turbines – as they are now called – are being used to supply electricity directly to small users such as farms and remote villages while the large 3 megawatt turbine at Burgar Hill in Orkney is providing electricity to the national grid.

Wind we get in plenty! In a modern system about 30% of the wind's kinetic energy can be changed into mechanical energy. This in turn, produces electricity.

The LS1 Wind Turbine Generator on Orkney has a diameter of 60 m and can generate 3000 kW.

When the sun shines!

In Britain our cloudy climate limits the use that can be made of radiant energy from the sun. We get most sunlight in the summer when our need for fuel is least!

Although large scale production of solar electricity may not be possible we can use solar panels to provide useful amounts of domestic hot water.

Sunlight falls on the black absorbing panels containing pipes through which water flows on its way to the hot water tank.

The transparent cover acts like the glass on a greenhouse and helps to 'trap' the collected solar energy.

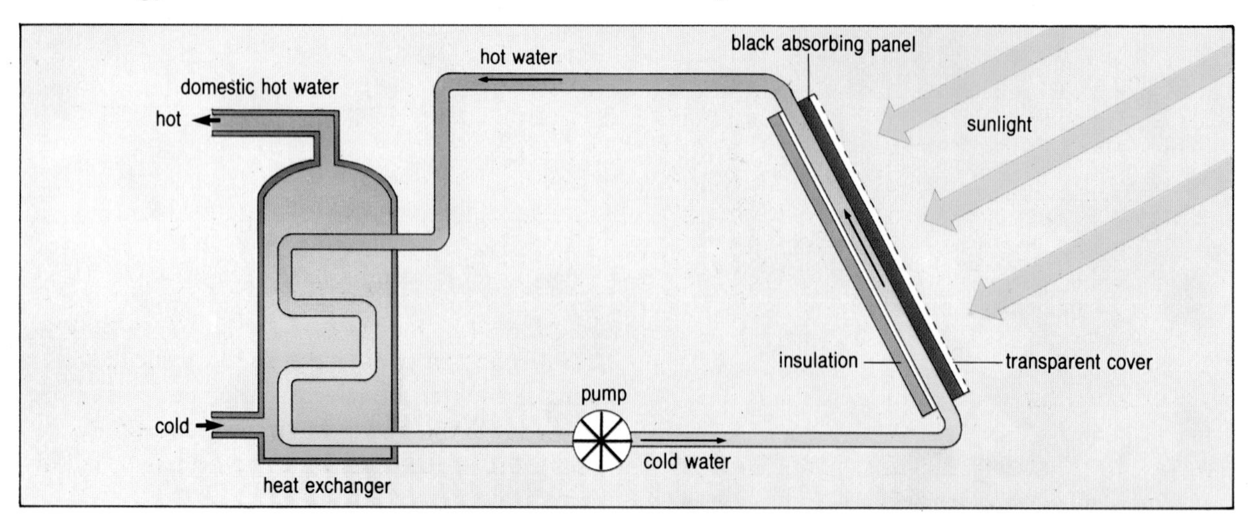

Where the sun shines brightly

In countries which get more sun collectors can take the form of solar furnaces and boilers. These focus the sun's rays with a lens or concave mirror to produce intense heat. This can be used to produce steam which drives turbines and generators. There are plans to build power stations based on this idea in the USA, USSR and the EEC.

Electricity can also be produced directly from sunlight by the use of solar cells which were first developed during the space programmes in the 1960's. The cells are made from semiconductor materials such as silicon and can change solar radiation into electricity. A 16.5 megawatt power plant based on solar cells has been built at Carrina Plain in California while in the EEC fifteen smaller plants of up to 300 kW are now in use.

Solar power that works when the sun's gone down

AN experimental type of solar power cell has reached record efficiency levels – and one version has been able to provide power at night.

The results, with "photo-electrochemical" cells, appeared in the latest edition of the journal, *Nature*.

Photoelectrochemical cells generate power when light strikes a solid semiconductor, that touches the liquid. The semiconductor then draws electrons out of the solution, and the electron flow – which is electrical current – runs through a circuit to do its work before returning to the solution.

'The Scotsman' May 1987.

» *Find out about the 'greenhouse effect'. Build a model to demonstrate this effect.*

A solar panel installation in California.

1 In the UK energy from the sun arrives on average at 200 W per square metre. A solar power station is required to produce power at 20 MW. What area of land surface would require to be covered by solar cells if they are only 10% efficient in transforming solar energy to electricity?

2 The angle of tilt of a solar panel greatly affects the amount of energy it receives at different times of the year. The diagram shows what is meant by the angle of tilt. The table of data shows the effect of different angles of tilt for the summer months.

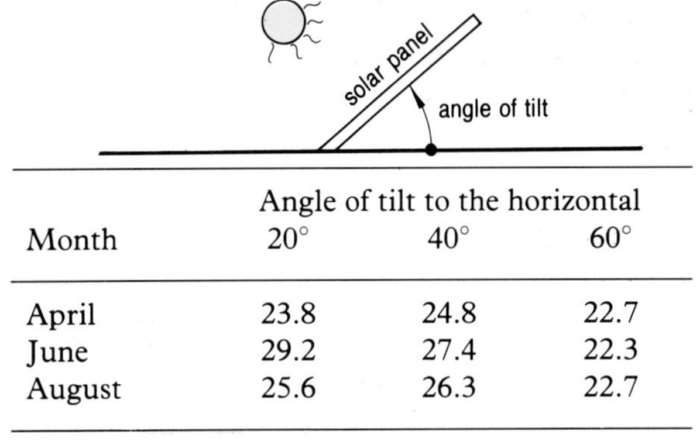

Month	Angle of tilt to the horizontal 20°	40°	60°
April	23.8	24.8	22.7
June	29.2	27.4	22.3
August	25.6	26.3	22.7

Maximum daily input of energy in megajoules to a 1 m³ panel.

a) What angle of tilt would be ideal for a solar panel in April?
b) Is it better to have the solar panel tilted at an angle of 40° or at an angle of 60° for all the months shown in the table? Give reasons for your answer.
c) What is the maximum amount of energy that a 4 m² panel could receive during a day in June?

More possibilities

Energy beneath our feet!

Geothermal energy is literally 'heat from the Earth'. It comes both from the heat which has remained since the Earth was created, and also from the natural radioactive decay of rocks. In some parts of the world large volumes of rock or water at temperatures of 200°C or more have been found within the Earth. These are close enough to the Earth's surface to provide useful sources of energy.

The 'Old Faithful' geyser in Yellowstone National Park, Wyoming, USA. Geysers are found where there are cracks in the Earth's crust.

Hot water test drill a success

Britain's first attempt to tap a reservoir of hot water deep down in the earth's crust to use as an energy source has proved a success, the Junior Energy Minister, Mr John Moore, announced yesterday.

The £1.8 million well, at Marchwood, near Southampton, found water at 5,500 ft. Allowing for heat loss in transit, the water will come to the surface at a temperature of 65–67 degrees centigrade. There is sufficient to heat about 1,000 homes for several decades.

'The Guardian' 25 April 1980.

Hot water on tap

Hot springs, or geysers, are found where the heated water can burst through cracks in the Earth's crust. More often the water is trapped between layers of rock several kilometres down and can only be released by drilling wells.

In Paris more than 100 000 homes are heated in this way. In places where the water temperatures are high enough electricity can be produced in commercial quantities.

Drilling for heat

Ways are also being developed of using the energy in the hot dry rocks. At its simplest this involves pumping cold water down through bore holes where it is heated by the rocks. The water is then brought back to the surface and passed through a heat exchanger before being returned underground.

At Camborne in Cornwall trial wells have been drilled to a depth of 2 km where the temperature is 70°C. But if electricity is to be produced, depths of 8km and temperatures of about 200°C will be needed.

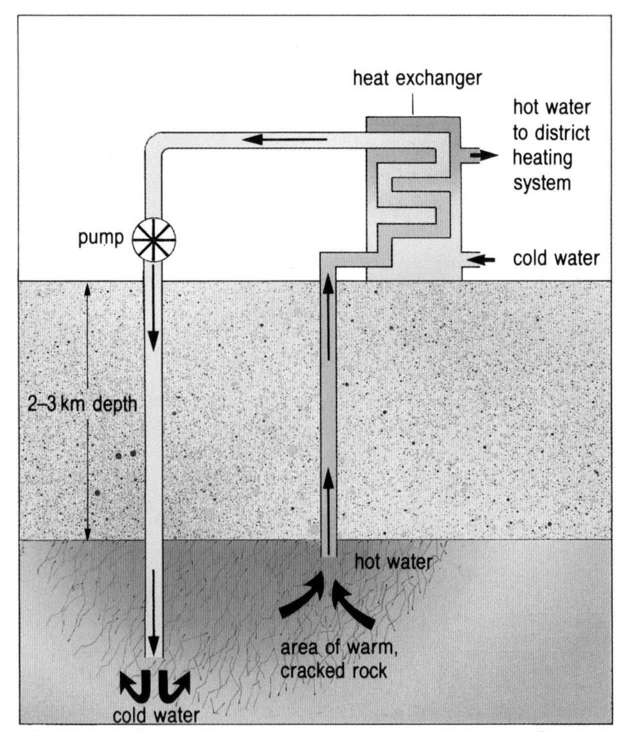

Deep inside the earth the temperature can be as high as 200°C. Heat in the rocks can be used to produce hot water for a district heating scheme.

Government backs plans to make oil from rubbish

By Mary Fagan
Technology Correspondent

THE GOVERNMENT is to back a scheme for turning cellulose from household and agricultural waste into fuel oil.

The idea is to simulate the way nature turns biomass — such as plant life — into fossil fuel by using pressure and heat in the absence of air, condensing what nature does in millions of years into minutes. The result is high-quality oil with an energy content of up to 41 million joules per kilogramme, compared with 44 million joules for North Sea crude oil.

The process would not be economical as a source of fuel at current oil prices (it would cost about $24 per barrel to produce, about $10 more then crude oil), but has major potential for getting rid of the increasingly large amounts of household waste.

About 40 per cent of household waste is made up of cellulose, which is tipped into "landfill sites" — holes in the ground.

Finding somewhere for the rubbish is already a major problem in the United States.

Although the project began as a way of getting rid of waste, it also has huge potential as a source of fuel for poor countries. As well as household waste, it can convert residue from sugar beet and grape harvesting, for example.

An added advantage is that the oil contains negligible amounts of sulphur or nitrogen and so when burnt should not cause acid rain. In tests, more than 100 gallons of oil have been produced from every ton of waste fed in.

The oil cannot be used to fire a petrol engine, although it can be readily used as boiler fuel. Recent experiments, however, show that it could be further processed to provide a substitute for gasoline.

Energy from bio-fuels can be as simple as burning wood or peat for heat. Other more unusual possibilities lie in the huge resources that we throw away every day – our rubbish.

One large plant at Edmonton in London has been working since 1974 burning 400 000 tonnes of refuse every year to produce electricity.

Even the methane gas which builds up under huge rubbish dumps can be harnessed for our use. For example, at Stewartby in Bedfordshire ten gas wells are continously removing gas from a site. The gas is then compressed and piped to a tank. It is then burned to generate electricity thus, replacing coal or oil as a fuel.

Scientists are now experimenting with micro-organisms which will speed up the decomposition of waste and so produce the gas more quickly and efficiently.

In Brazil, many cars now run on alcohol rather than petrol. The alcohol can be produced from the fermentation of cane sugar, and small changes to the design of the car engine make it easy for the car to use alcohol as a fuel.

1 Why are high underground rock and water temperatures needed if we are to produce electricity from geothermal sources?

2 How can the use of bio-fuels help us to conserve our fuel resources?

›› *Find out about the use of geothermal energy in Iceland. Why do many people in that country rely on geothermal energy for their domestic heating?*

Electrical generators

Cheap a.c.

Did you know that you can generate alternating current for nothing? Well ... maybe you *do* need a plate of porridge to give you enough energy to pedal your bike! You can then produce alternating current from your bicycle dynamo.

Inside the dynamo, a permanent magnet rotates close to a coil of wire. This generates alternating current in the coil. The current is small but it is enough to light the bicycle lamp.

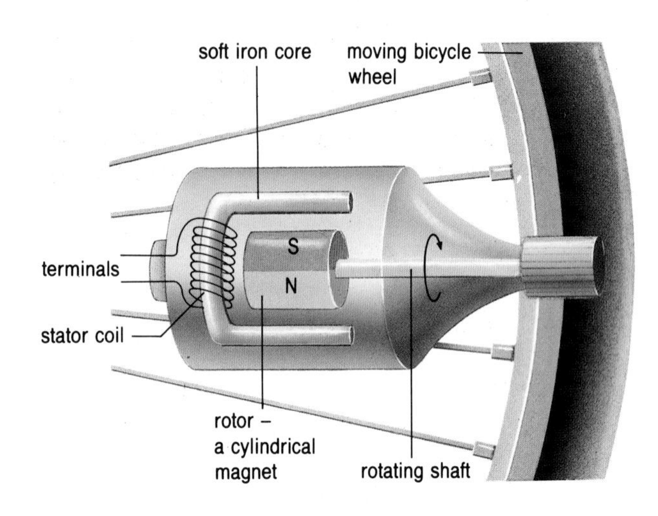

Expensive a.c.

In a commercial a.c. generator (*alternator*) the rotating magnet is an electro-magnet supplied with direct current. The alternating current is generated in a number of fixed coils (stator coils).

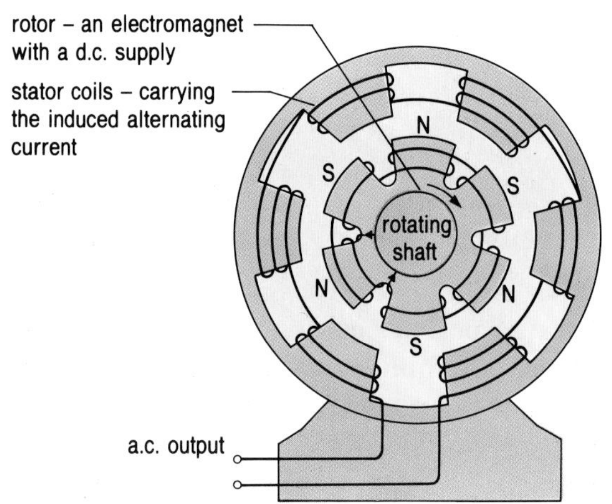

From heat to electrical energy

In power stations huge turbines are spun round by the action of high pressure steam. The heat needed to produce it comes from burning fossil fuels in a boiler or from a nuclear reactor.

Each turbine has a rotating shaft fitted with a series of bladed wheels. Another set of blades is fixed to the casing to guide the steam on to the wheels.

The shafts of the turbines are linked up to 25 000 V electrical generators (alternators) which rotate to produce up to 20 000 A of alternating current (a.c.).

The photograph shows a 660 MW generator stripped for maintenance. The large rotor is seen in the foreground and the spoked wheels are the blades of the steam turbine.

Electromagnetic induction

If there is a current in a conductor in a magnetic field, a force is produced which can make the conductor move.

If instead, a conductor is moved across a magnetic field, electrons will flow through the conductor. This effect is called *electromagnetic induction.* A current is said to have been *induced.* By rotating a coil in the magnetic field the induced current will change in size and direction – an *alternating current* (a.c.) has been generated.

Experiments show that this induced current can be increased by:

1 using a coil with more turns;

2 using a stronger magnet;

3 winding the coil on a soft-iron core;

4 rotating the coil at a higher speed.

conductor moved across magnetic field
N
S
the induced current is read on a sensitive meter

maximum current in the forward direction
current
time
maximum current in reverse direction

Spreading the load

The demand for electricity varies from hour to hour, month to month and place to place. By feeding the electricity from all the country's power stations into a national network, called the *grid*, the supply can be adapted to suit the demand.

For much of the day the demand is low and all the country's power stations are not needed. However, there must always be enough generating capacity (power station output) available to meet the peak demand when it occurs. The graph shows a typical daily pattern of demand.

The National Grid Control Centre in London which coordinates all power transmission on the supergrid.

1 List some of the reasons for changes in demand **a)** during the day, **b)** during the year.

2 Why do Electricity Boards require an even greater generating capacity than is needed to meet peak demand?

3 Use the graph on the right to find the number of hours each day the demand is greater than **a)** 3500 MW, **b)** 2500 MW.

4 Describe two simple experiments to show how induced current can be increased.

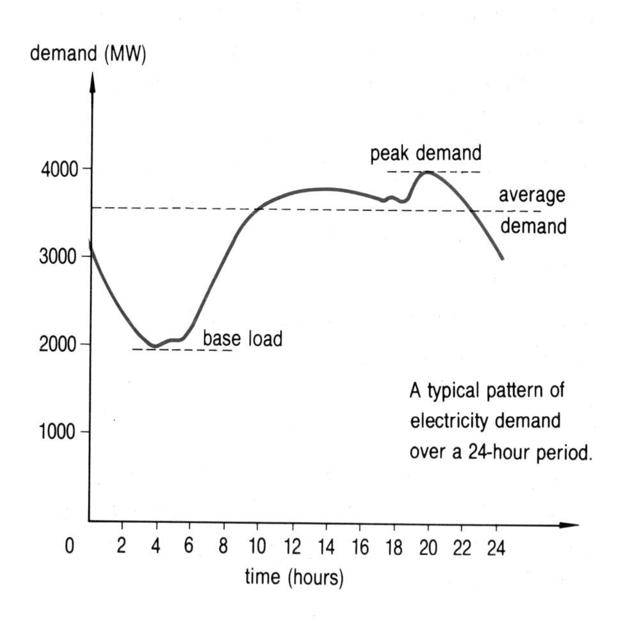

A typical pattern of electricity demand over a 24-hour period.

›› *Take a bicycle dynamo apart and identify the coil and the rotating magnet.*

Transformers

Step-up, step-down

Your 'radio cassette' works on 9 V. Door bells need 6 V and a car battery is charged at 12 V. Yet all of these can be powered from the 240 V mains supply. How is this possible? What is needed is an iron ring and a couple of coils of wire! It is called a transfomer.

In the transformer shown one coil has 1000 turns and the other 100 turns. It can be used to transform 1000 V to 100 V, 630 V to 63 V, 240 V to 24 V and so on. You can see that the ratio of the number of turns is the same as that of the alternating voltages. In the case shown here we have what is called a *step-down transformer*. It changes the voltage to a lower one – but the current is stepped up.

A transformer can also work the other way round. If the input is wired to the coil with smaller number of turns we have a *step-up transformer*. This transforms a low voltage to a higher one – but the current is stepped down.

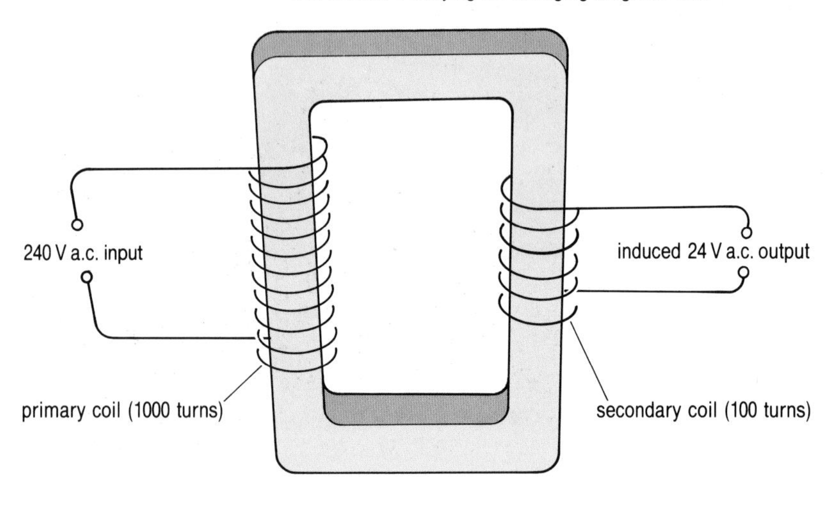

Transformer Equations

Electrical power supplied to the primary

$$= V_p \times I_p$$

Electrical power delivered by the secondary

$$= V_s \times I_s$$

where V_p = voltage across the primary
I_p = current in the primary
V_S = voltage across the secondary
I_s = current in the secondary

The efficiency of the transformer will be given by,

$$\frac{(V_s \times I_s)}{(V_p \times I_p)} \times 100\%$$

Since large commercial transformers are almost 100% efficent we can often approximate by writing.

$$V_p \times I_p = V_s \times I_s$$

The alternating voltages and currents in the primary and secondary coils of a transformer are related to the number of turns in each.

$V_p/V_s = N_p/N_s$ and $I_p/I_s = N_s/N_p$

where N_p = number of turns in primary
and N_s = number of turns in secondary.

You win some, you lose some

A transformer is not 100% efficient – some energy is lost by heating the coils, some by heating the iron core and some by magnetizing and demagnetizing the iron core.

This means that more electrical energy has to be supplied to the primary circuit than is delivered from the secondary. However, by having the voltage and current transformed to suit your needs the energy that you do get out is in a more useful form.

bar magnet moving in and out of coil

S

N

The relative movement of a magnet and coil produces a changing magnetic field around the coil. This results in a current being induced in the coil.

induced current varies in size and direction

coil

soft-iron core

electromagnet

When the stationary electromagnet is switched on and off the magnetic field around the coil changes. This produces an induced current which varies in size and direction.

coil

Power transmission-keeping costs down

A current in a wire produces heat because of the resistance in the wire. The electricity from our power stations comes to us mainly through long cables carried overhead on pylons. Because of their resistance we could end up with a warmer atmosphere and very little energy arriving in our homes, shops and offices.

The power wasted as heat (I^2R) can be greatly reduced if the current(I) is kept low. However, a better solution is to use alternating current instead of direct current.

At the beginning of the transmission lines, a transformer steps up the voltage to 275 000 V or 400 000 V, and so reduces the current in the lines. At the other end of the lines another transformer steps down the voltage and so increases the current again.

›› *Why don't the Electricity Boards use extremely thick cables to transmit energy as this would cut down the power wasted as heat? Find out the solution used by the Boards.*

In power transmission this 100 MW transformer requires to be oil cooled to remove the waste heat. You can see the tanks that hold the oil.

1 A transformer is used in a battery charger. If the input is 240 V, 0.5 A and the output 14 V, 7 A find the input power, the output power and the efficiency of the transformer. If the mains (primary) coil has 600 turns how many turns would you expect the secondary coil to have? In practice, will this be the actual ratio of turns used? Comment on your answer.

2 There are certain advantages in using underground cables rather than carrying them overhead on pylons. Write down what you think these might be.

From source to consumer

The national grid

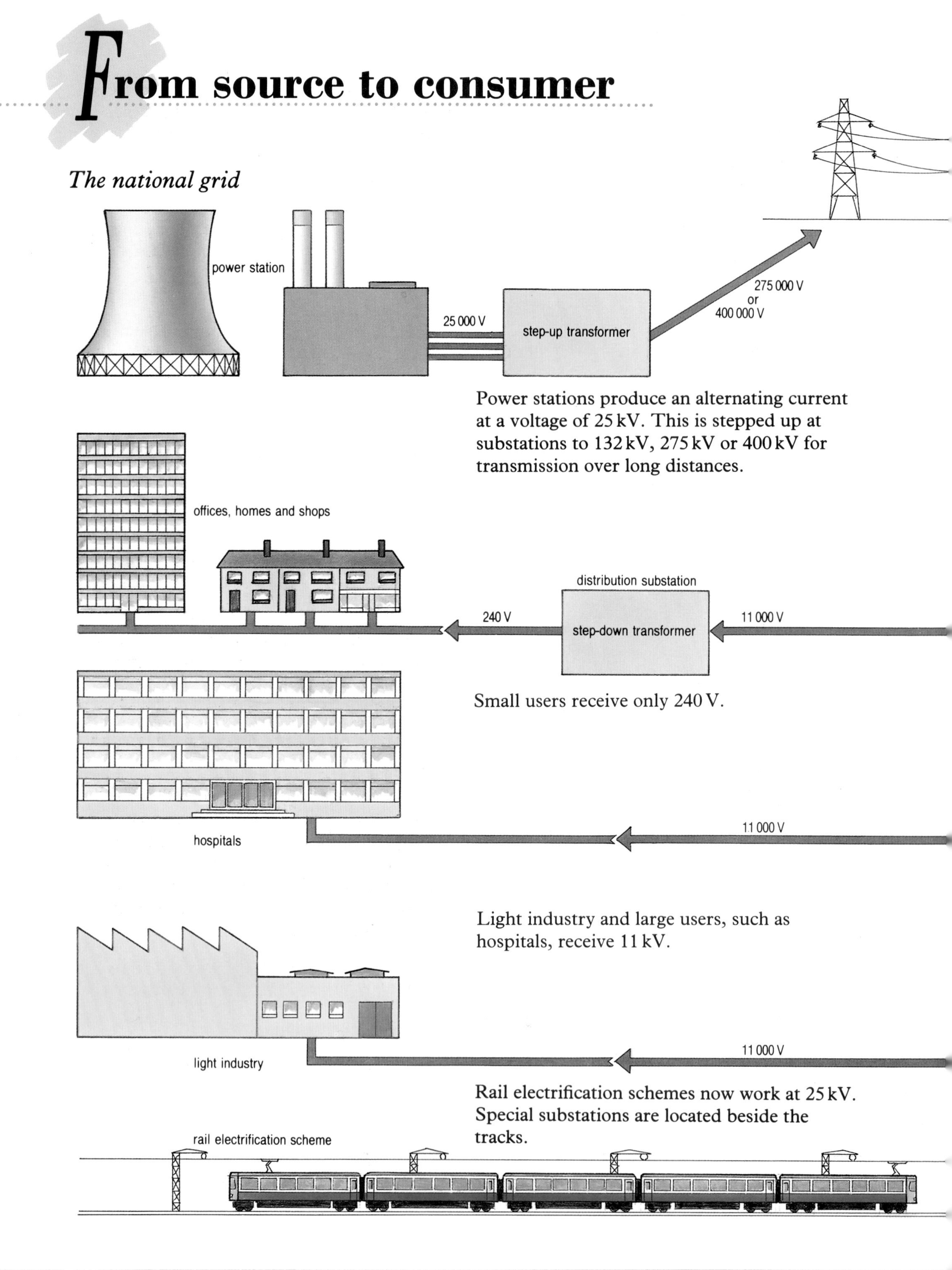

Power stations produce an alternating current at a voltage of 25 kV. This is stepped up at substations to 132 kV, 275 kV or 400 kV for transmission over long distances.

Small users receive only 240 V.

Light industry and large users, such as hospitals, receive 11 kV.

Rail electrification schemes now work at 25 kV. Special substations are located beside the tracks.

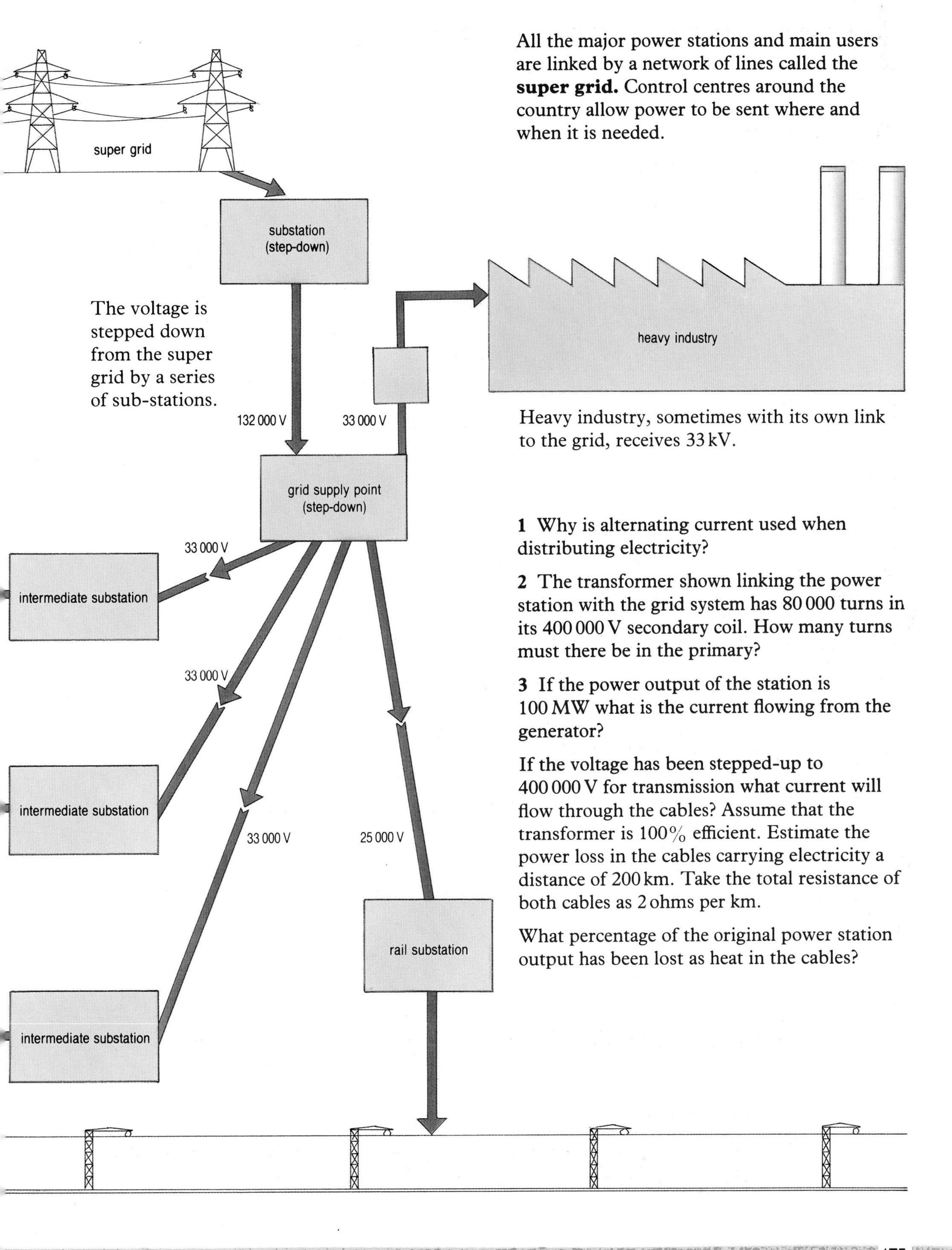

All the major power stations and main users are linked by a network of lines called the **super grid.** Control centres around the country allow power to be sent where and when it is needed.

The voltage is stepped down from the super grid by a series of sub-stations.

Heavy industry, sometimes with its own link to the grid, receives 33 kV.

1 Why is alternating current used when distributing electricity?

2 The transformer shown linking the power station with the grid system has 80 000 turns in its 400 000 V secondary coil. How many turns must there be in the primary?

3 If the power output of the station is 100 MW what is the current flowing from the generator?

If the voltage has been stepped-up to 400 000 V for transmission what current will flow through the cables? Assume that the transformer is 100% efficient. Estimate the power loss in the cables carrying electricity a distance of 200 km. Take the total resistance of both cables as 2 ohms per km.

What percentage of the original power station output has been lost as heat in the cables?

Moving heat around

Conduction, convection and radiation

For the average family 80p in every pound spent on energy goes on heating the house and heating water. Of course heat is also produced in a range of household appliances which are designed to 'move heat around' to where it will be most useful. The transfer of heat can take place in three ways – by **conduction, convection** and **radiation**.

Good heat transfer is also important for machines which have to *lose* heat if they are to work properly. For example large transformers and refrigerators are designed with cooling fins which are usually painted a dull, black colour. The large surface area of the fins allows more air molecules to come into contact with the hot surface. This increases the heat loss by *convection*. With the fins set vertically convection currents can easily rise. The surface is painted black to *radiate* away as much energy as possible.

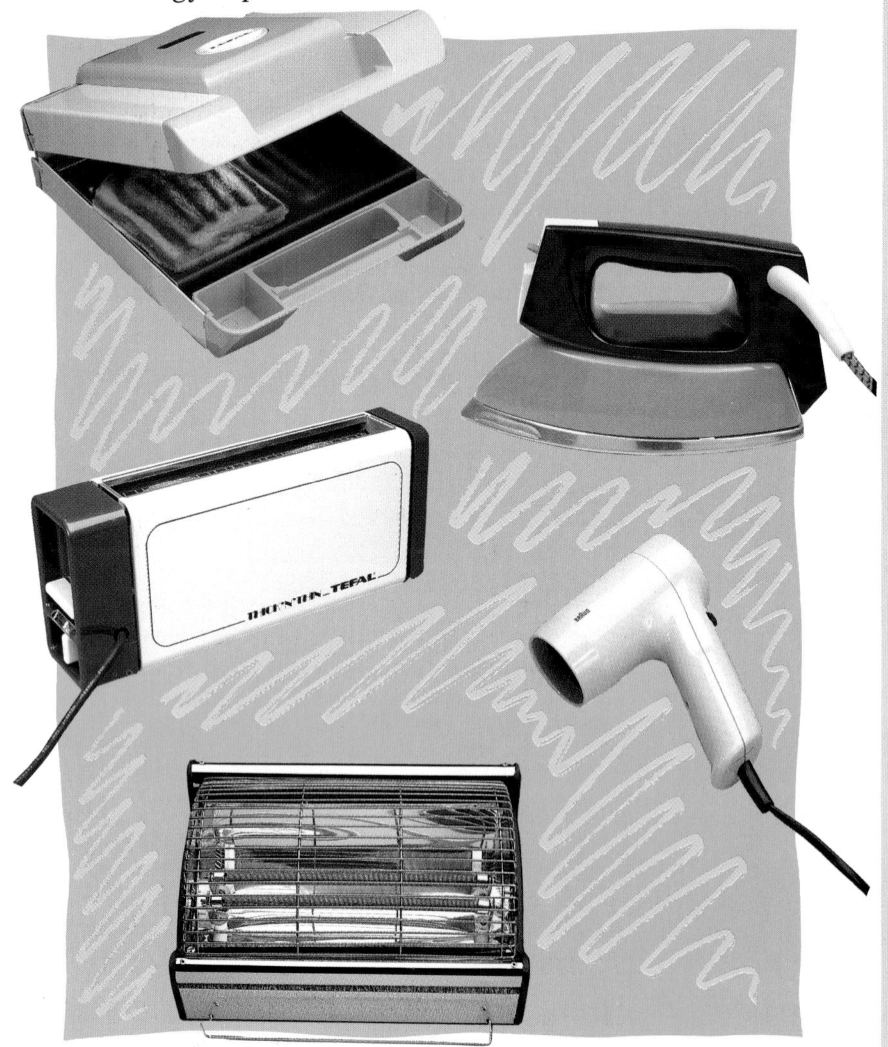

How an average family uses energy.

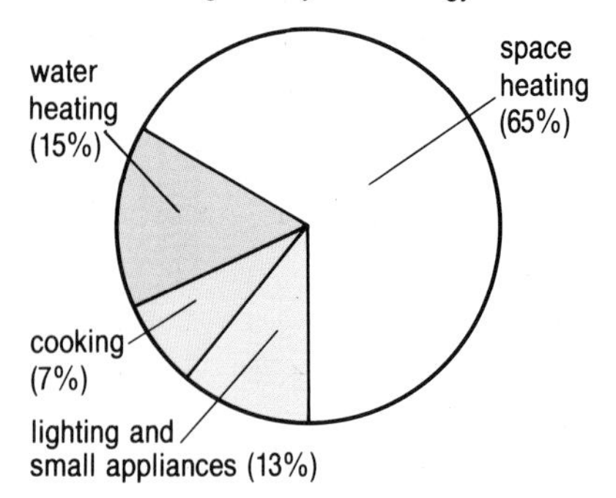

Conduction

If a source of heat is brought close to one end of a solid the molecules there vibrate more quickly. This energy is passed on to neighbouring molecules until eventually the molecules at the other end also vibrate more quickly. This process is called *conduction.*

Some materials are much better at conducting heat than others. Metals are the best conductors while non-metals and most fluids (liquids and gases) are poor conductors. Bad conductors of heat are called *insulators.*

Convection

If part of a fluid is heated the molecules gain energy, vibrate more violently and move further apart. This means that the fluid has *expanded* and become less dense. The colder, denser fluid above is pulled down more strongly by gravity and pushes the hot, less dense fluid upwards.

Radiation

A heater element gives off both visible light (the red glow) and radiant energy (which is invisible). The radiant energy or *infrared radiation* is part of the electromagnetic spectrum. The radiation does not need a medium in which to travel (as in conduction) nor molecules to carry energy from place to place (as in convection).

Heat for the home

The current in the element of an **electric fire** produces heat. Modern heaters have elements which have silica glass sleeves, and the hot element heats the glass which then gives out *infrared radiation*. Most heaters also have polished reflectors which direct most of the radiant energy out into the room.

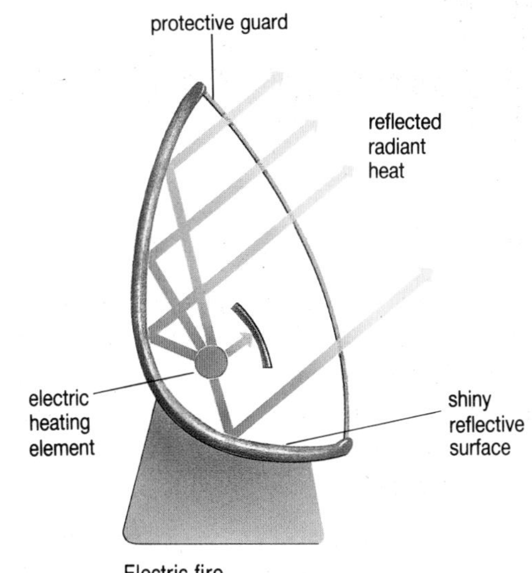

Electric fire.

In a **convector heater** a heating element warms the air around it. This rises and passes out into the room as it is replaced by cold air which flows in at the bottom. So eventually all the air in the room passes through the heater. Rather than just relying on convection currents to circulate the air, a fan is often used to speed up the process.

The **oil-filled 'radiator'** actually works mainly by *convection*! The oil in the panel is heated from below by an electric element and heated air around the panel rises so that convection currents are set up in the room.

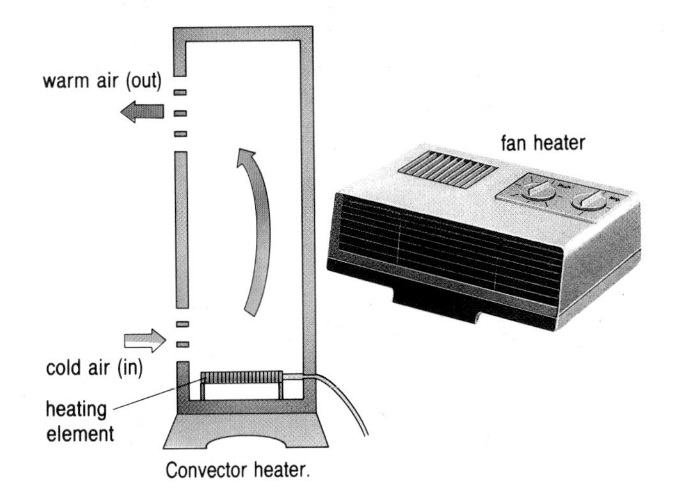

Convector heater.

In a **domestic hot water system** water is heated in the boiler and rises by convection into the top of the storage tank. Meanwhile the cooler water in the tank sinks down to the bottom of the tank and is heated. Normally the hot taps in the house are supplied with water taken from the top of the tank.

Central heating systems which use narrow bore pipes to supply many radiators need a water pump to help circulate the hot water through the system.

1 Is it sensible to put 'radiators' beneath windows? Give your reasons.

2 a) Why do you think that the water storage tank in the diagram on the right has an expansion pipe?
b) What might happen if the cold water tank ran dry?

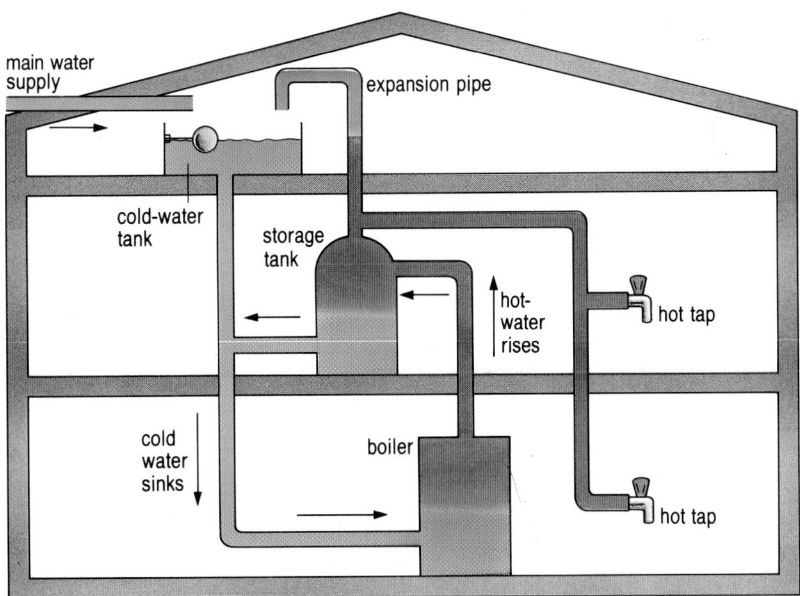

the water heated in the boiler is circulated by convection currents

Conserving energy in the home

Preventing heat transfer

The movement of heat from place to place can be reduced by making use of materials which are poor conductors – i.e. good *insulators*. Air, for example, is a particularly good insulator and pockets of air trapped in materials such as wool, fibre glass and expanded polystyrene are widely used for insulation.

The walls of ovens and refrigerators are lined with fibre glass to reduce the movement of heat from higher to lower temperature regions. The air in fibre glass keeps the oven hot – and the refrigerator cold.

Cutting costs

There's certainly no shortage of advice on the different ways of saving energy in our homes, but preventing heat from escaping is likely to give the greatest savings.

Compared with houses built around 1920 houses built today require only about two-thirds of the heating fuel. This is mainly because of their improved insulation.

You can see from the diagram how different parts of a house allow heat to escape. Although it cannot be stopped completely there are many ways of slowing down this heat loss.

Unfortunately we often have to spend quite a lot of money before we get any benefits. So before making a decision it's well worth calculating the **pay-back time** – this is the time taken to pay for the insulation from the expected savings.

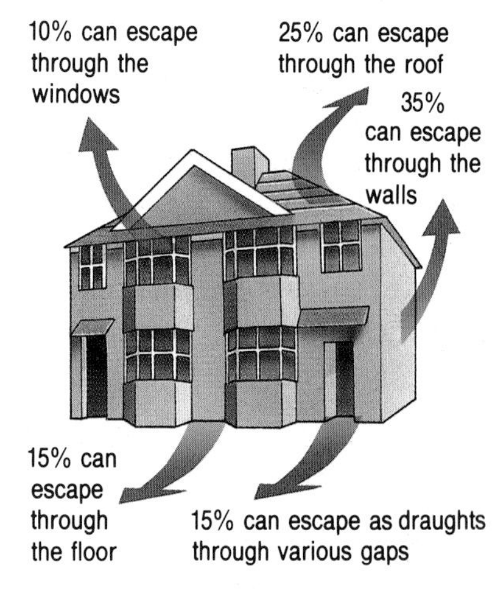

How heat escapes from a house.

The great escape – and how to reduce it

Preventing draughts This is not really a method of insulation because it works by reducing the escape of warm air, and keeping cold air out. Some fresh air will always be needed for ventilation, but draught excluders stop about 60% of heat losses at doors and through the window by about 50%.

Lagging The hot water tank and metal pipes are good conductors of heat. Fitting an insulating jacket to the tank and lagging the pipes will soon repay the cost. Cold water pipes are also lagged but for a rather different reason. They only lose heat when the surrounding temperature drops very low in winter.

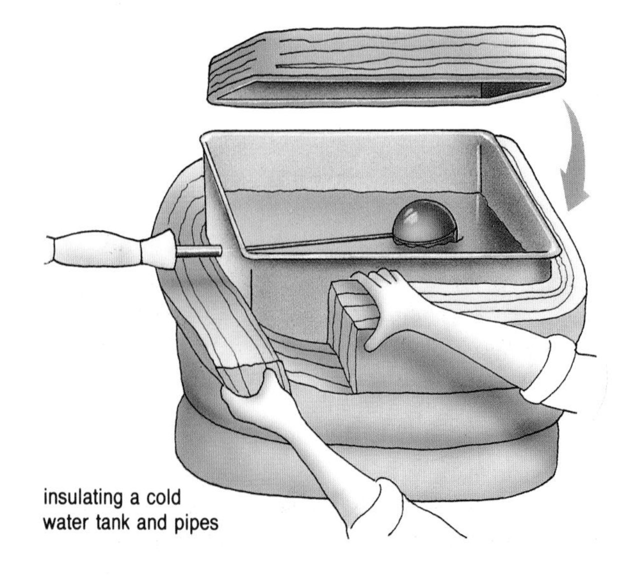

insulating a cold water tank and pipes

Water is not like other liquids which always expand when heated and contract when cooled. When water cools it does contract but then it expands slightly as the temperature approaches 0°C. As the water freezes to form ice it expands even more before starting to contract again with further cooling. The effect of this expansion on water pipes can be dramatic – and expensive.

Roof insulation Heated air in the house rises and can be prevented from passing into the roof space by insulating material laid between the ceiling joists. The thicker the insulation the more heat is saved. At least 75 mm should be used and this will cut the losses from an uninsulated roof by about 80%.

Wall insulation Cavity walls in modern houses have a double layer of bricks with an air space in between them. Convection currents in the cavity can be prevented by filling the space with an insulating material such as foamed plastic, polystyrene beads or mineral wool. This can save more than half the heat losses through the external walls of a house.

Floor insulation Fully fitted carpeting on top of a good insulating material (underlay) will reduce heat loss through the floor by about one third.

Double glazing A double-glazed window has two panes of glass and a layer of air in between. As air is a good insulator this trapped layer reduces the heat lost by conduction through the window by about 50%.

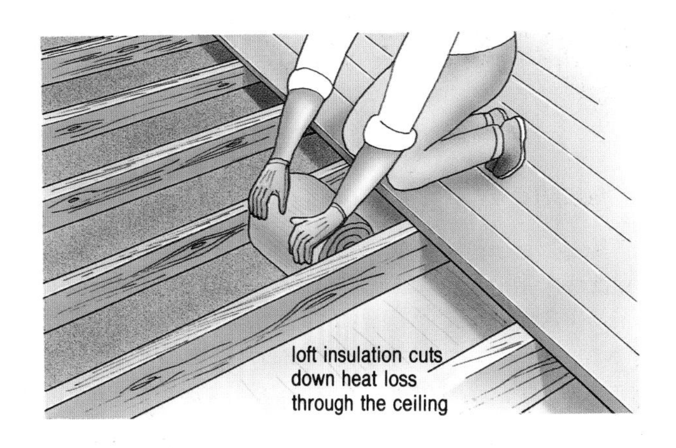

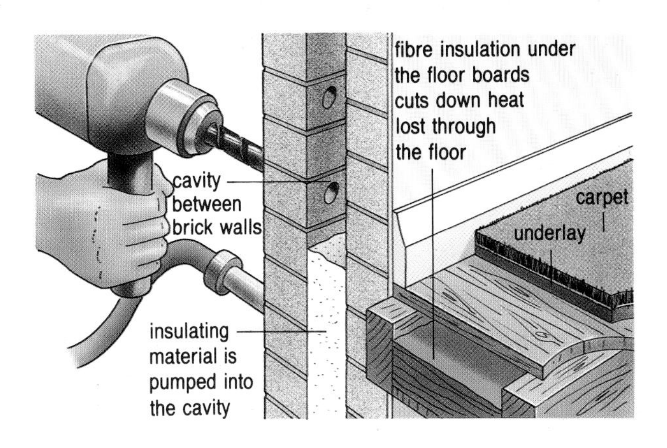

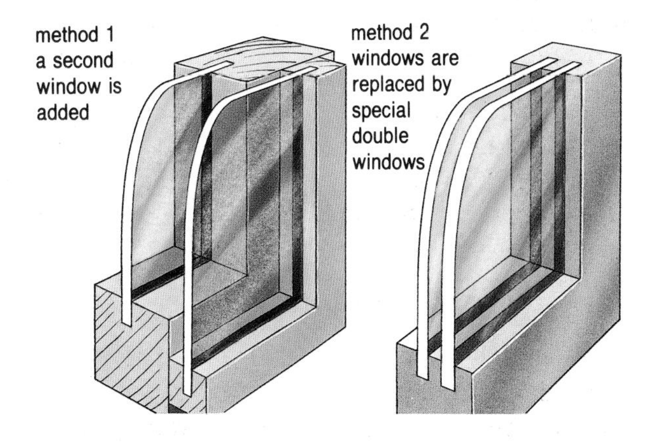

1 The table opposite shows the costs and savings for various methods of insulation. How many years would it take to cover the cost of each method from the annual savings? Draw up a new table showing this pay-back time for each of the five methods.

2 Which of the methods of insulation mentioned above help to reduce heat losses due to **a)** conduction; **b)** convection; and **c)** radiation?

Method of insulation	Heat loss reduction	Cost estimate	Annual saving
draught proofing	50–60%	from £30	£20–£40
roof insulation	80%	£250	£100
lagging (tank, pipes)	80%	£40	£20
wall insulation	60–70%	£250–£450	£50–£125
double glazing	50%	£300–£1500	£25–£80

Using heat absorbers

Specific heat capacity

When thermal energy is absorbed there is an increase in the temperature of the material that has been heated. We can recover this 'stored' energy when the temperature of the material falls.

Some materials need larger amounts of energy than others to produce an increase in temperature. They are much better at storing thermal energy and so have much more to get rid of when they cool down. Such materials are said to have large **specific heat capacities.**

The specific heat capacity (c) of a material is the heat required to raise the temperature of 1 kg by 1°C. The table allows you to compare the values for a few materials.

Material	Specific heat capacity
water	4200 J/kg °C
concrete	850 J/kg °C
copper	400 J/kg °C
glass	670 J/kg °C
air	1200 J/m³ °C

Notice that water has a particularly high specific heat capacity. This makes it a useful substance for storing and for carrying energy. For example, water can be heated in a boiler and then moved through pipes to taps, or to radiators placed in rooms where the heat is required.

Although concrete has a lower specific heat capacity than water, it is more dense and so the same mass of concrete takes up less space. This makes it very suitable for storage heaters which contain a number of closely stacked blocks. They are made of a special mixture of concrete which absorbs heat well and are heated by electric elements which use cheaper night-time electricity. This energy is absorbed during the night and then released slowly when it is required during the day. Such heaters warm the room mainly by convection, but a small amount of energy is also radiated.

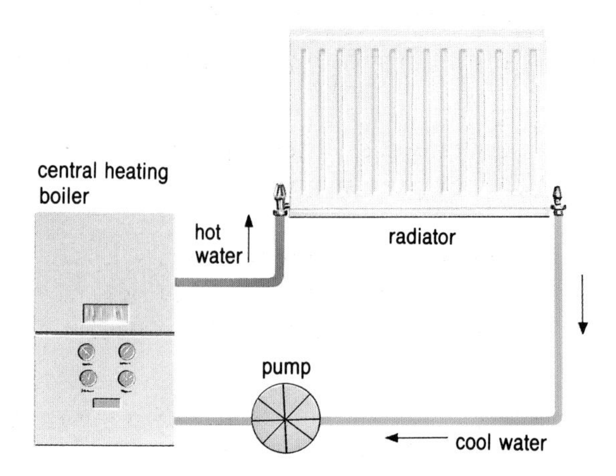

Water is used to carry heat around the house.

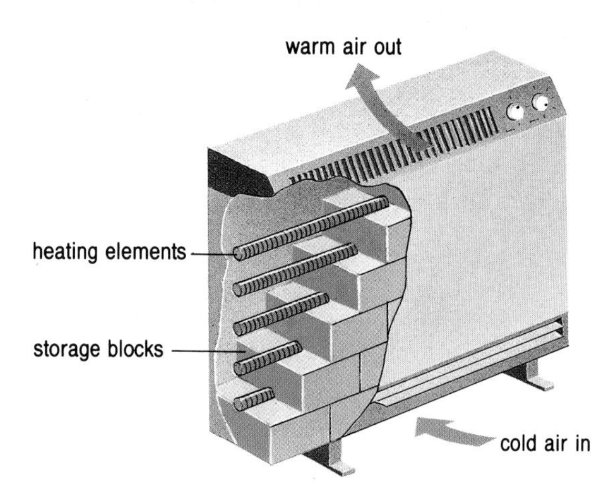

A storage heater contains concrete blocks which store thermal energy.

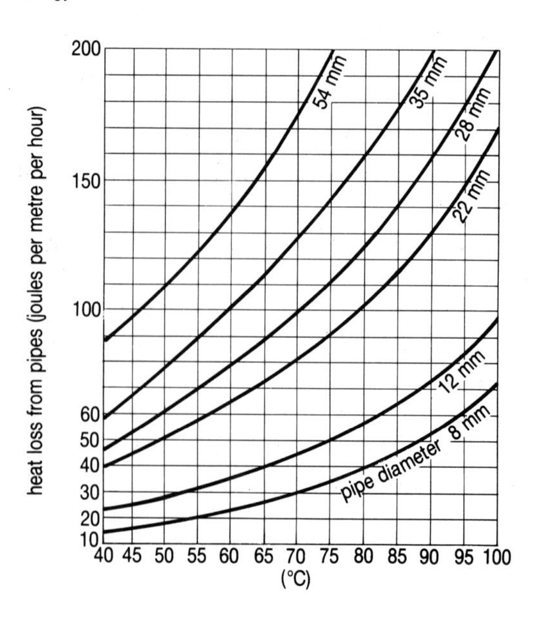

Boiling away

The boilers in our homes are normally used to heat water for two purposes:

1 to supply domestic hot water; and

2 to transmit heat through a central heating system.

The energy needed to heat up some water can be calculated by using the equation shown in the panel.

Energy will also be required to make up for loss of heat from the tank and pipes to the surroundings. Even a tank with a reasonable amount of lagging loses heat at a rate of about 145 W.

The energy lost, per hour, from each metre of piping is found from graphs such as the one shown here. You can see that the rate of heat loss depends both on the diameter of piping and the temperature difference between the pipes and the surrounding air.

The **Specific Heat Capacity** (c) of a material is the energy needed to raise the temperature of 1 kg of it by 1°C. Then, the total energy (E_h) is given by:

$$E_h = c.m.\Delta T$$

J J/kg°C kg °C

The same equation can be used to calculate the energy given out when the temperature of an object falls.

When substances at different temperatures are brought together, or 'mixed', heat will pass from the hotter to the colder. This continues until both are at the same final temperature. If no energy is lost to the surroundings then:

Energy lost by the hotter object = Energy gained by the colder object

1 A 400 litre tank of water has to be heated up from 5°C to 60°C with a heating-up time of one hour.
a) How much energy has to be supplied in one hour if each litre of water has a mass of 1 kg?
b) What must be the output power rating of the boiler?
c) If the boiler is 55% efficient what input power rating would you have to quote to a manufacturer?
d) If instead, a two hour heating up time was thought to be acceptable, what input power rating would then be needed?

2 If a 3 kW immersion heater takes 40 minutes to heat 30 kg of water from 10°C to 60°C, how much heat is lost to the tank holding the water and to its surroundings?

3 A hot water tank is kept at a temperature 50°C higher than its surroundings. The 28 mm connecting pipes carrying the hot water away are 10 m long. If the tank is reasonably well lagged, estimate the total energy lost each hour from the tank and the pipes.

4 The hot water tap of a bath delivers water at 80°C at a rate of 10 kg/min. The cold water tap of the bath delivers water at 20°C at a rate of 20 kg/min. Assuming that both taps are left on for three minutes, calculate the final temperature of the bath water, ignoring heat losses.

Calculating heat loss

Just passing through!

Even after all steps have been taken to insulate a home some heat will continue to be lost through the fabric of the building. Energy will be required from the heating system to replace this. The rate at which heat escapes from a room depends on three factors:

1 The type of material used. In construction work special tables are prepared to show how easily different materials allow heat to pass through. These are called *U-values* and tell us how good a conductor a material is. A good conductor has a high U-value.

2 The surface areas (A) of the different materials making up the room. For example the more windows there are the greater will be the heat loss through glass.

3 The difference in temperature between the room (T_2) and outside (T_1). The greater the difference, the greater will be the rate of heat loss. So in winter a room loses heat more quickly than it does in summer.

Putting these ideas together leads to a formula for P, the rate of heat loss, i.e. heat loss per second.

$$P = U.A.\Delta T$$

W W/m²°C m² °C

In practice the walls, floor and ceiling of a room are made of different materials. The rate of heat loss (P) must be calculated for each, and then a total found at the end.

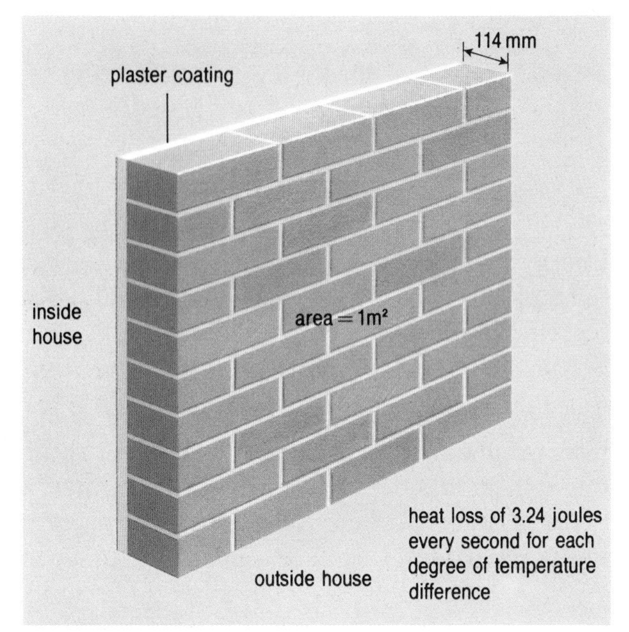

A single brick wall with plaster has a U-value of 3.24 W/m².°C.

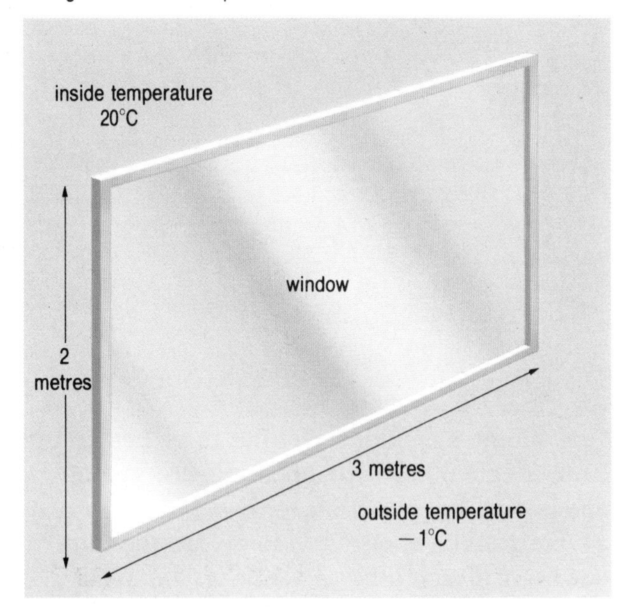

In-out, in-out

Energy will also be required to heat the air that has to pass through a house for ventilation. Although we want our houses to be well insulated there has to be a regular change of air. This prevents rooms from becoming stuffy and avoids problems of condensation and dampness. You can see from the table that different rooms require different amounts of ventilation.

Room	Air exchange rate (changes/hour)	Normal room temperature (°C)
Bathroom	2	21
Bedroom	1	16
Living rooms	1.5	20
Stairs/Hall	2	16
Kitchen	2	18
Toilet	1.5	16

Type of Fabric	U-Value (W/m².°C)
Tiled roof with no insulation	2.2
Tiled roof with 75 mm of insulating material	0.45
Single glazed window with 6 mm glass	5.6
Double glazed window with 20 mm gap	2.8
Single brick wall with plaster (114 mm)	3.24
Double brick wall with 280 mm air cavity/plaster	1.7
Double brick wall with cavity insulation/plaster	0.6
Timber floor set on joists	1.7
Timber floor on joists with carpet and underlay	0.7

There has to be some ventilation but this is ridiculous.

1 From the table of U-values find which material gives the lowest rate of heat loss.

2 Less energy is lost through the walls between rooms inside a house than through outside walls. Suggest a reason for this.

3 How much energy would be saved in a day if the single glazed window shown on p.182 were double glazed with a 20 mm air gap?

4 Calculate the rate of heat loss to the outside from the room shown here through **a)** the walls; **b)** the window; **c)** the floor. What is the total ignoring the ceiling?

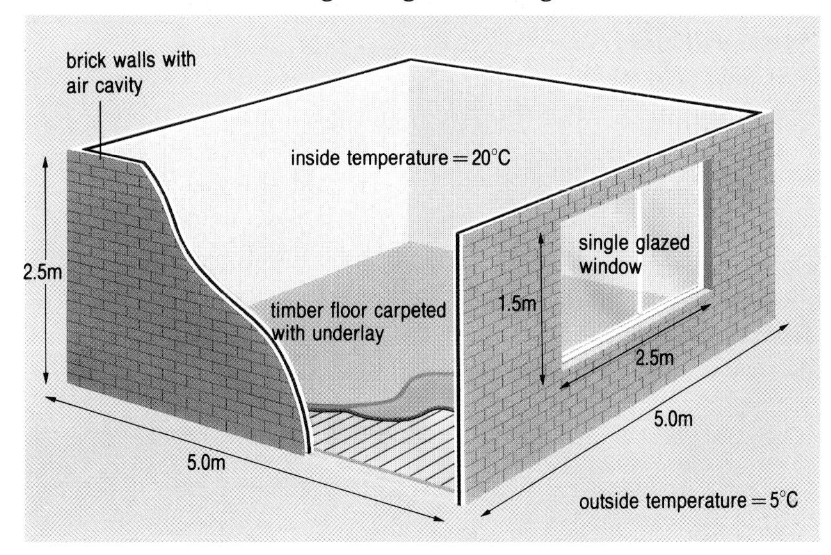

5 The room shown in question 4 is a living room.

a) Calculate the volume of air that should pass through the room each hour to provide ventilation.

b) How much energy will be used each hour to heat this air up to 20°C? (1200 J of energy is needed to raise the temperature of 1 m³ by 1°C.)

c) Work out a formula for the energy required/hour in terms of: (i) the energy needed to raise the temperature of 1 m³ by 1°C (1200 J/°C);
(ii) the number of air changes per hour (N);
(iii) the volume of the room (V);
(iv) the inside temperature (T_2); and
(v) the outside temperature (T_1).

»Where is the biggest heat energy loss in your home – is it through windows, floor, walls or roof? You will need to estimate the area of the outside walls of your home, the floors, the roof and the areas of all your windows. You can ignore any losses through surfaces that are shared with neighbouring homes.

Complete a table of results which includes areas, U-values and energy loss/year. Make a bar-chart to show how the different amounts of energy loss compare. Discuss how your results compare with those of others in your class.

Pumping heat

High to low

Every day we come across lots of examples of heat flowing from hot objects to cooler ones. You pour a hot drink into a cold cup and almost immediately you have a cooler drink and a hotter cup!

A material which has a low temperature still contains some energy. But can we make use of it? This energy is in the form of what engineers call **low-grade heat** – and there's a lot of it around. We have rivers used as cooling water for factories; warm air is given off buildings; and warm bath water is simply poured down the drain. The ground around warm buildings holds energy which has escaped from inside the buildings. Fortunately it is possible to convert some of this 'low-grade heat' into useful energy. To do this we use a heat pump.

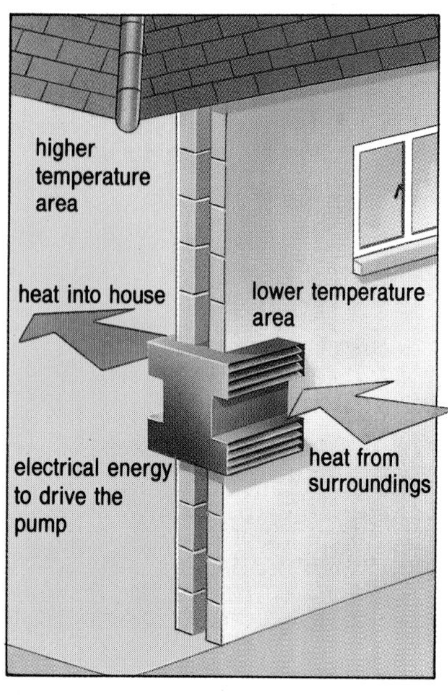

Energy flow using a heat pump.

Heat pumps

When a liquid evaporates (i.e. changes to a vapour) it takes in energy which we call **latent heat**. If the vapour is compressed it changes back to a liquid and a latent heat is given out again. So if we pump a liquid round a closed circuit of pipes heat can be taken in at one place and given out at another.

A heat pump is not a heat source – it simply transfers energy from a lower temperature region (which then becomes even colder) to a higher temperature region – and this requires energy! But it is energy well spent because about three times as much useful energy is transferred from the low to the high temperature region as is used in electrical energy to drive the pump!

The energy required to melt 1 kg of a solid at its melting point is called the **Specific Latent Heat of Fusion** (L_f) of the substance. For ice, $L_f = 3.34 \times 10^5$ J/kg.

If 1 kg of water at 0°C changes into ice at 0°C a similar amount of energy will be given out.

The energy required to change 1 kg of a liquid to a vapour (or gas) at its boiling point is called the **Specific Latent Heat of Vapourisation** (L_v). For water $L_v = 2.26 \times 10^6$ J/kg

When 1 kg of steam at 100°C changes into water at 100°C the same amount of energy is given out. The energy (E_h) that must be taken in or given out for a change of state to take place can be calculated from:

$$E_h = mL$$

J   kg   J/kg

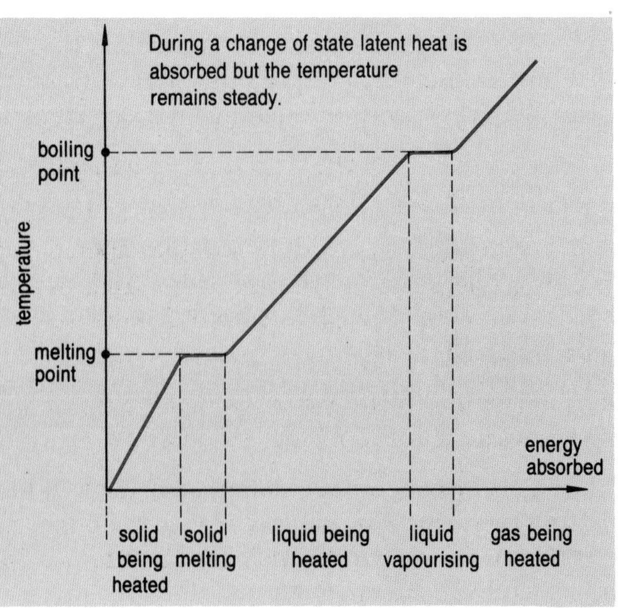

Heat pumps in practice

Freon, a liquid which boils at −29°C, is pumped round the circuit shown here. The freon changes to a vapour outside the house getting the necessary latent heat to do this from the surrounding air. The vapour is then pumped to a condenser inside the house where it is compressed back to a liquid, and gives out latent heat.

Heat pumps are often combined with solar panels which are used to warm up some water. This then is used as the source of heat for the pump.

In hot summers the heat pump can be used in reverse to cool the air in the house.

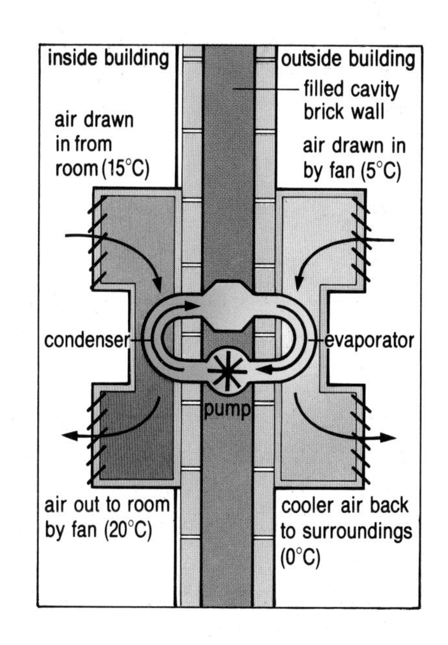

The refrigerator

A refrigerator (or freezer) uses a heat pump to remove heat from inside the food compartment and transfer it to the kitchen. The insulated walls of the refrigerator prevent most of the heat from re-entering.

Freon, is pumped round the closed system of pipes and when it is passing through the freezer compartment it is allowed to evaporate. The latent heat needed for this evaporation is taken from inside the refrigerator.

On leaving the compartment the vapour is put under pressure in the condenser pipes that run up the back of the cabinet. This pressurising makes the vapour condense and give up its latent heat to the surrounding air. The metal cooling fins at the back help to disperse this heat into the room.

A valve controls the flow of liquid in the pipe and, adjusting this to increase the flow, allows the refrigerator to become colder.

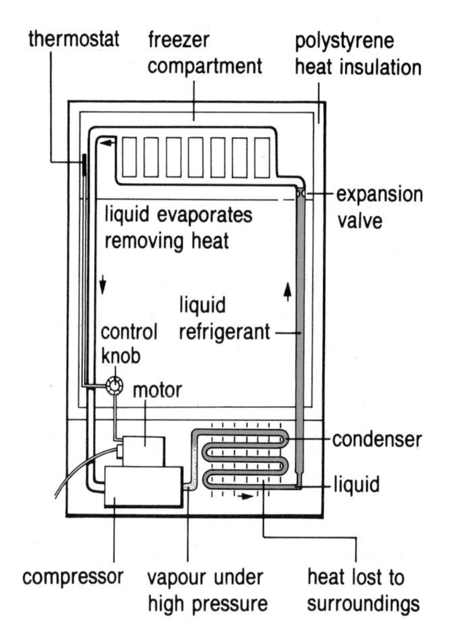

1 Why does the pump of a refrigerator keep switching on from time to time? What would happen if it stayed on all the time?

2 How do you think the switching on and off of the refrigerator is controlled?

3 The pump (compressor) in a freezer operates at 250 W. Estimate the amount of heat removed from the compartment in one hour. What mass of water at 0°C would be changed to ice during this time?

4 What mass of water would be produced if 1.67×10^6 J of energy were added to a block of ice at 0°C?

5 A steam heating system delivers steam at 100°C to radiators. When the water eventually returns to the boiler its temperature is 85°C. What mass of steam is required to produce 10^8 J of heat with this system?

LEISURE

7

In camera

The lens system

Setting the controls

Larger than life

Action replay

Newton's disco

Music in the air

Why different sounds sound different

The heart of hi-fi

His Master's Voice

Tape it

Compact discs

The first and last links

Gravity and games

Physics and fitness

In camera

The pinhole camera

People may argue about whether the chicken or the egg came first. But there is no doubt that the camera came before photography! A thousand years ago Arab astronomers were projecting images of well-lit objects on to the walls of darkened rooms. They were using what was really just a giant pinhole camera. As with every other kind of camera, the images were formed by the diffuse reflection of light from the objects themselves.

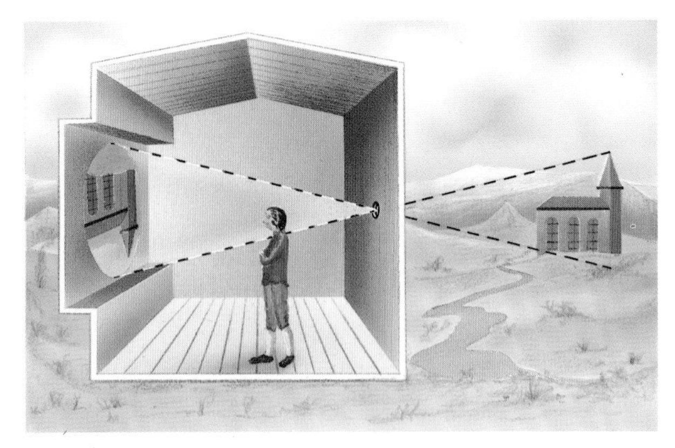

Ancient camera obscura (pinhole camera).

Writing with light

It was not, however, until the nineteenth century that a way was found of making a permanent picture from the image projected on to the camera's screen. Many systems were suggested, but it is the process patented by Fox Talbot in 1841 that has survived.

The early photographers had to carry a heavy camera and a tripod around with them, together with a tent to work in and a selection of chemicals and processing equipment.

Regular reflection of light from a mirror or a shiny metai surface.

Diffuse reflection of light from a non-shiny surface.

Camera equipment c1860.

A century of amateur photographers

Photography became popular when George Eastman produced his Kodak box camera in 1888. The camera contained a roll of paper coated with light-sensitive emulsion. Each roll took 100 photographs but the camera and film had to be returned to the factory for developing, printing, and reloading.

The worlds first Kodak.

A *Draw a diagram showing how the image of a candle flame is formed in a pinhole camera.*

B *Fit a lens to a pinhole camera and focus it on near and distant objects.*

C *Examine a commercial camera and identify the basic features listed on the diagram.*

Basic features of a camera

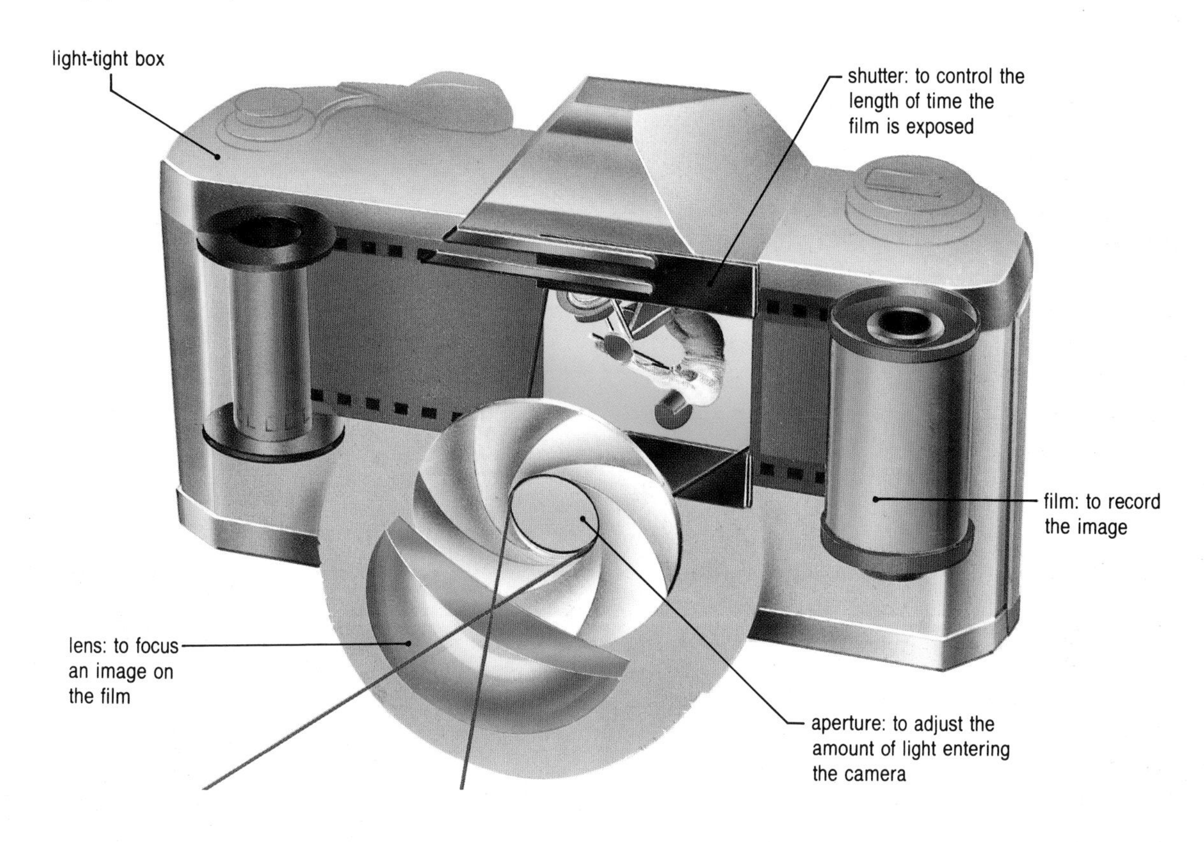

Basic controls on a camera

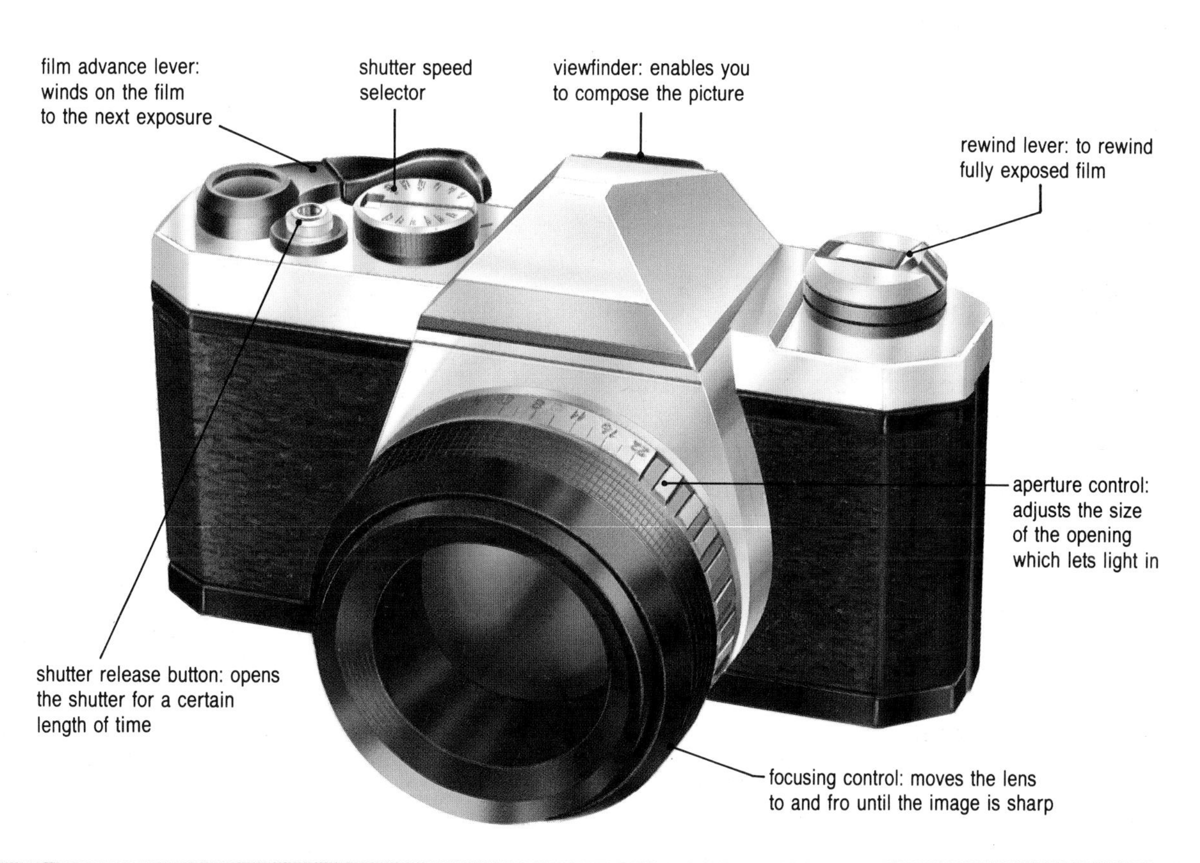

The lens system

Far and near

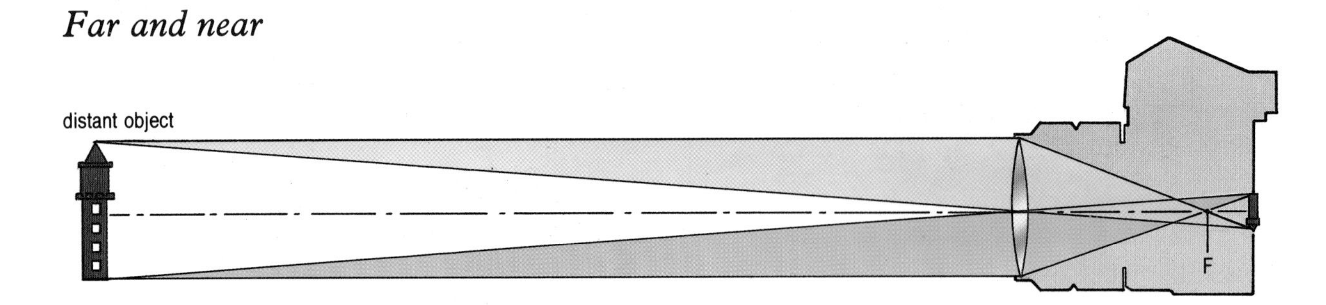

If you are photographing a distant tower, a small inverted image is formed at, or very near, the **principal focus** (F) of the lens. For close-up work the camera lens has to be moved further from the film to bring the image into focus. A large inverted image is then produced.

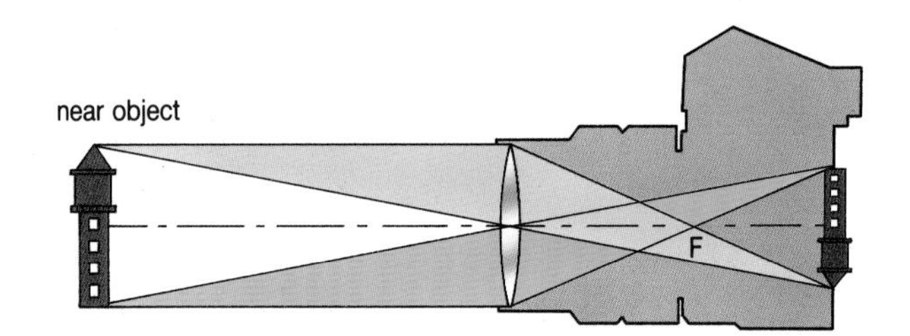

Aperture and shutter speed

The total amount of light passing through a camera lens affects the **exposure**. It depends on:

1 The time the shutter is open (the **shutter speed**).

2 The size of the opening or aperture (the **f number**).

No matter what size or type of lens you use, the same f number will always let the same amount of light reach the film.

To appreciate how the f number is defined, look at the diagram. If the **focal length** of a convex lens is f, then an aperture of diameter $\frac{1}{2}$f will be marked f/2 or simply f2. An aperture of diameter $\frac{1}{4}$f will be marked f/4 or f4 and so on. Halving the aperture diameter reduces its area – and the light passing through – to a quarter. So for an f4 aperture to give the same exposure as an f2 aperture the shutter would have to be open four times as long.

For practical purposes, it is useful to have intermediate f number settings – or f 'stops' as they are often called. Each 'stop' then changes the area, and thus the exposure, by two (i.e. it halves or doubles it). The range includes f2, f2.8, f4, f5.6, f8, f11, f16.

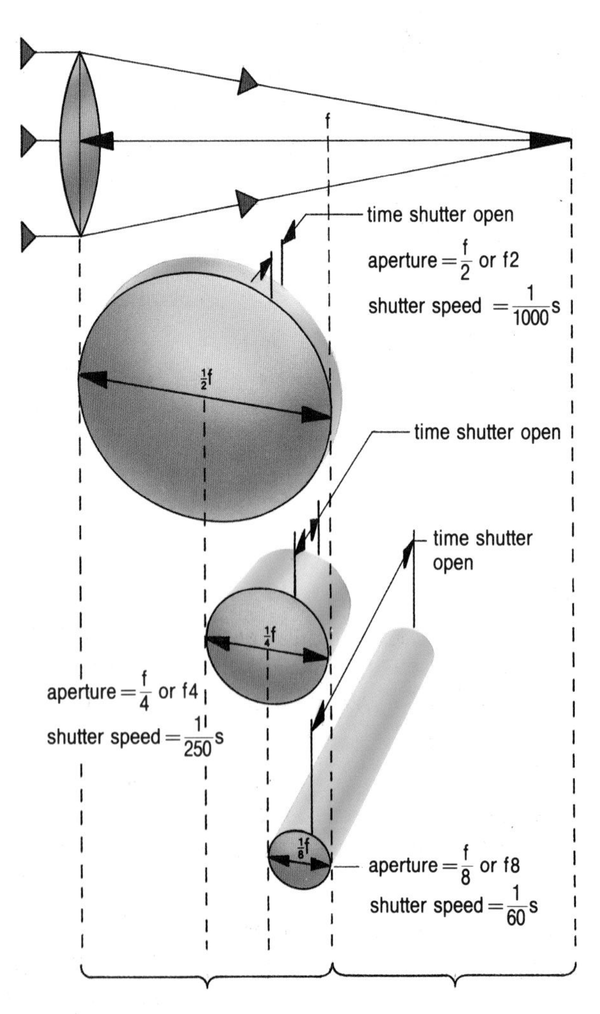

in each case the 'volume' of the cylinder is the same

Wide angle, telephoto and zoom

Lenses come in a variety of different focal lengths. Short focal length lenses are called **wide angle** (e.g. 28 mm with a 75° angle of view). Long focal length lenses are called **telephoto** (e.g. 500 mm with a 5° angle of view). The difference is illustrated in the two photographs taken from the same camera position.

Zoom lenses can be adjusted to a variety of different focal lengths (e.g. 60–300 mm).

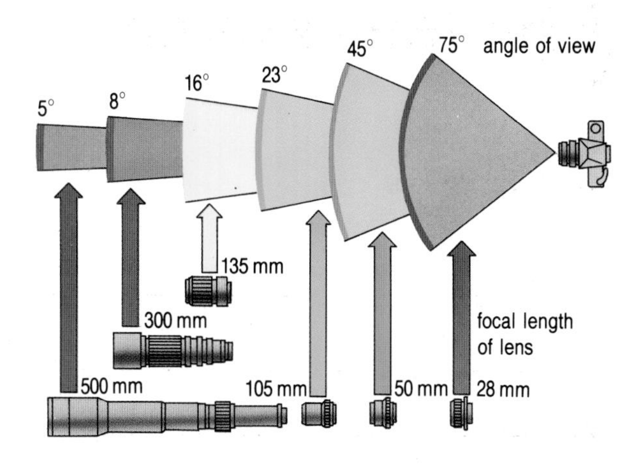

Photograph taken with a 28 mm lens.

Photograph taken with a 500 mm lens.

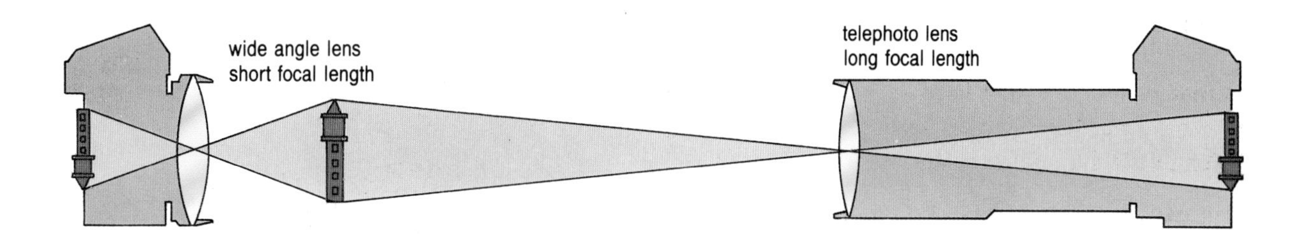

Image

A camera lens produces, on the film, an image which is inverted, laterally inverted, and smaller than the object.

Aperture

If the aperture diameter is $\frac{1}{4}$ of the focal length (f) of the lens it is written as f/4 or f4. This is called the **f number**.

The greater the f number the smaller the aperture.

1 What is the aperture diameter of an f2 28 mm lens?

2 Draw a diagram to show how the image is formed on the film of a camera fitted with a 50 mm lens, when the object viewed is 2 m from the lens.

3 Suggest when you might use a telephoto lens and when a wide angle lens.

4 The amount of light entering a lens is proportional to the area of the aperture. Use this fact to explain the numbers chosen for the aperture scale.

›› *Devise and conduct an experiment to measure the amount of light reaching the film of a camera when the aperture is set at f16, then at f8, and then at f4. Record and discuss your results.*

Setting the controls

Selecting a film

Some films react quickly to light. They are called fast films and normally need a fast shutter speed (short exposure). Other films react slowly to light – i.e. they require a slow shutter speed (long exposure).

The **speed** of a film is its sentitivity to light and is stated on the film box as an **ISO number** (International Standards Organisation). The higher the number the faster the film.

For colour prints and slides, ISO 100 is a normal general purpose rating. With a lens aperture of f5.6 the exposure time needed on a *cloudy bright* day is 1/125 second. The exposure chart shows other aperture and shutter speed settings that can be used with this film to give the same exposure on that day.

If your camera has a built-in exposure meter you will have to set the film speed on the camera dial.

Using a large aperture.

Using a small aperture.

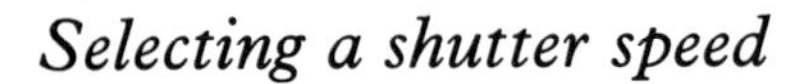

Selecting a shutter speed

If a camera moves while the shutter is open the photograph will be blurred. To avoid this you should always use a shutter speed of at least 1/60 with a hand-held camera.

If you want to 'stop' a fast moving object you will need a much faster speed.

Selecting the aperture

The aperture controls the amount of light which enters the camera. But it has another important use. A large aperture enables you to focus sharply on one thing while leaving everything else fuzzy. A small aperture, on the other hand, gives a photograph with everything sharply in focus as in the second example. We say it gives a greater **depth of field.**

To understand why a small aperture gives this greater depth of field look at the two diagrams. Imagine that in each case the bird's beak is focused on a point *behind* the film. Rays of light from the beak will then form a large patch of light on the film and so produce a fuzzy image of the beak. If a small aperture is used the rays form a smaller patch of light and so give a sharper image of the bird's beak. So if a small aperture is used objects which are not accurately focused on the film still produce reasonably sharp images.

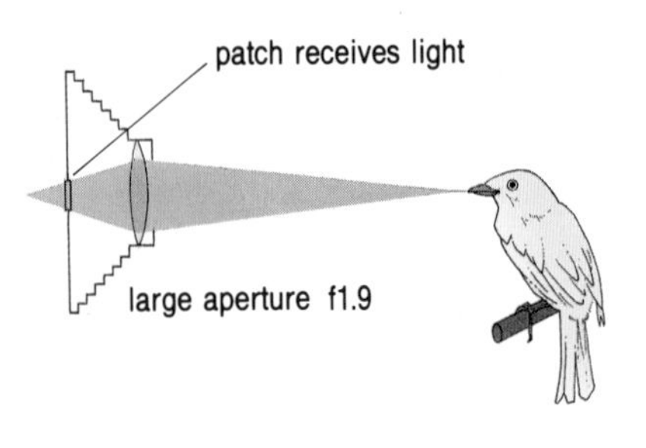

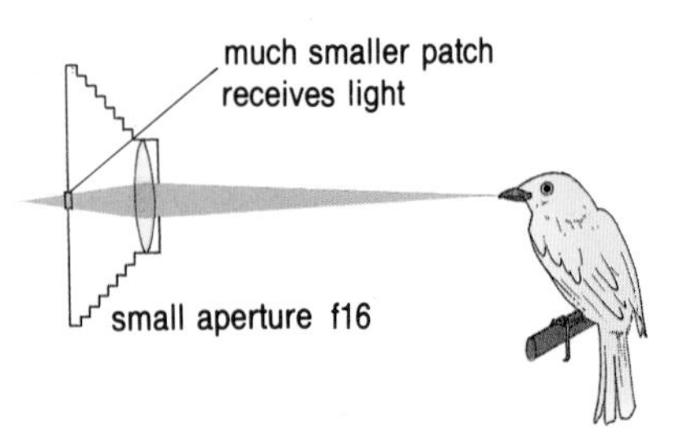

When choosing the aperture setting you must decide how much of the scene you want to be sharply in focus.

Decision time

There are five things which you have to take into account when preparing to take a photograph.

1 Film speed

2 Lighting conditions

3 Shutter speed

4 Aperture

5 Focus control setting

When you have bought the film, decided on the subject and looked at the lighting conditions, you must decide on the best pair of shutter speed:aperture settings before you focus on your subject. Use the exposure chart and the information given above to make the selection. To read the chart, first note the kind of weather and the film speed you are using. Then, follow the heavy black lines to find the coloured symbol at the point where they cross. The symbol will enable you to find the correct pair of settings. However, the same symbol appears at many points on the sloping line and each of these indicates another pair of settings which give the correct exposure. You must decide which pair suits your purpose.

Finally you will have to set the focus control before pressing the button. Happy shooting!

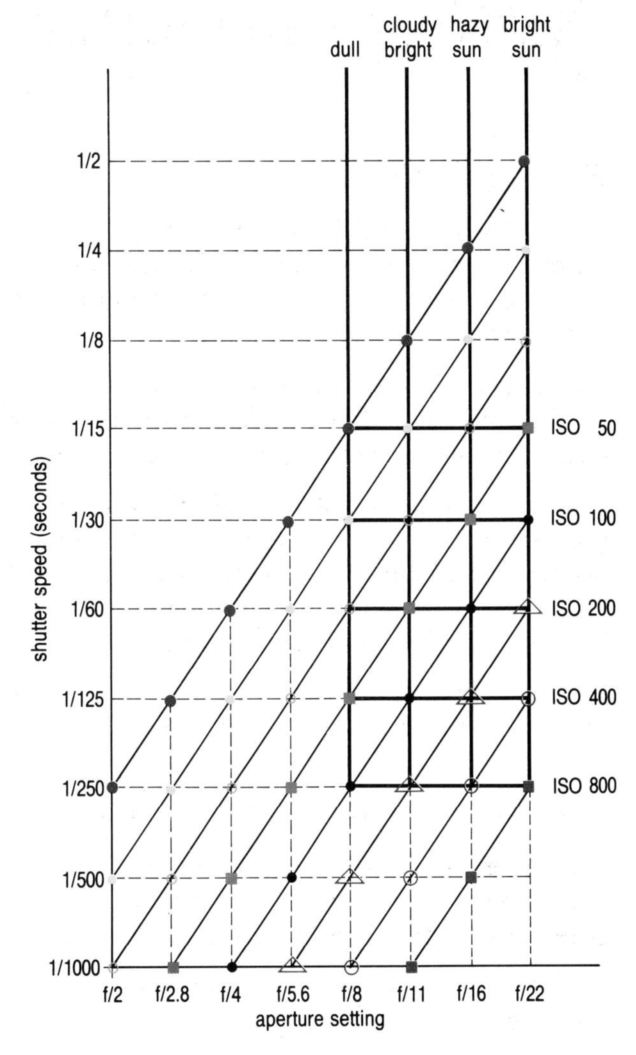

Exposure chart.

1 Copy out this table and fill in the gaps to give the same exposure each time.

f no.	5.6	2		4	
Shutter speed	$\frac{1}{250}$		$\frac{1}{1000}$		$\frac{1}{125}$

2 On a dull day you are using ISO 100 film to photograph:
a) a passing car;
b) a country scene in which you want everything to be in focus;
c) a close up of a child's face.
What shutter speed:aperture settings would you select for each? Explain your choice.

3 Your camera has a fixed-focus f5.6–f16 lens and only one shutter speed of 1/125 second. What speed of film would you buy to take photographs:
a) only on bright sunny days;
b) on bright *and* dull days?

4 Discuss the advantages and disadvantages of using:
a) slow film;
b) fast film.

5 Where would:
a) the shutter speed;
b) the aperture setting scales,
be zero if the axes on the exposure chart were extended?

Larger than life

The slide projector

In a projector a small slide can produce an image a hundred times larger than the slide itself. A convex lens is used in the projector and the slide is placed just beyond the principal focus. The image projected on the screen is then real, magnified, and inverted. The slide must therefore go in upside down!

You can see what is happening by drawing a scale diagram. Remember that rays from the slide which are parallel to the principal axis pass through the focus, and those which pass through the centre of the lens are not bent.

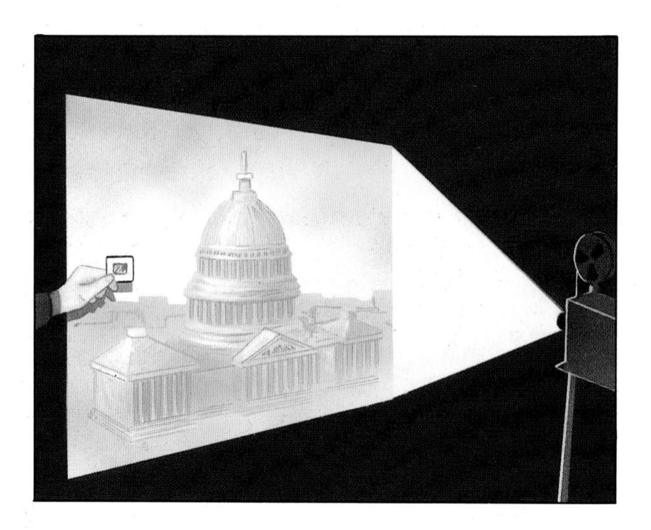

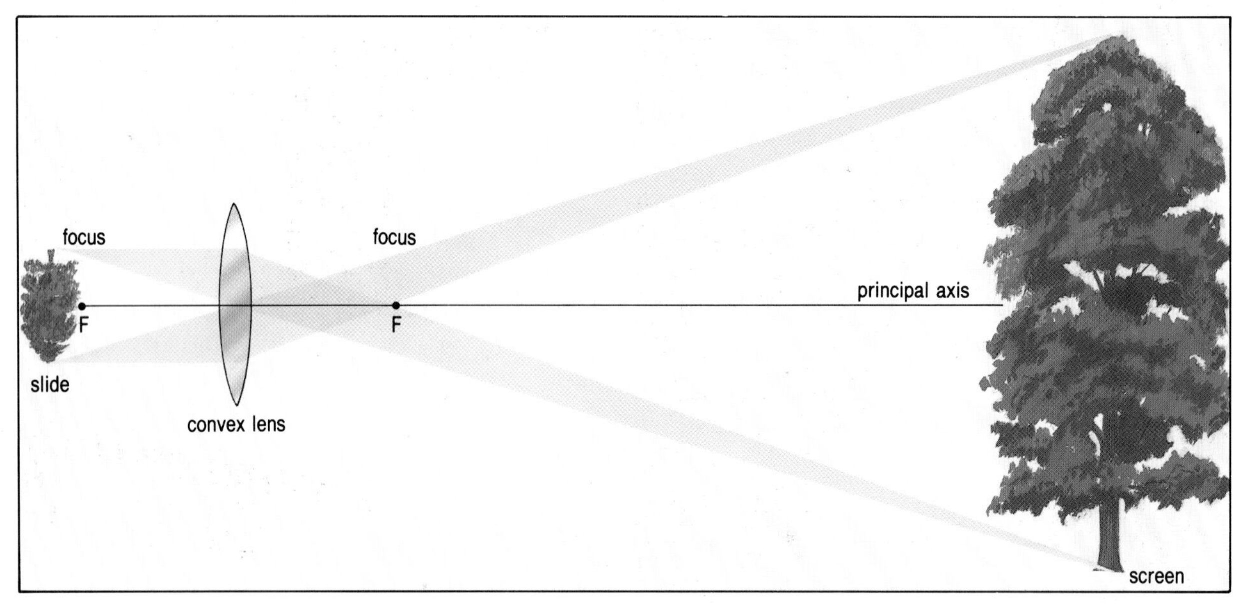

Illumination

To illuminate the slide a **clear** lamp with a small bright filament is used. But only a very small part of the light emitted from the filament actually passes through the slide. However, this situation can be improved in two ways.

1 Two large condenser lenses are placed between the lamp and the slide. Much more light can then be directed through the slide.

2 A concave mirror behind the lamp can reflect some of the 'stray' light back through the lamp.

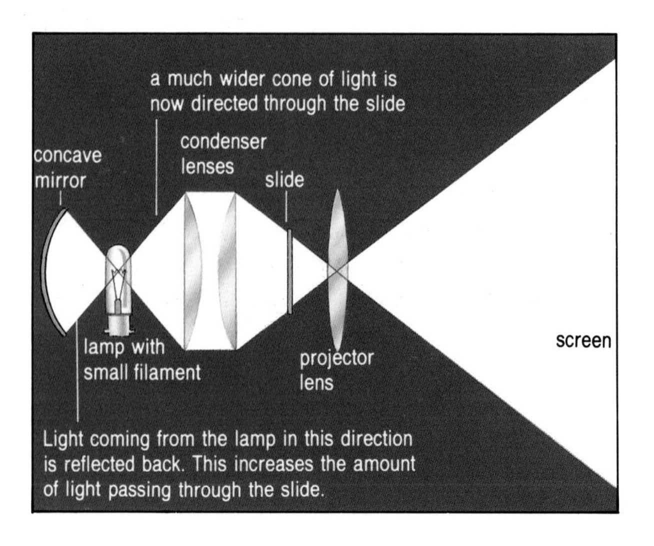

Enlarger

An enlarger works on exactly the same principle as the slide projector but it usually has an **opal lamp** instead of a clear lamp and reflector. The opal lamp gives a diffuse light which illuminates the negative uniformly. As the enlarger usually projects a much smaller image than the slide projector there is no need for very brilliant lighting. Moreover, photographic paper is very sensitive to light.

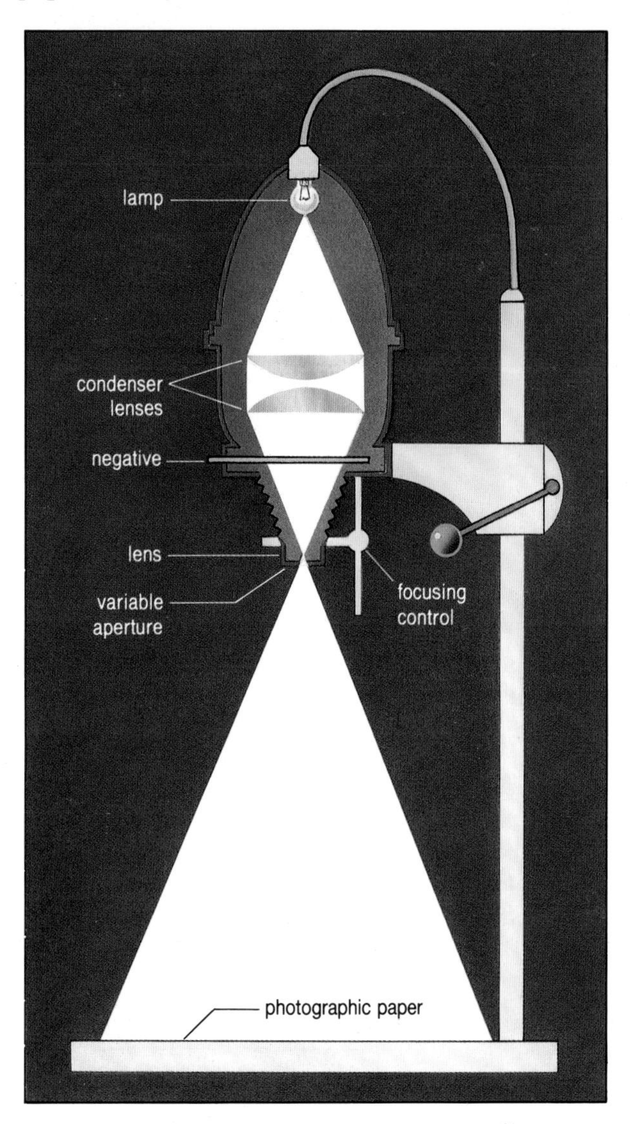

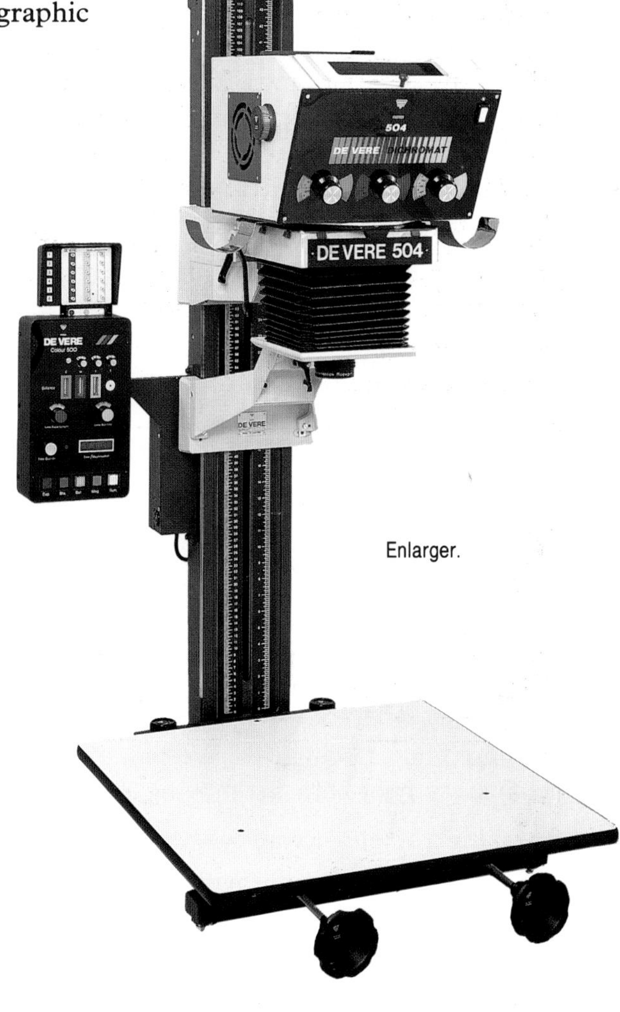

Enlarger.

›› *Use a mushroom type pearl lamp, a 35 mm transparency and a magnifying glass to make a model of a slide projector or an enlarger. Cleaning powder will remove print from the end of the lamp.*

Demonstrate and explain to your partner the purpose of the aperture in an enlarger.

Use a negative to make an enlargement on photographic paper and then process it.

1 Why are condenser lenses used in a slide projector?

2 Would you move a projector lens towards or away from the slide to focus on a more distant screen?

3 Find out why the lens in a slide projector often consists of several lens elements.

4 An enlargement 200 mm wide is produced from a negative 32 mm wide. If the distance from the enlarger lens to the paper is 500 mm, use a scale drawing to find the focal length of the lens.

Action replay

Time and motion

Moving things fascinate us. We can use photography to make a detailed study of both fast moving and slow moving objects.

By taking a single picture every minute, every hour or even every day, scientists can examine in detail the gradual growth of plants or crystals, the germination of seeds, or the development of flowers and fruits. This is called **time-lapse photography**. The pictures here were taken at one hour intervals and show a tulip gradually opening up.

Movie magic

The cinema, television, and video would soon lose their popularity if they showed only still pictures. But this is exactly what they do show! They have never displayed a single moving picture. It's really all a great trick. What you see, every second, is a series of 24 or 25 still pictures. These still pictures are called frames.

We interpret these slightly different still pictures as one moving picture because we go on 'seeing' things for a fraction of a second after they've disappeared. This ability to go on perceiving things after they've gone is called **persistence of vision**. You can test this out by making your own 'movie' in the activity described later.

Movie film of a humming bird.

Cine camera and projector

The world's first movie camera was produced about a hundred years ago by the French scientist Etienne-Jules Marey. One of his early films, taken at 50 frames a second, shows that a cat always falls on its feet.

When a movie film is being projected, one frame is seen for, say, 1/48th of a second. A shutter then covers the light beam while a claw jerks the film down to the next frame. The shutter then opens again so that the new picture can be projected.

The world's first successful movie camera (1890).

Fast and slow

Normally the camera speed and the projector speed are the same, but it is possible to use a movie camera for time-lapse photography. An opening flower could be filmed at one frame a minute and then projected at 24 frames a second. One hour's action would then be seen in $2\frac{1}{2}$ seconds!

The opposite effect can also be useful for studying motion. A high-jumper may want to improve his technique or a golfer her swing. By shooting at 240 frames a second and projecting at the 'normal' speed of 24 frames a second, people can see themselves in slow-motion. This enables them to analyse their performance and, with luck, improve it.

Very high-speed cameras are also used to examine industrial processes in slow motion. Often, single frames are enlarged and examined, one by one, to measure the changes which have taken place in, say, 1/1000 second.

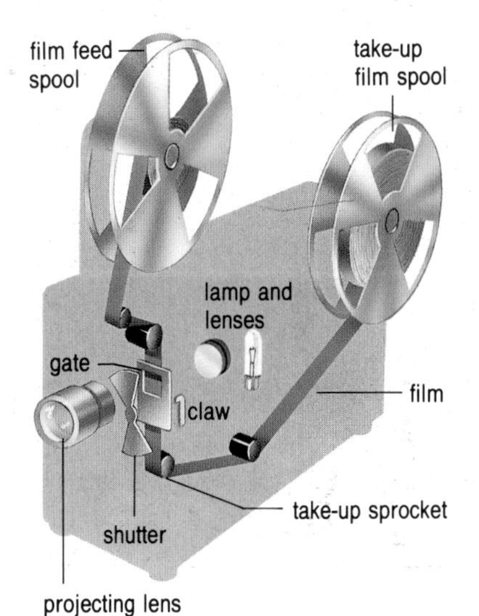

Early movie film showing that a cat always lands on its feet.

1 Explain why a high-speed camera is used to produce slow-motion pictures.

2 A movie film is set up to film a rose at 720 frames a day. Calculate the number of times by which the action is speeded up when the film is projected at the normal speed of 24 frames per second.

3 Why does a movie projector have a rotating shutter?

›› *Demonstrate persistence of vision to a young friend by making your own movie. Draw four little figures on squares of paper, fold each of them in the middle and glue them together round a pencil. Now spin them and see the action.*

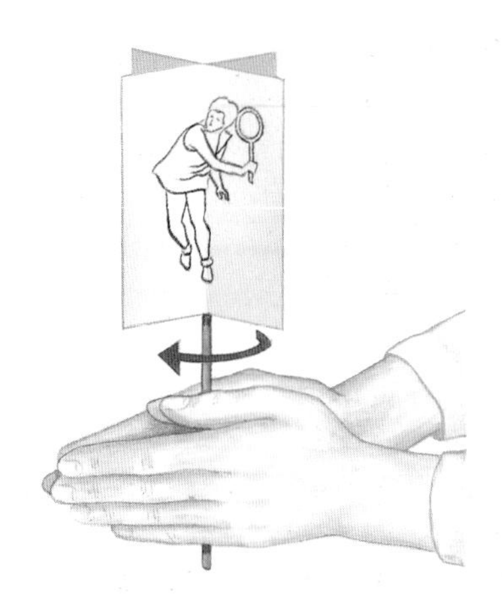

Newton's disco

Multiflash photography

A stroboscope lamp.

Flashing lights illuminate the world of entertainment, industry, and science. The electronic flash lamps which produce bright, brief light pulses at regular time intervals are called **strobe lamps**. We use them to take **multiflash** pictures which enable us to study the motion of all kinds of moving things from tennis players to trolleys.

For good results, total darkness is needed and it helps if the background is black. To record the gymnast performing a somersault, a camera is set up on a tripod, the strobe lamp is switched on and the room lights are switched off. Just as the gymnast starts to jump, the camera shutter is opened and kept open. The final print enables us to study the somersault.

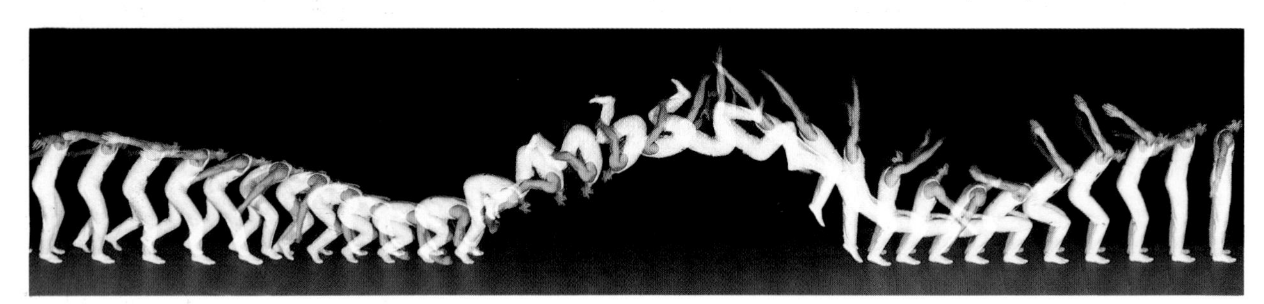

Newton's First Law

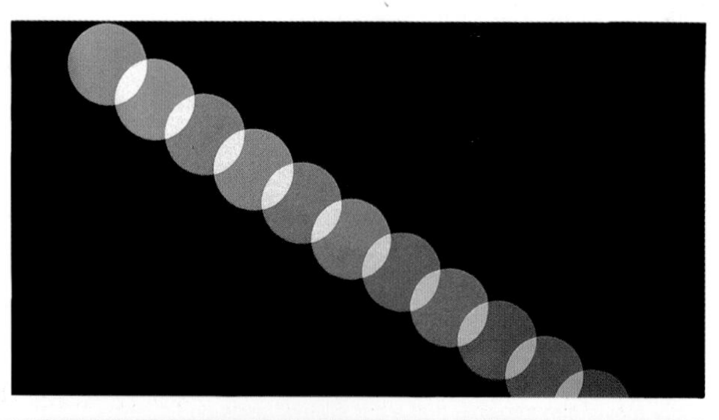

Here are two multiflash photographs in which the images are equally spaced. The dry ice puck is floating on gas so that there is practically no friction and therefore no horizontal forces acting on it. The diameter of the puck is 8 cm and the strobe frequency 15 Hz. The model car is running down a gentle slope.

1 What do these photographs tell you about the speed of each object?

2 From the information given find the speed of the puck.

3 Explain how it is possible for the model car to run down a hill at a constant speed.

Newton's Second Law

Two spheres, one bigger and heavier than the other, are dropped at the same instant. A multiflash photograph shows that the acceleration is the same for each. Strobe frequency = 25 Hz, scale separation = 80 mm.

4 Using this result and Newton's Second Law, deduce a relationship between mass and weight for bodies in the Earth's gravitational field.

5 Give two descriptions for the force/mass ratio and state the units for each.

6 Plot a graph showing how the speed of the small ball varies with time. From the graph calculate the acceleration due to gravity.

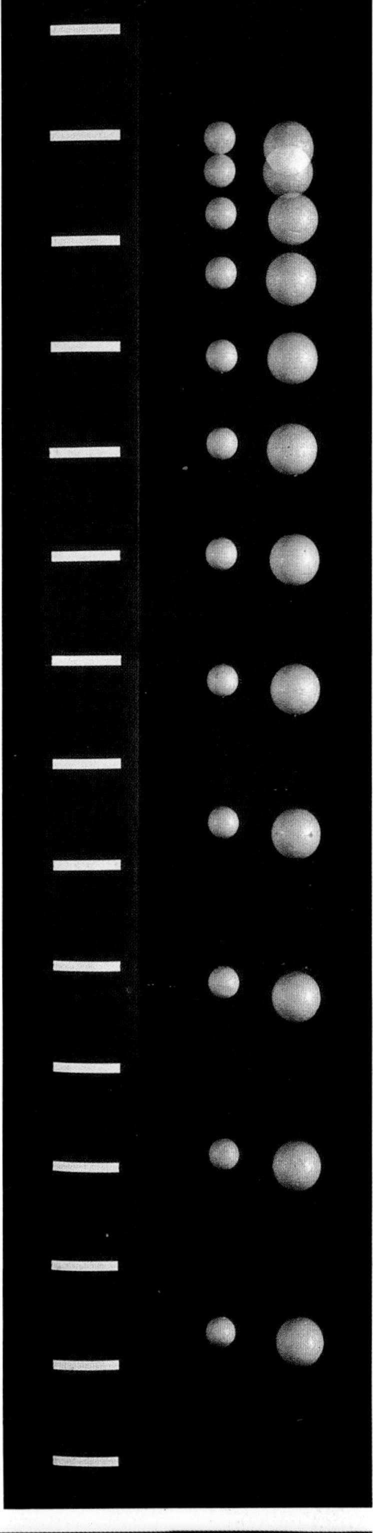

Newton's Third Law

In the photograph of an air-track collision a moving vehicle (A) collides with a stationary vehicle (B) of the same mass. They stick together and move off at a reduced speed. Strobe frequency = 5 Hz, scale separation = 10 cm.

7 Find the speed (v_1) of the single vehicle before the collision, and the speed (v_2) of the two together when they move off after the collision.

8 Now calculate the total 'mass × speed' for both vehicles before the collision, and the total 'mass × speed' for them after the collision.

If we call the product 'mass × speed' 'momentum' what can you say about the total amount of momentum before and after the collision?

9 Find how much speed (Δv_1) the first vehicle (A) lost during the collision. Now find how much speed (Δv_2) the second vehicle (B) gained during the time (Δt) of the collision.

10 $$\text{acceleration} = \frac{\text{change of speed } (\Delta v)}{\text{time for the change of speed to take place } (\Delta t)}$$

Compare the accelerations of the two vehicles during the collision. Assume that $\Delta t = x$ seconds.

11 Now use Newton's Second Law to calculate the average force acting on each vehicle during the collision. State your results in terms of the force which A exerts on B, and the force which B exerts on A.

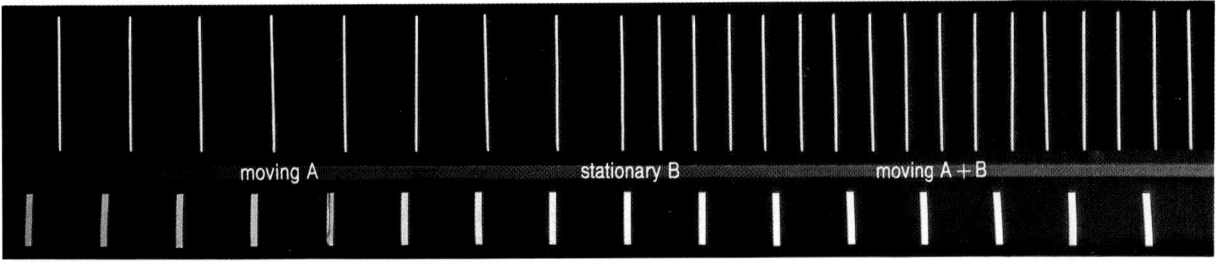

Music in the air

The amazing paper cone

What sort of sounds do you hear every day? People speaking? Machines clanging? Traffic roaring? Instruments playing? Cones waggling? 'All but the last' you may say. Yet every time you listen to a radio, television, video or audio recorder – or hear an announcement over a public address system you are listening to a paper cone vibrating to and fro!

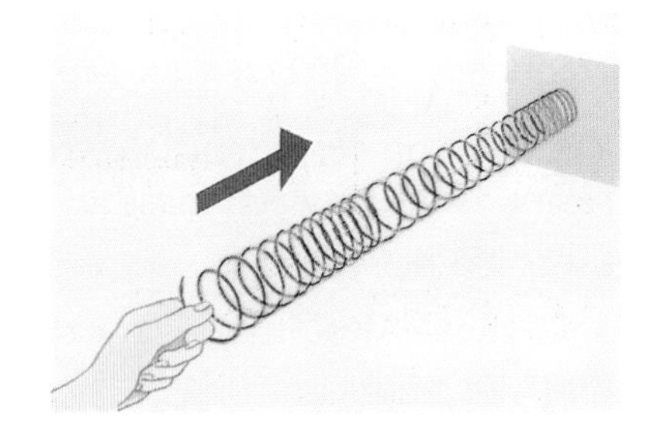

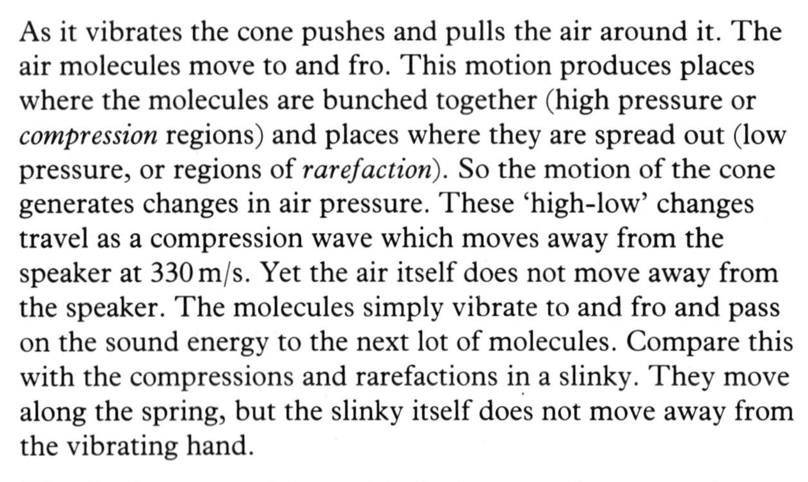

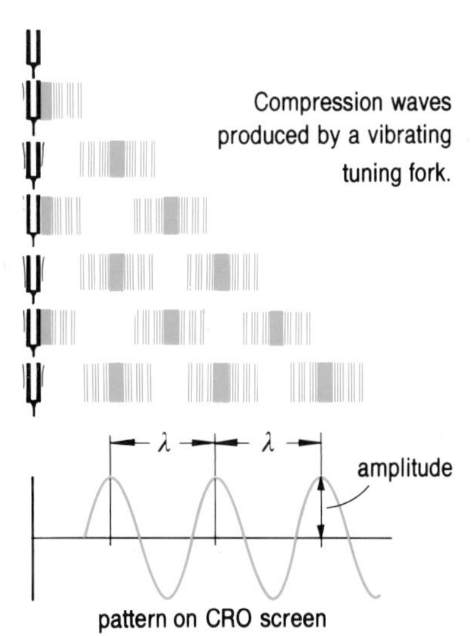

As it vibrates the cone pushes and pulls the air around it. The air molecules move to and fro. This motion produces places where the molecules are bunched together (high pressure or *compression* regions) and places where they are spread out (low pressure, or regions of *rarefaction*). So the motion of the cone generates changes in air pressure. These 'high-low' changes travel as a compression wave which moves away from the speaker at 330 m/s. Yet the air itself does not move away from the speaker. The molecules simply vibrate to and fro and pass on the sound energy to the next lot of molecules. Compare this with the compressions and rarefactions in a slinky. They move along the spring, but the slinky itself does not move away from the vibrating hand.

The displacement of the molecules in a sound wave can be represented on a graph. Don't confuse this graph (or oscillo-scope pattern) with the compression wave it is representing.

The illustration shows regions of compressed and rarefied air moving away from a tuning fork. The distance between successive compressions (or rarefactions) is called the *wave-length*. The distance from the centre line to the crest of the wave is called the *amplitude*. And, as with all types of wave, speed (v), frequency (f) and wavelength (λ) are related by

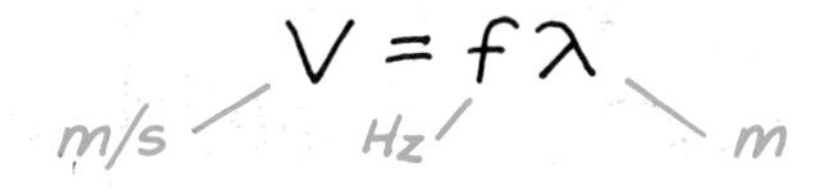

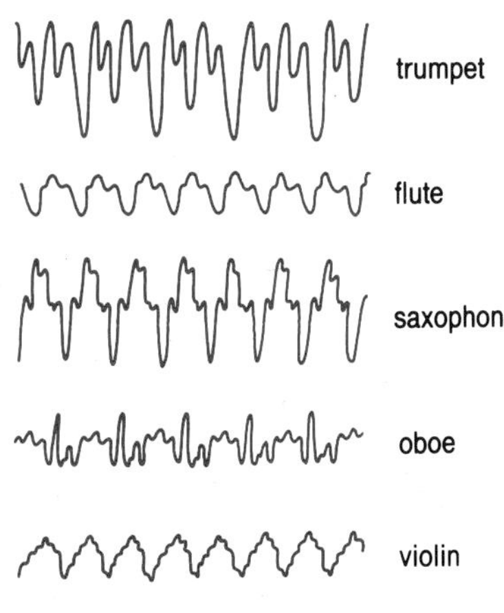

Oscilloscope traces of musical sounds.

Instruments of music

A range of sound sources from tambourines to tubas is shown here. In each case something is vibrating: a reed, a skin, a rod, a string – or a tubeful of air!

An oscilloscope, or a computer monitor, can be used to show the waveform produced by each of these instruments. You can see that different instruments produce different waveforms.

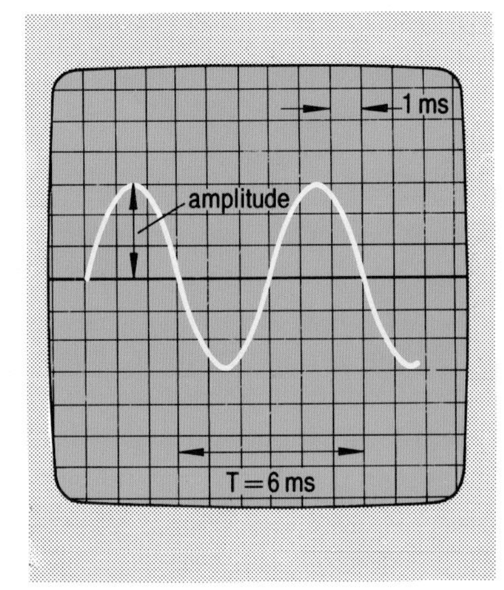

Electronic graphs

In addition to investigating various waveforms an oscilloscope can be used to compare the amplitudes of different signals and to measure the period of a sound wave. The period (T) is the time for one complete vibration. If it takes $\frac{1}{80}$ second to produce one vibration there will be 80 vibrations each second. Thus

$$f = \frac{1}{T}$$

Hz s

In the example shown here the time control of the oscilloscope has been set so that the electron beam takes 1 millisecond to move between the vertical lines on the screen. The period (T) for a complete wave must therefore be 6 ms. This makes the frequency

$$f = \frac{1}{T} = \frac{1}{6\,\text{ms}} = \frac{1000}{6\,\text{s}} = \frac{1000}{6}\,\text{Hz} = 167\,\text{Hz}$$

1 State what is vibrating to produce sounds from each of the instruments shown here.

2 The same wave pattern as the one shown on the oscilloscope screen is produced on another oscilloscope which has its input connected to a signal generator.
a) Is the output control on the signal generator set at 'square', 'triangular', or 'sine' wave?
b) If the oscilloscope time control is set at '100 μs per cm' what is the signal generator frequency?

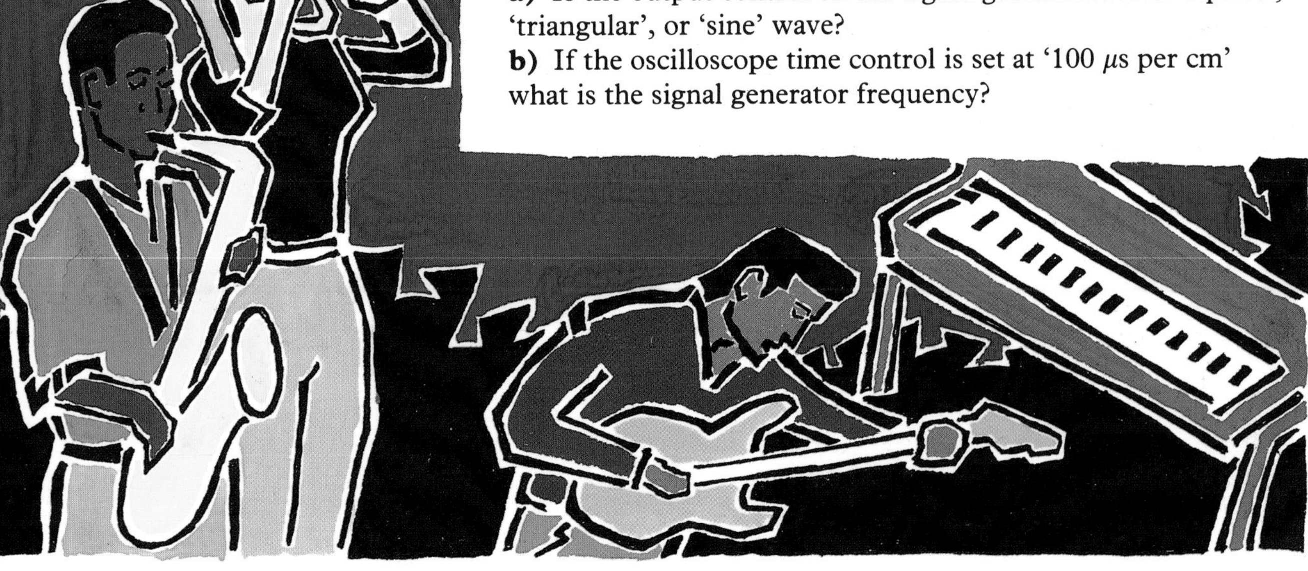

Why different sounds sound different

Quality and quantity

To the question 'When will you be ready?' there are two types of answer. 'I'll be ready shortly' is a qualitative answer. It gives only general information. 'I'll be ready in 20 minutes' is a quantitative answer. It gives more precise and more useful numerical information.

Sound can also be described qualitatively – that is, without measurements. For example we can speak vaguely about its loudness, its pitch or its quality. But if we are going to accept Lord Kelvin's advice and 'advance to the stage of science' we have to find ways of describing the sound with *numbers*. We have to measure it.

"When you can measure what you are speaking about, and express it in numbers, you know something about it; but when you cannot measure it, when you cannot express it in numbers, your knowledge is of a meagre and unsatisfactory kind: it may be the beginning of knowledge, but you have scarcely, in your thoughts, advanced to the stage of science."

Lord Kelvin.

Pitch and frequency

A note can have a high pitch or a low pitch. The quantity used to measure it is called the *frequency*, that is, the number of vibrations each second. It is measured in hertz (Hz). A young person can normally hear frequencies from about 20 hertz to 20 000 hertz. Unfortunately as we get older we find it more and more difficult to hear the higher frequencies. So, by the time we are old enough to afford a high quality audio system – we can't hear the higher pitched notes!

The frequencies of the notes on a piano and the frequency range of various other instruments are shown here. Notice that from middle C to C′, an octave higher, the frequency is doubled. The frequencies on the diagram are all in hertz.

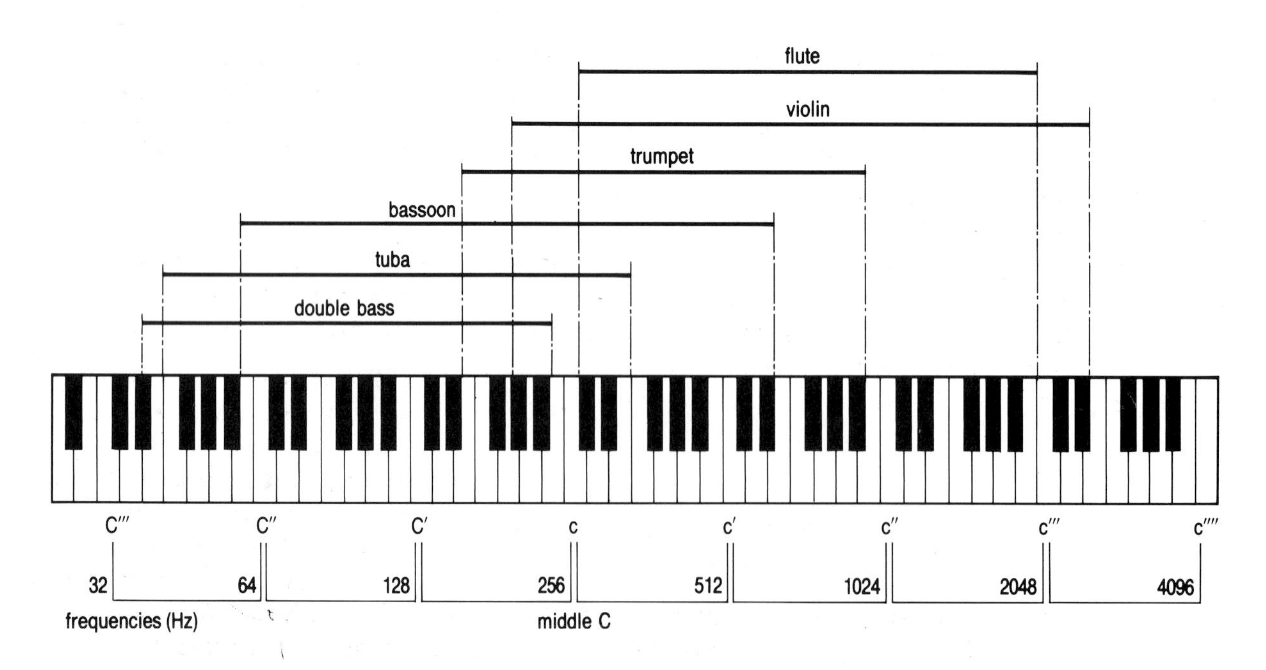

Loudness and sound levels

The loudness or volume of a sound can be measured using a rather complicated scale. For our purposes we need only know how to use its units – decibels (dB). Just as 0°C is not the coldest possible temperature, 0 dB does not mean 'no sound'. What it indicates is the lowest sound level which can just be detected under very quiet conditions. It is called the threshold of hearing.

At the other end of the scale we start to feel uncomfortable with sound levels of 120 dB, and we experience severe pain at 140 dB. So the range is from the undetectable (0 dB) to the unbearable (140 dB). To measure sound levels we need a sound level meter. In it a microphone detects the small variations of air pressure in the compression waves and changes them into variations of voltage. These are then amplified and displayed in decibels on the meter.

0 dB corresponds to an air pressure change of 20 millionths of a newton per square metre (20 μPa) and 120 dB is a change of 20 Pa. As normal atmospheric pressure is 100 000 Pa you can see just how sensitive our ears are to sound. In fact a few milliwatts of sound power is all you need to fill a room with sound at a comfortable level. Compare this with the power needed to light or heat a room!

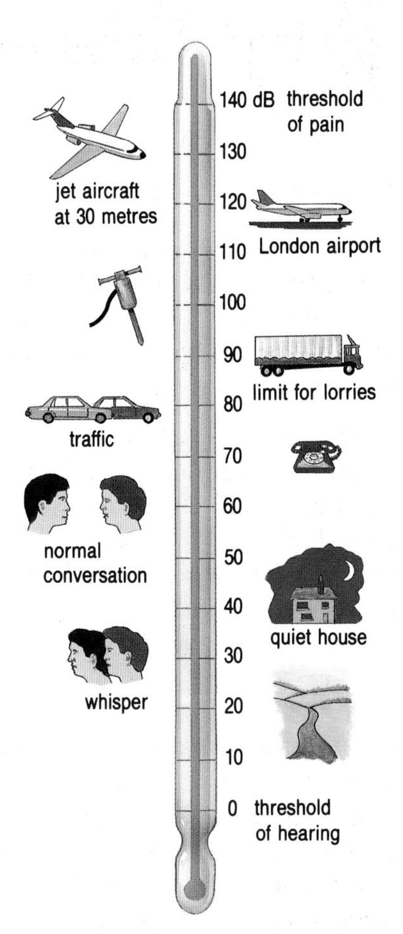

Sound levels in dB.

Device	Power	Approximate cost to run for 24 hours
loudspeaker	0.1 W	0.01p
lamp	100 W	15p
heater	3 kW	£4.50

Relative costs of sound, light and sound.

Quality and harmonics

A tuning fork produces the simplest kind of sound and it can be studied using a microphone and an oscilloscope. The waveform produced is a sine wave. As the tuning fork produces only *one* frequency the note is called a pure tone, and the same note can be obtained from a signal generator switched to the 'sine wave' output.

If you look at the oscilloscope patterns on page 201 you will see that each waveform is different. Each instrument has its own special *quality* or *timbre*. Again these are qualitative terms. However, a French mathematician called Fourier found that these, and any other repeating wave patterns, can be made up from a sine wave – called the fundamental – and a number of other sine waves whose frequencies are multiples (harmonics) of the fundamental. It is these harmonics that give each instrument its distinctive quality even when the instruments are playing the same note. Good audio equipment should accurately reproduce all the harmonics and the fundamental frequency of each note.

›› *Examine the notes from various musical instruments with a computer or an oscilloscope.*

Find out, approximately, what power is needed to operate a burglar alarm bell, a disco amplifier and speaker system, and an immersion heater. Now calculate the cost of running each of these for 24 hours and display your results in a table similar to the one shown above.

1 Use the diagram of the piano keyboard opposite to estimate the frequency range of **a)** a tuba, and **b)** a flute.

2 Estimate the power needed **a)** to light a large living room **b)** to heat the room **c)** to fill the room with sound.

The heart of hi-fi

Inputs

Audio systems are designed to use a number of different sources. These include a microphone, pickup cartridge, radio tuner, compact disc and cassette player. The required input is selected by a switch on the amplifier.

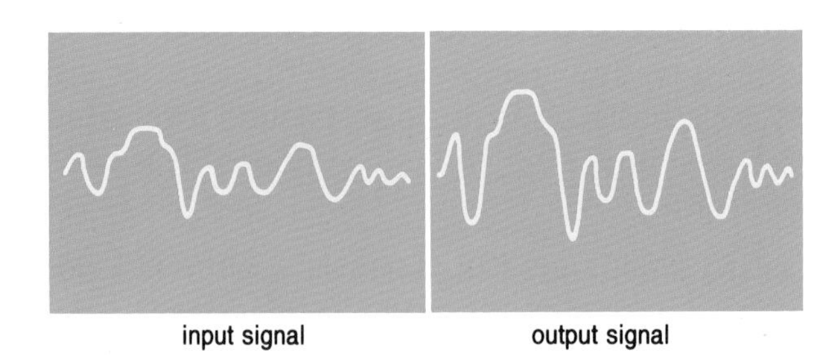

Amplifier

The heart of every system is the amplifier. It must be able to increase the size of the input signal with as little distortion as possible. This means that the output waveform should simply be a bigger version of the input waveform. To achieve this the gain of the amplifier should be the same for all frequencies. A graph of the gain against the frequency will therefore be a straight line (solid line). It is called a linear frequency response curve.

In practice an absolutely flat response curve is difficult to achieve, and the amplifier gain usually falls off a little at the lowest and the highest frequencies (dotted lines).

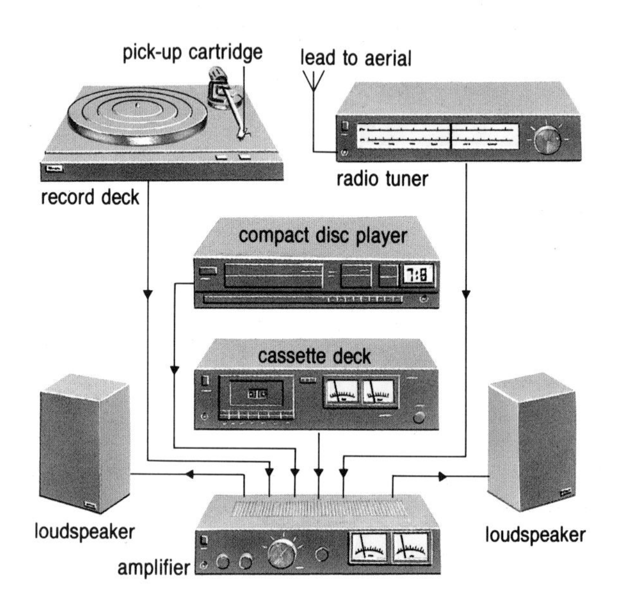

Tone controls

If the low frequencies in a signal are weak the bass control should be turned up. This increases the low frequency gain of the amplifier (red curve) and so strengthens the weak part of the signal. Similarly if the high frequencies are weak the treble control should be turned up to increase the high frequency gain of the amplifier (blue curve).

If the low frequencies in the signal are too strong the bass control should be turned down to reduce the low frequency gain (orange curve). Similarly if the high frequencies are too strong the treble control should be turned down (green curve).

In each case the adjustment should make the reproduced sound more like the original.

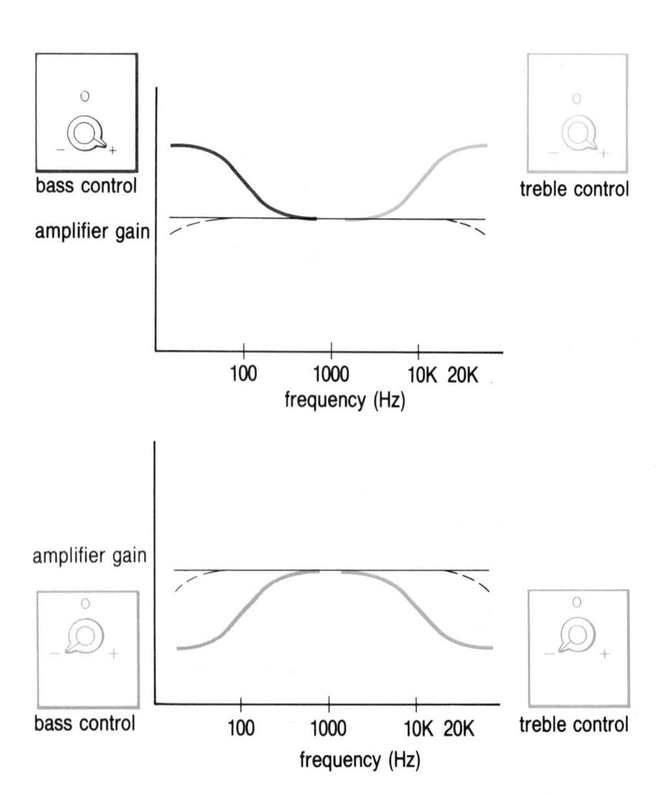

Voltage gain

Some audio systems have two separate amplifiers: a preamplifier (preamp) and a power amplifier. Both may, however, be housed in the same box.

The preamp uses transistors (in chips) to increase the small voltages available from a source. A pickup cartridge, for example, could produce a 0.3 mV signal which is amplified to 15 V. The voltage gain is defined as the number of times the output of the preamp is bigger than the input. In this case

$$\text{voltage gain} = \frac{\text{output voltage}}{\text{input voltage}}$$

$$= \frac{15}{0.3 \times 10^{-3}} = 50\,000$$

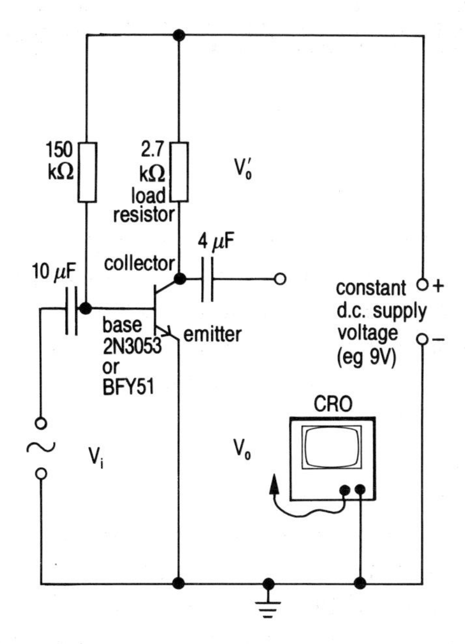

Circuit for finding voltage amplification.

You can measure the voltage gain of a transistor amplifier using an oscilloscope in the circuit shown here. The input voltage is fed into the base-emitter circuit and the output voltage is produced across the load resistor in the collector circuit.

In practice it is usually more convenient to measure all voltages with respect to the zero (earth) line. In this case we measure the output voltage (V_0) between the collector and the common negative line.

To see that these two alternating voltages (V_0 and V_0') are equal, remember that the supply voltage remains constant throughout. So as V_0' increases V_0 must decrease by the same amount. Consequently the amplitude and the frequency of the changing (alternating) voltages (V_0' and V_0) are the same.

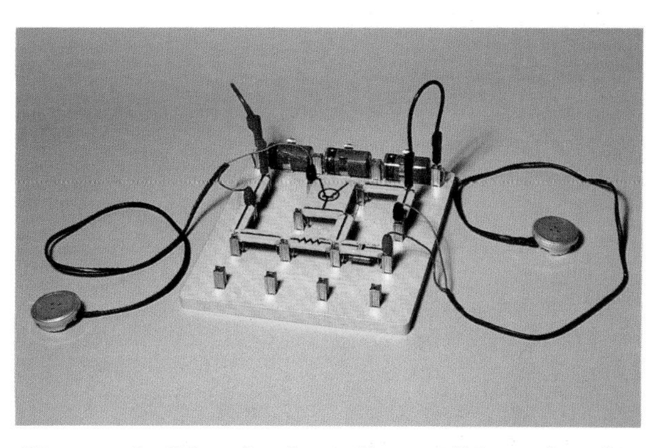

›› *You can build a simple audio amplifier using the transistor circuit shown here. In place of the signal generator insert a microphone and replace the load resistor with an earphone. A high resistance magnetic earphone is suitable both as a microphone and as an earphone.*

Power gain

Modern stereo amplifiers often produce a total of 100 watts; that is 50 watts per channel. If the output from the preamp is 15 V and the input resistance to the power amplifier is 1000 ohms, the power input is

$$\frac{V^2}{R} = \frac{15^2}{1000} = 0.225\ \text{W}$$

The power gain can be calculated as follows:

$$\text{power gain} = \frac{\text{output power}}{\text{input power}} = \frac{100}{0.225} = 444$$

1 If a microphone has a poor low-frequency response and too great a high-frequency response how would you adjust the amplifier tone controls?

2 If the input voltage to a preamp is 2 mV and the output voltage is 12 V find the voltage gain.

3 A power amplifier delivers 60 W and has a power gain of 300. Find the input power.

His Master's Voice

'Mary had a little lamb'

Thomas Alva Edison was determined to find a way of transmitting messages quickly, and so in 1877 he devised a machine to send Morse Code on punched paper tape. As the tape was running through the machine at high speed he heard a sound which he described as 'resembling human talk'. This gave him an idea! Later the same year Edison launched the world's first recording machine – the phonograph. He shouted 'Mary had a little lamb . . .' into a small horn which had a needle attached to it. The needle pressed the pattern of sound vibrations into the tinfoil. A recognisable voice was heard on playback.

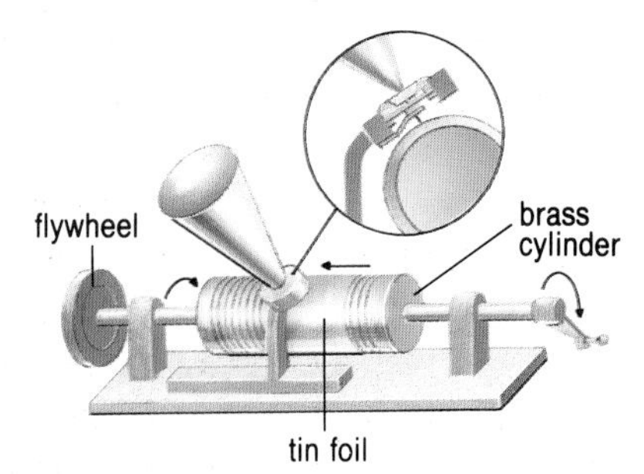

Thomas Edison's phonograph.

Ten years later, in 1887, the gramophone was born. Invented by Emile Berliner it used flat discs instead of cylinders and was the forerunner of today's record players. Nipper, the famous fox terrier, is listening to His Master's Voice on a gramophone which was built around the turn of the century.

The gramophone of 1887.

Mono

We have seen (page 201) that a complex sound, containing many different frequencies, can be represented by a single line graph. Similarly a single groove in a gramophone record contains information about a whole range of sound frequencies. These can be reproduced by a needle vibrating jerkily, to and fro, in this groove while the record is turning. However, the early 78 r.p.m. (revolutions per minute) records could reproduce only frequencies between about 200 and 2000 Hz – a very small part of the complete range of human hearing.

His Master's Voice.

Stereo

The introduction of electrical recording techniques in the 1920's revolutionised the gramophone. And when the stereo L.P. disc was unveiled in 1958 high-fidelity audio had arrived. The frequency range of disc recordings dramatically increased until it now covers practically all the frequencies we are able to hear (20 Hz–20 000 Hz).

A stereo disc carries two quite separate soundtracks, each of which is recorded by a separate microphone. Each is then played back through a separate amplifier and loudspeaker. If the equipment is set up properly the two loudspeakers can produce the illusion of a continuous row of sources extending from one loudspeaker to the other. This gives realistic reproduction of a large sound source such as a rock group, a band or an orchestra.

Grooves on a stereo LP disc.

Moving coil cartridge

On a stereo L.P. disc the two separate soundtracks are built on the sides of the groove. The angle between them is 90°. As the disc turns the stylus is pushed in two directions perpendicular to each other. This simplified picture of a moving coil cartridge shows that the stylus is connected to two small coils of wire. Each coil vibrates in a strong magnetic field and so an alternating voltage is induced in it. The amplitude and the frequency of the voltage depend on the shape of the soundtrack on the groove. These variations in voltage, which correspond to the variations in the original sound, are then amplified and the output fed to a loudspeaker.

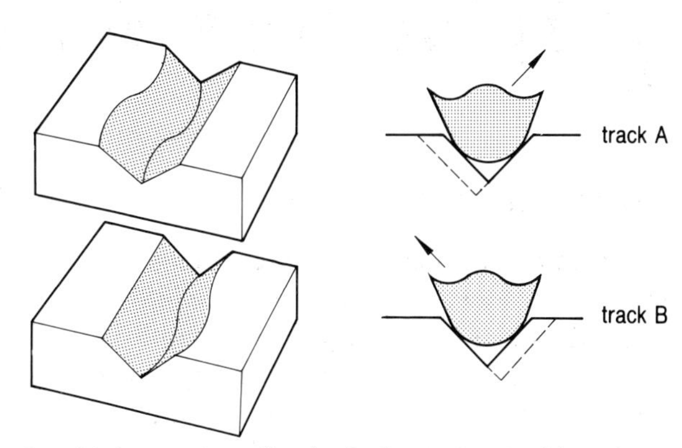

Soundtracks on a stereo disc showing the movements of the stylus on each track.

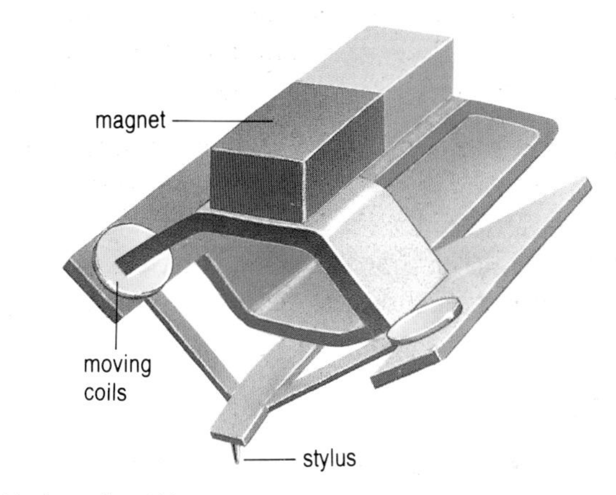

Moving coil cartridge.

Turntables and tonearms

If good quality sound reproduction is to be achieved it is important that the disc turns at a constant speed; otherwise music will keep changing pitch with disastrous effects. Irregular fluctuations below 10 Hz are referred to as 'wow'. Above 10 Hz they are called 'flutter'. A heavy turntable acts as a flywheel which smooths out these unwanted fluctuations in speed.

The tone arm holds the cartridge in place as it travels across the disc, and there is usually a counter-balance adjustment at the pivot to allow you to select the recommended downwards force exerted by the stylus on the groove.

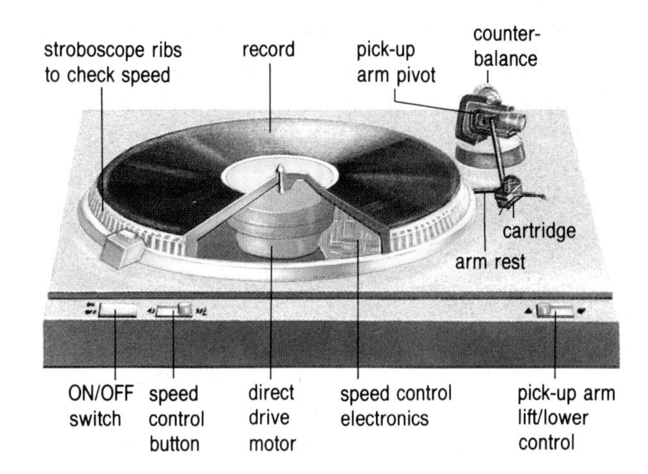

1 Why are the two sides of the groove on a stereo disc set at 90° to each other?

2 Construct a chart to show the frequency range of a modern L.P. disc compared with the range of early recordings.

›› *Use a long pin or needle and a yoghurt carton to play back an old 78 r.p.m. record.*

Tape it

Portable high tech

The miracle of the microchip has brought modern marvels to millions. For a few pence we can have a digital watch which tells the exact time, a calculator giving accurate answers or a personal stereo producing fantastic fidelity.

The cassette tape used in the personal stereo carries four tracks and is only 3.8 mm wide. Running at 4.8 cm/s a C90 tape gives ninety minutes of good quality stereo music.

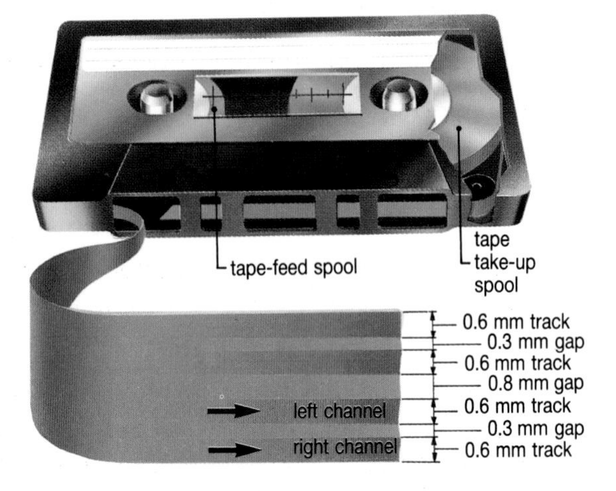

Recording

Plastic recording tape is coated with a layer of magnetic material such as ferric oxide. This contains millions of 'magnetic particles' which behave like tiny magnets. Normally these are all jumbled up so that their effects cancel each other. In this state the tape is unmagnetized. However, should part of the tape lie in a magnetic field the picture changes. An invisible magnetic pattern will then be imprinted on that part of the tape. Moreover this pattern will remain on the tape after the magnetic field has disappeared. So a permanent recording can be made as follows.

1 A microphone changes sound signals into electrical signals.

2 The electrical signals are then amplified.

3 The amplified signal energizes an electromagnet – called the **record head**. It consists of a ring magnet with a narrow gap of a few micrometres.

4 The strong magnetic field produced in this gap magnetizes the tape as it passes over it.

5 Since the magnetic field is controlled by the sound signal the magnetic pattern on the tape forms a permanent record of the sound.

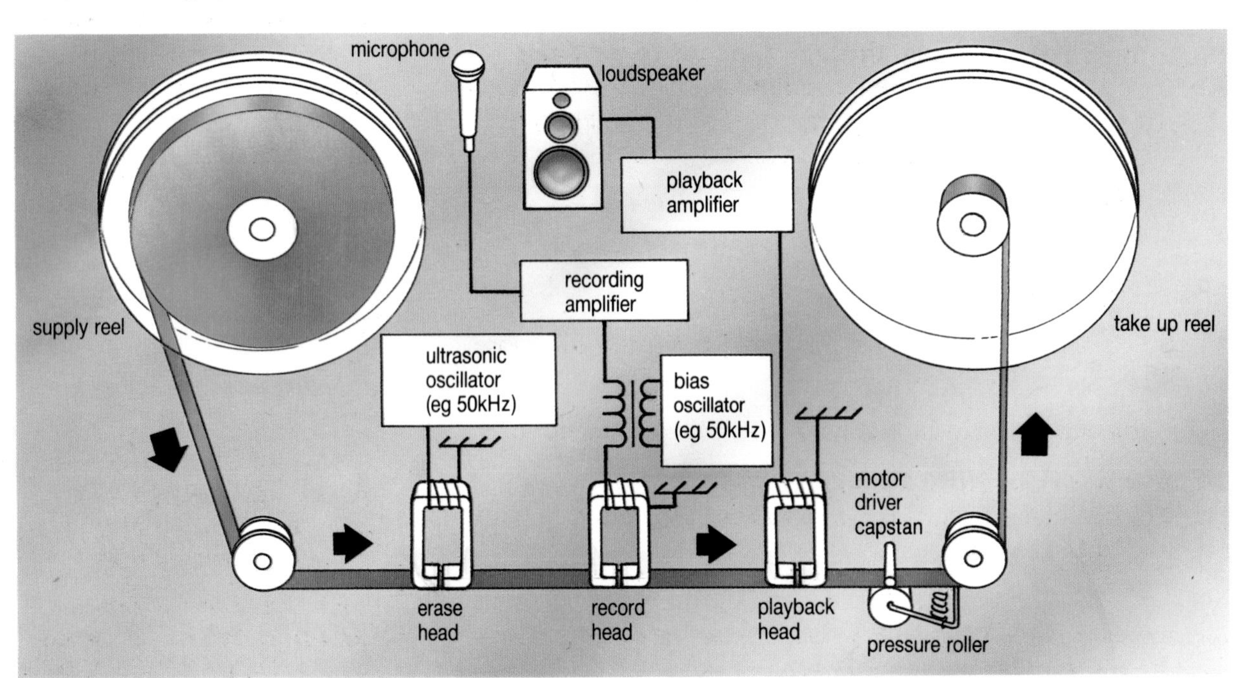

Playback

Playback is just recording in reverse. The tape carrying the magnetic pattern moves across the playback head. This head is similar to the record head but, for high frequency reproduction, it should have a smaller gap of about 1 micrometre.

1 The moving magnetic pattern produces a changing magnetic field.

2 This changing magnetic field produces a changing voltage in the coils of the playback head.

3 This changing voltage is then fed to an amplifier and then a loudspeaker.

In some cheaper tape recorders the same head is used for both recording and playback. In others the two heads are combined in a single unit. Two heads are always better than one.

Were you surprised the first time you heard your own recorded voice? Even if the reproduction were perfect your voice would still sound 'unnatural' to you. This is because you normally hear your voice through the bones in your head as well as through the air. The combined sound has a different quality from the sound which travels only through the air.

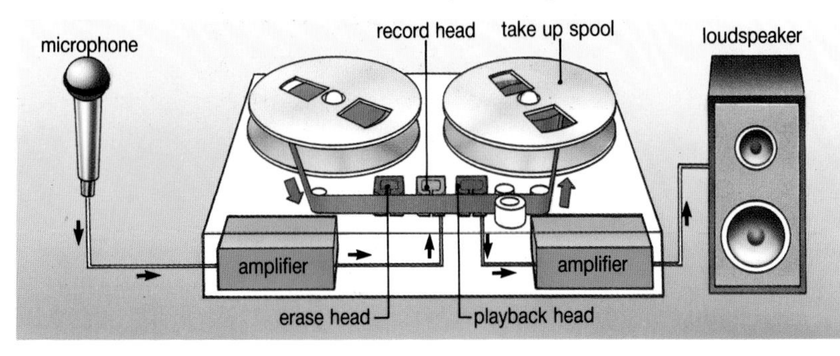

Erase

Before making a recording, the tape must be magnetically 'clean'. The tape is therefore passed over an erase head which is situated just before the record head. A high frequency (ultrasonic) signal is fed to the erase head to jumble up the magnetic particles on the tape.

1 What is the length of tape on a C90 cassette?

2 Explain why, in Oersted's experiment, the compass needle turns.

3 Distinguish between induced voltage and induced current.

4 A friend is listening to a good quality recording of your voice. Will she think it sounds the same as your natural voice? Explain your answer.

›› *Set up some simple apparatus to demonstrate the discoveries of Oersted and Faraday mentioned in this section.*

Important Dates in the Development of Tape Recording

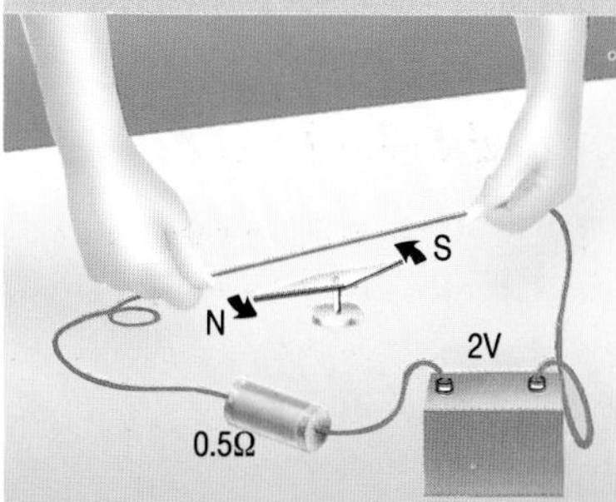

Oersted's experiment

1819 Hans Oersted discovered that when a wire carrying an electric current is placed parallel to a magnetic compass needle, the needle turns. This shows that a current produces a magnetic field. The field lines form concentric circles round the wire.

1831 Michael Faraday discovered that if the strength of a magnetic field near a wire is changing a voltage is produced across the wire. This is called an **induced e.m.f.** (electromotive force). If the wire is part of a complete circuit there will be an **induced current**.

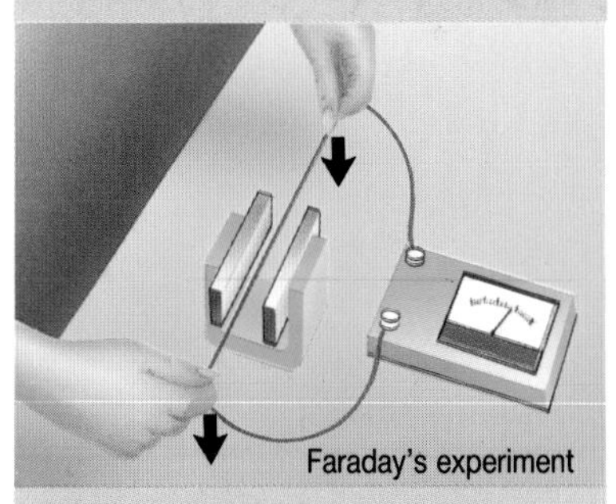

Faraday's experiment

1898 Valdemar Poulson invented the magnetic sound recorder. In it he stored the sound information on a length of magnetized iron wire.

1935 The first magnetic tape recorder was built in Germany and demonstrated at the Radio Show in Berlin.

Compact discs

From gramophone to compact disc

The first gramophone record was produced just over 100 years ago (see page 206). Forty years later electrical recordings were being made on 78 r.p.m. discs. But it took another 30 years (1958) before the $33\frac{1}{3}$ r.p.m. stereo L.P. disc arrived. This gives extremely good quality reproduction – but there's always room for improvement. Even the cleanest disc produces some background noises.

Selection of discs.

C.D. player

The digital transmission of information in telecommunication (page 19) has been followed by digital recording in the world of hi-fi. It works like this.

A microphone produces a continuously changing (analogue) electrical signal. The strength (amplitude) of this signal is automatically measured 44 100 times a second. Each measurement is then coded as a series of pulses in binary form: '1' represents 'a pulse' and '0' represents 'no pulse'.

On the analogue graph, the amplitude is shown sampled to the nearest decimal number. These samples are then shown in binary code on the digital graph.

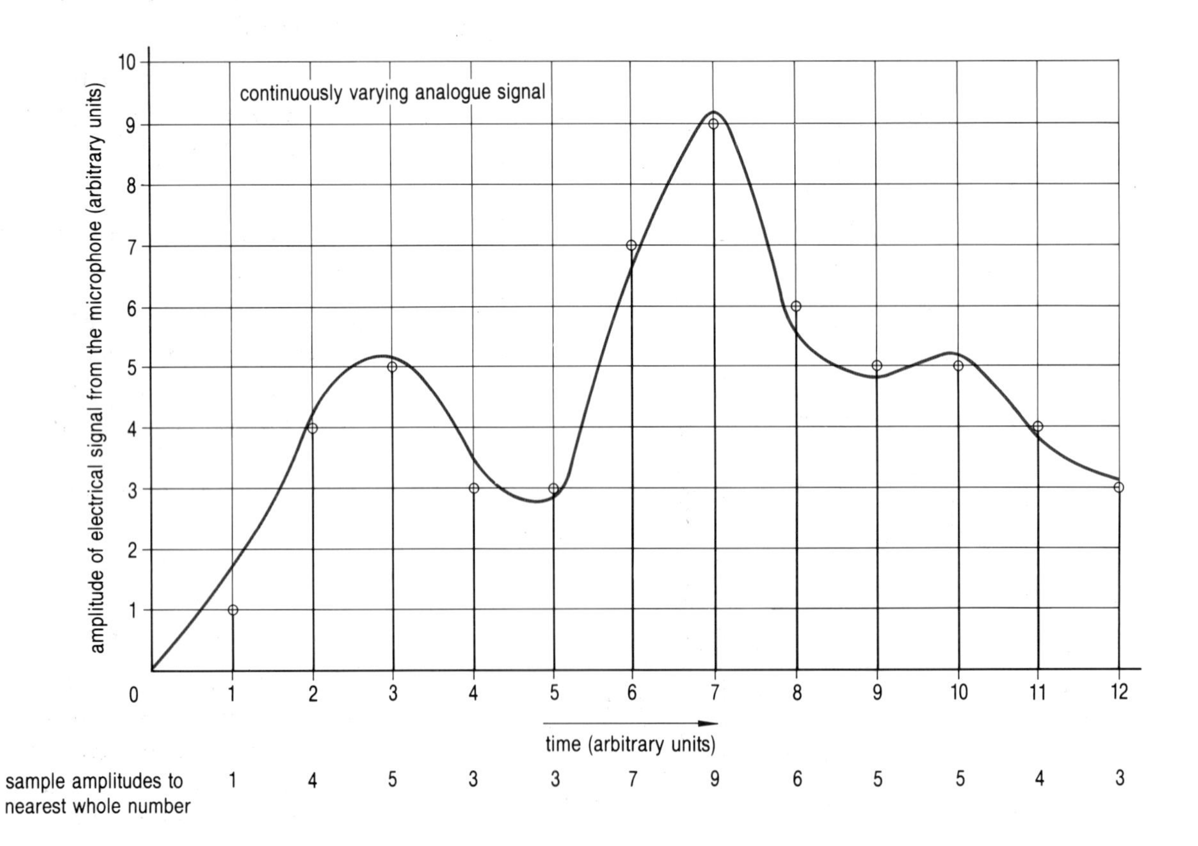

On the compact disc this digital information is stored as a series of bumps on the flat reflective surface. These bumps are formed by making depressions or 'pits' on the other side of the reflective surface. To confuse you the bumps are often called pits!

When the disc is being played back it spins round at several hundred r.p.m. and a laser beam is used to 'read' the recording. The beam is sharply focused on the under side of the spinning disc and reflections from it are then picked up by a photodiode. As different amounts of light are reflected from the flat areas and from the bumps the electrical signal from the photodiode gives a series of '0's and '1's. This digital output from the photodiode is fed to a digital-to-analogue converter which produces a high quality replica of the original signal from the microphone. After amplification the analogue signal is fed to a loudspeaker.

On a compact disc about one a half million pieces of information are recorded every second. By keeping the tracks very close together (600 per millimetre) a 12cm disc will play for over an hour. As only the laser beam touches the disc surface there is no wear and tear. There is therefore no deterioration with use. In fact you can handle or even drop the disc (gently!) without damaging the recording. Slight scratches and dust on the disc do not affect the sound in the same way they do on an ordinary L.P. disc.

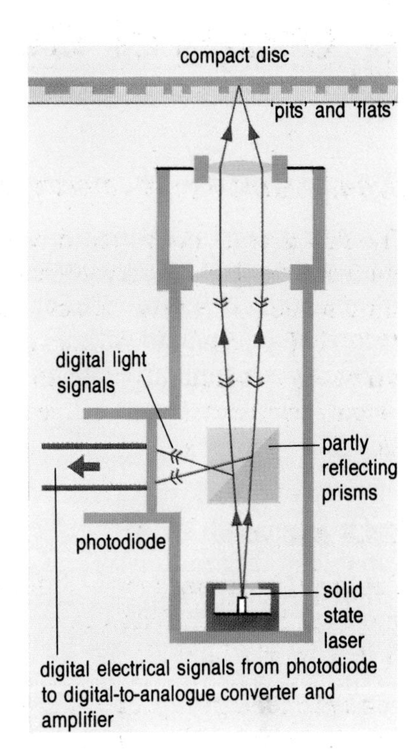

1 Discuss the advantages of a compact disc compared with other recording systems.

2 Why is there so little background noise with a compact disc?

3 What kind of recording system is used in a telephone answering machine?

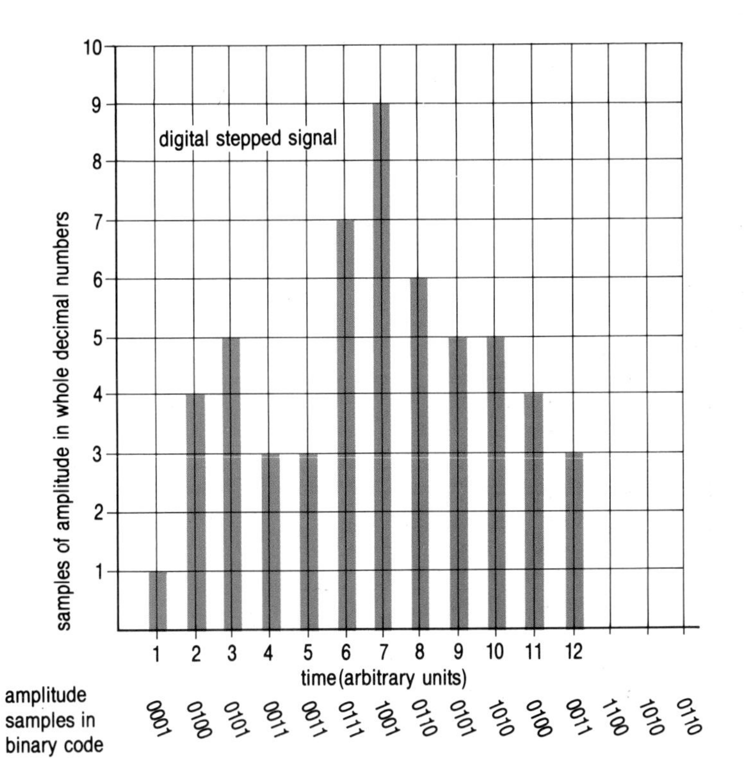

›› ***A*** *Find out other uses for solid state lasers.*

B *Copy, on blank graph paper, the x and y axes scales of the analogue and digital graphs shown here. On the analogue graph draw a sine wave which starts at 5 and swings between 9 and 1 arbitrary units on the y axis and has a period of 12 arbitrary units on the x axis.*

Now choose the nearest whole-number amplitudes for each unit of time, and transfer them to the digital graph. Indicate each value in binary code.

The first and last links

The first link – the microphone

A chain is only as strong as its weakest link. This is also true in any audio system. There is, for example, no point in using high-fidelity recording equipment with a carbon microphone which cannot respond to frequencies above 4 kHz (see page 12).

A wide range of microphones is now available, but one of the most popular for high-quality recording is still the moving-coil type. It is capable of responding well to sounds in the whole frequency range of human hearing.

A small diaphragm is attached to a coil of wire as shown. When someone speaks into the microphone the diaphragm moves up and down. This causes the coil to move between the poles of a permanent magnet. This, in turn, generates a voltage which varies at the frequency and amplitude of the sound waves. The principle on which this microphone is based was originally discovered by Michael Faraday (see page 209).

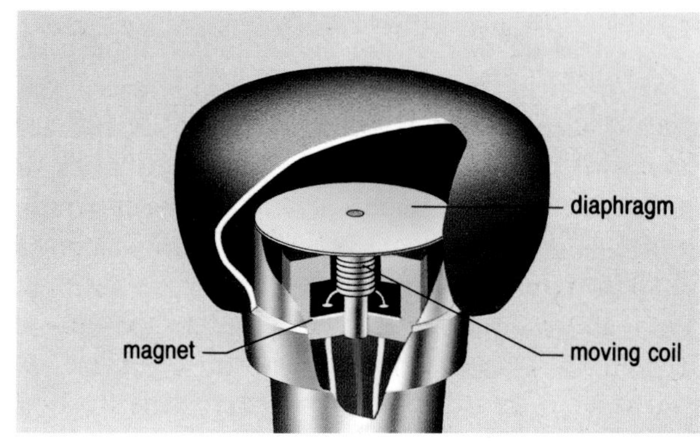

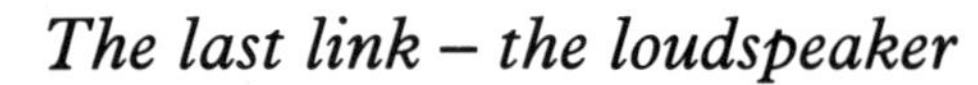

The last link – the loudspeaker

Although they look very different a moving coil microphone and a moving coil loudspeaker are very similar. Each has a paper cone fixed to a coil of wire which moves in a magnetic field.

The principle on which the loudspeaker is based was discovered by Hans Oersted (see page 209).

If a current through the coil in one direction makes the coil move to the left, reversing the current will make it move to the right. An alternating current in the coil will therefore make the cone move to and fro.

In practice the simple bar magnet shown in the panel is surrounded by a ring magnet. The coil then vibrates in a strong radial field. As you can see the outer edge of the cone is attached by a springy suspension to a metal frame which also holds the magnet.

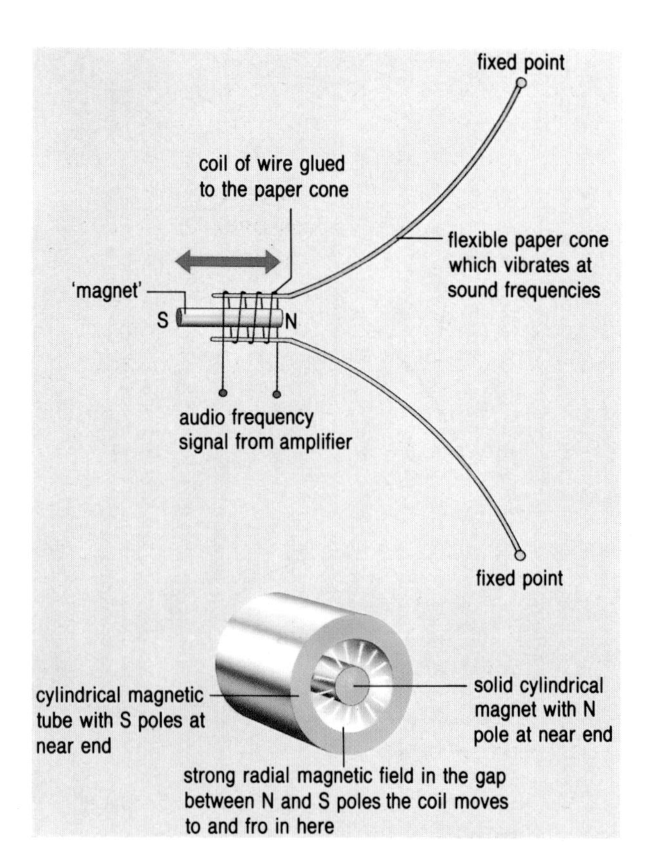

The enclosure

Today nearly all audio systems use moving-coil loudspeakers. They come in a variety of sizes:
large base units (woofers) for the lower frequencies (30–500 Hz);
mid-range units (500 Hz–4 kHz); and
small treble units (tweeters) for high frequency reproduction (4 kHz–20 kHz).

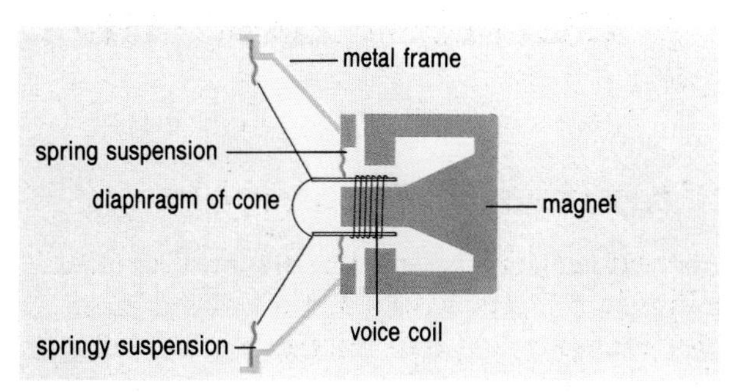

Moving coil loudspeaker.

When a speaker cone moves forward the front surface sends out a compression wave. But at the same time the rear surface sends out a rarefaction wave (i.e. reduced pressure).

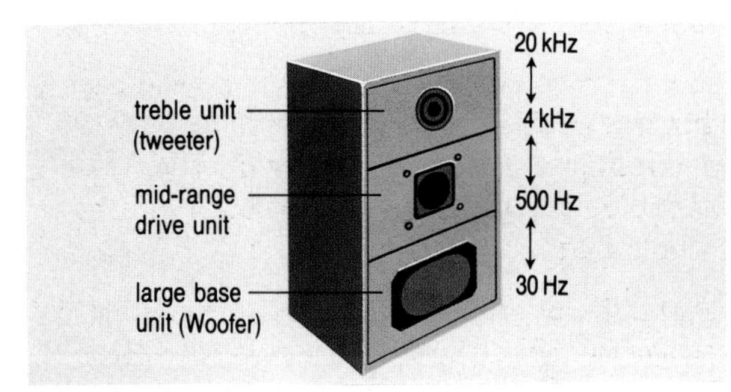

Hi-fi loudspeaker unit.

Low frequency (long wavelength) sounds bend readily round corners. You may have seen a similar *diffraction* effect with long waves in a ripple tank.

So, at low frequencies (less than 200 Hz) the sound waves bend round the outer rim of the loudspeaker and tend to cancel each other. The remedy is to mount the speaker in a box which is filled with some absorbent material. Sounds coming from the back surface of the cone are then absorbed, and so they cannot interfere with the sound radiated from the front surface of the cone.

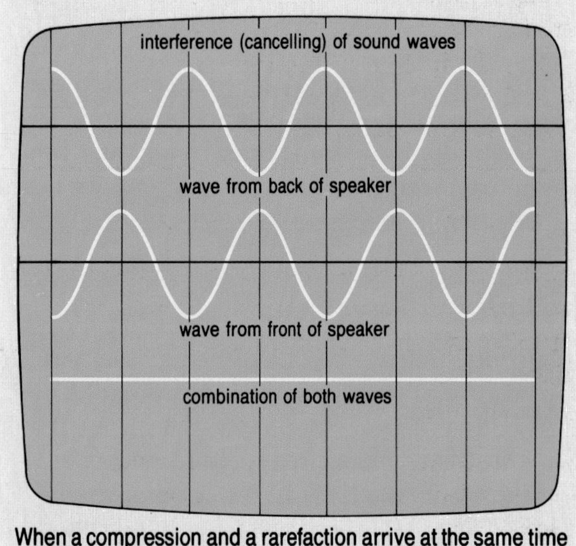

When a compression and a rarefaction arrive at the same time they cancel. Two sounds produce no sound!

It is usual to mount the other speakers in the same enclosure, although the higher frequencies do not suffer so much from diffraction effects. To ensure that each speaker is supplied with the appropriate range of frequencies a filter system, called a cross-over network, is fitted. It sends the high frequency signals to the tweeter and the low frequency signals to the woofer. Without such a filter too much energy could reach the tweeter and damage it.

Many smaller audio systems fit only two units in the enclosure: a bass/mid range unit and a tweeter.

1 What do you think are the greatest differences between a moving coil 'mike' and a moving coil speaker?

2 What size of moving-coil speaker would make the best microphone? Why?

3 Why are tweeters no use for low frequency reproduction?

›› ***A*** *Make tape recordings using different types of microphone. Rank them in order of sound quality.*

B *Investigate the difference an enclosure or a large wooden board (baffle) makes to the low-frequency output of a speaker.*

Gravity and games

Balanced bodies

In sport, balance is very important. Gymnasts can balance and even do somersaults on a beam only 10 cm wide. Ice skating also calls for great balancing skills and in sports such as judo and wrestling the aim is to keep your own balance and to catch your opponents off theirs!

Centre of gravity

Your body behaves as if your weight were concentrated at one point – your centre of gravity. When standing straight up with your hands by your side your centre of gravity is a few centimetres below your navel. Raising your arms above your head raises your centre of gravity by 5 cm or so.

In preparing for the spring board dive shown here the diver is delicately balanced. Just two forces are acting on her body – her weight (W) and the upward push (P) on her feet from the board. She can hold her balance as long as W and P act along the same line. In fact anything can be balanced when its centre of gravity lies directly above its base.

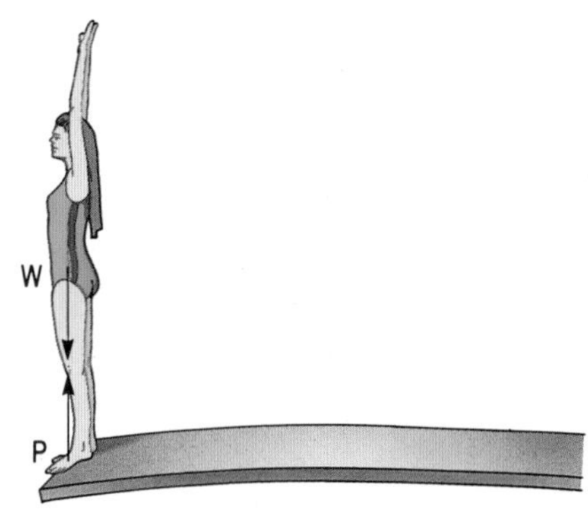

Two forces providing balance on a springboard.

Lift off

In a sport such as volley ball being able to raise your centre of gravity is very important. A high body position close to the net means that you are able to block an opponents shot. Or you can strike the ball, with a large downwards force, into your opponents court. However, once your feet are off the ground there is nothing you can do to change the path followed by your centre of gravity. In the air you are at the mercy of only two forces – the pull of gravity and air resistance. After take off in the long jump it is again these two forces alone which determine the length of the jump.

Jump judgement

In the Mexico City Olympic Games of 1968 Bob Beamon leaped an incredible 8.90 m. Measure this distance out for yourself and you cannot fail to be amazed. However, Beamon's effort has been judged by some to be partly due to the position of Mexico City, where the acceleration due to gravity is 9.7794 m/s². In addition the altitude is 2500 m and the air resistance is therefore smaller than at lower levels.

In Moscow, where the 1980 Games were held, 'g' was greater – 9.8155 m/s² – and at the altitude of only 150 m, the denser air produced more air resistance. The distance jumped at the Moscow Games was a mere 8.54 m, so we could argue that the reduced air resistance and the smaller gravitational pull made for a longer jump in Mexico City.

Bob Beamon in Mexico City, 1968 jumped 8.90 m in the long jump, a record that still stands.

Computer calculations

Experimental data, the rules of physics and a computer program have settled the argument on Beamon's jump. The computer simulated the jump and calculated that the reduced air resistance and smaller gravitational pull added about 4 cm to the distance. So even in Moscow the jump would have been 8.86 m – still an Olympic and World record. Bob Beamon simply got it right on the day!

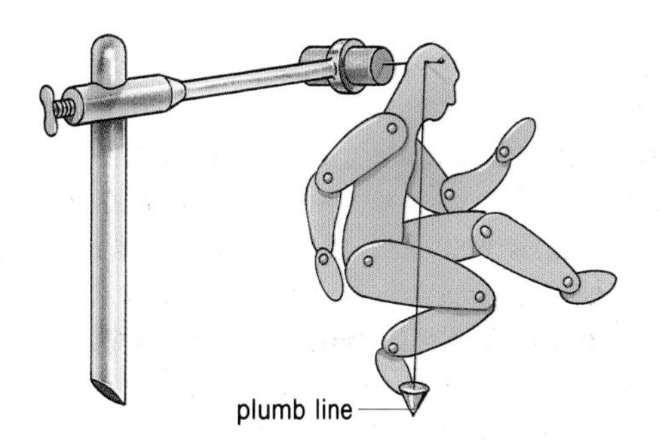

›› ***A*** *Make a cardboard manikin. Suspend the manikin and use a plumb line to find the position of its centre of gravity. Find where its centre of gravity lies with its arms and legs in different positions.*

B *Use the apparatus shown to find the position of your centre of gravity.*

C *Is the angle of take off or release important in the long jump or throwing events such as the shot-put? To answer this question you could use a spring to project a steel ball and then investigate what effect the angle of launch has on the horizontal distance travelled by the ball.*

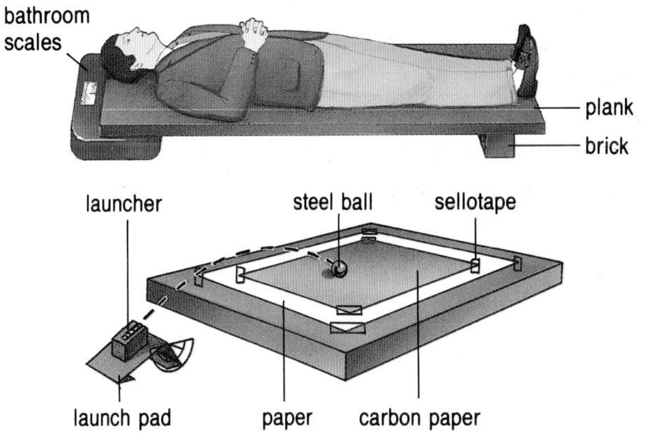

Physics and fitness

Work or play?

To enjoy a healthy life we must ensure that our bodies are exercised and do their fair share of work. Strange as it may seem a very enjoyable way of doing work is through play! Sports such as squash, tennis, football, athletics and swimming allow us to exercise the body machine and, at the same time, do some work! However, if we are to enjoy our work through play then our bodies must be fit.

A fitness formula

When someone is really fit they are able to keep their body working for quite a long time without getting too tired. One way of measuring this kind of fitness is to carry out a 'step' test.

This test involves stepping up and down for 5 minutes on to a box or bench about 0.4 m high. The 'step-up and down' frequency is kept constant at 30 per minute throughout the test.

One minute after completing this exercise the pulse is taken and beats counted over 30 s. After another 30 s the pulse beats are counted again for 30 s. Finally after a further half minute rest the person's pulse beats are counted over 30 s.

These three pulse measurements are substituted into a 'fitness formula' and the person's 'fitness factor' calculated. This fitness factor is then compared against a scale of fitness. The fitness factor for those who cannot complete the test is calculated by inserting into the fitness formula the time that was taken to carry out part of the test.

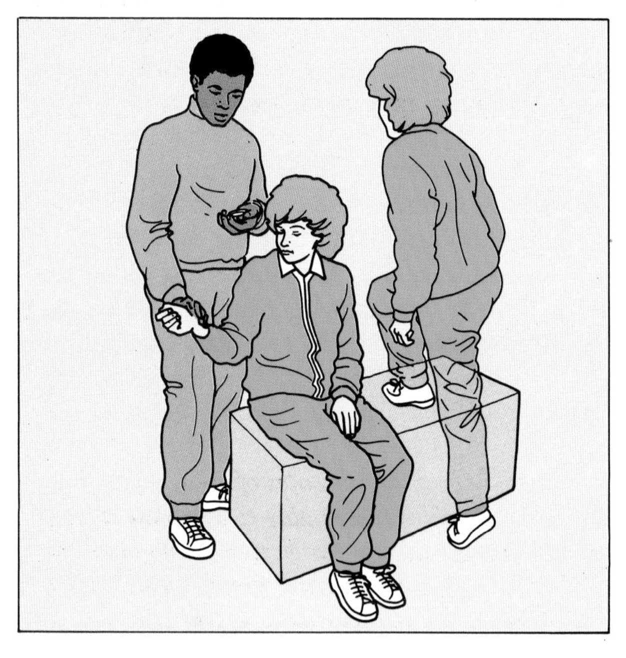

Fitness formula:

$$\text{fitness factor} = \frac{\text{time for exercise in seconds} \times 100}{2 \times (\text{sum of the three pulse counts})}$$

Fitness factor	level of fitness
90+	mega-fit
80–89	super-fit
70–79	fit
60–69	fairly fit
50–59	lacking fitness

A fitness scale

Jettison the joules

When carrying out a step test you are doing work and transforming energy. The rate at which you are transforming this energy is called your **power**. The energy comes, of course, from the food you eat. A 100 g chocolate bar, for example, will provide you with about 2 800 kJ.

If you do not transform this energy through doing work then your body will store the energy in the form of 75 g of fat! To fight the flab you have to jettison the joules. This takes longer than you might think but sporting activities use up these stored joules. Look at the table and compare the energy used in different activities.

Activity	Energy used per second of activity (J)
Driving a car	190
Walking	200
Golf	350
Cycling	350
Soccer	600
Squash	700
Cross-country running	740
Swimming	770

Power and performance

Sports scientists at Loughborough University of Technology have devised a self-propelled treadmill to investigate power output in sprinting. Look at the diagram of the equipment used in the investigation. A strain gauge attached to the tether on the sprinter detects the force pushing the sprinter forward. A small dynamo attached to the treadmill gives the track speed. The outputs from the strain gauge (force) and the dynamo (speed) are fed to a computer. Information like this could help to improve sprinting performance.

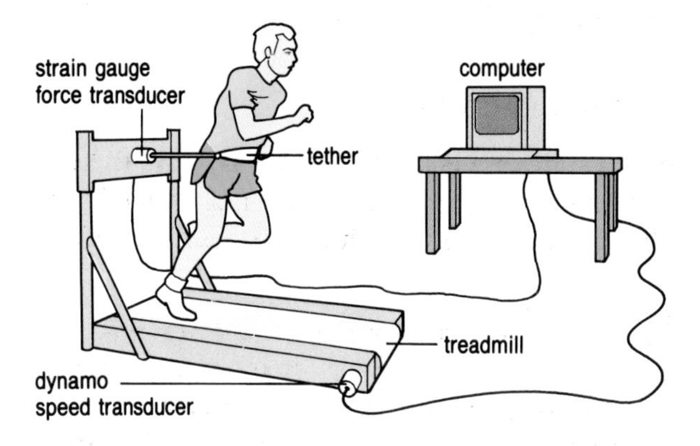

1 Some data from an experiment on a sprinter are given here. Use these results to display how his sprinting force, speed and power varied during the run.

Time (s)	Force (N)	Speed (m/s)
1	160	4.5
2	122	6.5
3	99	7.0
4	100	7.2
5	99	6.4
6	101	7.0

Ernest Obeng's sprint data.

›› ***A*** *Estimate how many joules of energy you would use up if you were able to complete a 'step test'. What would be your power throughout the test? How long would it take you to 'work off' a bar of chocolate at this rate of working?*

B *Investigate how much power you can develop when running up a flight of stairs? This has in fact been attempted using the stairs of the Empire State Building! The climb took just over 12 minutes for a vertical rise of 362 m, 1575 steps and 97 flights! The average power output for the climb was about 400 W. You had better settle for one or two flights!*

C *How much pedal power is required to keep a bicycle moving at a reasonably steady speed along a level surface? Carry out this investigation using a bicycle, a rope and a spring balance to tow another bicycle along.*

SPACE PHYSICS

8

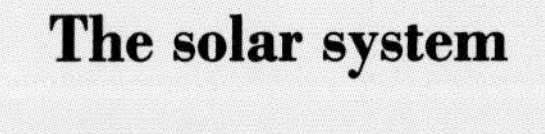

The solar system

The planets

The Sun and the nine planets in orbit around it make up the solar system.

Earth from space.

When seen from space, our planet Earth is a colourful globe with a spectacular pattern of brilliant white clouds. Because of its distance from the Sun it is neither too hot nor too cold and life as we know it flourishes on its surface. Its inhabitants are protected from the Sun's harmful radiation by the screening effect of the Earth's atmosphere.

Mercury and Venus are closer to the Sun and are considerably hotter than the Earth. The surface temperature of Venus has been found to be more than 400°C. Venus and Mercury have no atmosphere.

Mars has a thin atmosphere consisting mainly of carbon dioxide and was long considered to be the planet most likely to support extraterrestrial life. However, the information provided by the two Viking lander craft which touched down on Mars in 1976 has led scientists to conclude that Mars is a bleak desert. No evidence of life was found.

Jupiter is larger than all the other planets combined. The most prominent feature of the planet is the Great Red Spot which is thought to be a never-ending storm. Saturn is well-known for its rings which are considered to be one of the great mysteries of the solar system. Uranus, Neptune and Pluto are the planets about which astronomers have least information. Because of their extreme distances from the Sun they must be extremely cold and inhospitable.

Photograph of Saturn, taken by Voyager 2, showing the fine ring structure.

Planet	Mean distance from the Sun (10^6 km)	Diameter (km)	Time to complete one revolution in orbit round Sun	Mass as a fraction of the mass of the Earth	Number of satellites
Mercury	58	4 880	88 days	0.05	none
Venus	108	12 100	224.7 days	0.82	none
Earth	150	12 750	365.26 days	1	1
Mars	228	6 790	687 days	0.108	2 small
Jupiter	778	142 800	11.9 years	318	4 larger than the Moon and at least 12 smaller
Saturn	1427	120 000	29.5 years	95	at least 20 and the ring system
Uranus	2871	52 300	84 years	14.5	at least 15
Neptune	4492	48 600	164.8 years	17.2	2
Pluto	5913	3 300	248.5 years	0.002	1

The Sun

The Sun is just one of one hundred thousand million stars in our galaxy. In its hot core a continuous nuclear reaction produces an enormous amount of energy which is radiated in the form of electromagnetic waves.

The temperature of the core is about 20 million °C, although the surface temperature is a mere 6000°C. The fuel for this nuclear reaction is hydrogen. About 600 000 million kg are converted to helium every second, although there is no need for concern; there is enough hydrogen for the Sun to continue burning for another 5000 million years!

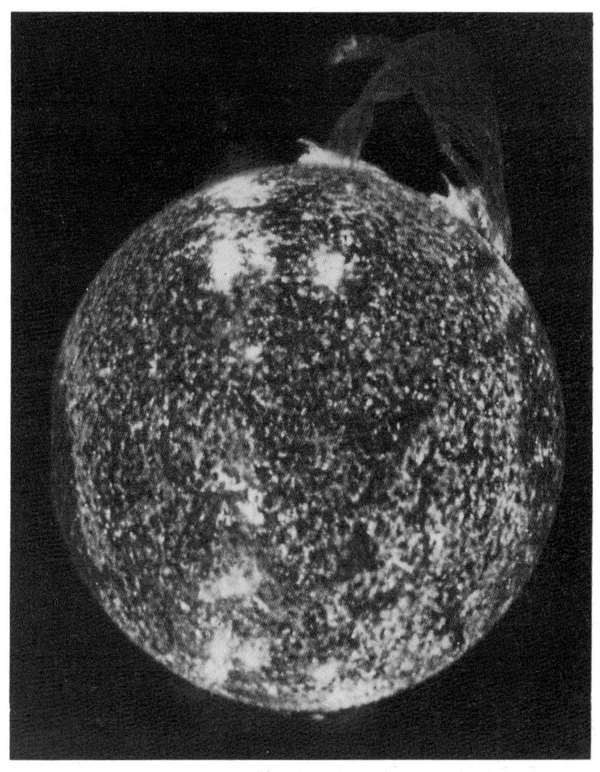

The Sun showing a spectacular solar eruption.

Natural satellites

Many of the planets have their own natural satellites. Earth, of course, has one – the Moon – which orbits the Earth with a period of approximately one month. The table above gives the number of satellites of the other planets. In between the orbits of Mars and Jupiter lies the asteroid belt. Comets and meteorites are also common in the solar system.

» *What are asteroids, comets, and meteorites? Draw a diagram of the solar system showing where each of these can be found.*

1 Which planets closer to the Sun than Jupiter have an atmosphere?

2 Use the table of planetary data to show that Jupiter has a mass greater than that of all the other planets combined.

3 The mass of Earth is 5.97×10^{24} kg. Use the table to find the mass of the other planets.

Beyond the solar system

The Milky Way

The distance from the Earth to the Sun is enormous: approximately 140 million kilometres. But this distance is small compared with the distance from the Earth to the nearest star – Proxima Centauri.

The speed of light is 3×10^8 m/s (300 million metres per second). So it takes approximately 8 minutes for light to travel from the Sun to the Earth. The distance that light travels in one year is called a **light year** (ly). This unit is used by astronomers to measure distance. The distance to Proxima Centauri from the Earth is 4.2 ly.

Astronomers have found that stars are grouped into galaxies. Our Sun and Proxima Centauri are two of approximately one hundred thousand million stars which make up our galaxy. Our galaxy is called the Milky Way.

The diameter of the Milky Way is an incredible 100 000 light years! Our Sun is a fairly typical star in this galaxy, but there are many stars that are much bigger (red giants) and many that are much smaller (white dwarfs).

Astronomers believe that stars change during their lifetime. Some, when they have burnt all their fuel, collapse and become very small and dense – these are called **neutron stars**. It is possible that some stars become *extremely* dense. The gravitational field around them is then so strong that not even light can escape this gravitational attraction – they become **black holes**.

A spiral galaxy. The Milky Way would look similar to an observer outside our galaxy.

The brightest stars

Of the 7000 stars visible to the unaided eye, the 20 which look the brightest are listed opposite in order of how bright they *appear*. Their brightness is compared with the Sun.

Name	True brightness (Sun: 1)	Distance from the Earth (light years)
1 Sirius	23	8.7
2 Canopus	1 500	100
3 Proxima Centauri	1.5	4.3
4 Arcturus	110	36
5 Vega	55	27
6 Capella	170	47
7 Rigel	40 000	800
8 Procyon	7.3	11.3
9 Betelgeuse	17 000	500
10 Achemar	200	65
11 Beta Centauri	5 000	300
12 Altair	11	16.5
13 Alpha Crucis	4 000	400
14 Aldebaran	100	53
15 Spica	2 800	260
16 Antares	5 000	400
17 Pollux	45	40
18 Fomalhaut	14	23
19 Deneb	60 000	1 400
20 Beta Crucis	6 000	500

To the edge of the Universe

With a telescope it is possible to see many other galaxies which contain as many stars as the Milky Way. Our nearest neighbour, called Andromeda (M31), can just about be seen with the naked eye, yet it is 2.2 million light years away from us. It is estimated that there are more galaxies in the Universe than there are people on the Earth!

Astronomers are still making exciting discoveries. The extreme range of their telescopes is now about ten thousand million light years. Mysterious phenomena called **quasars**, which give out rapidly changing radio signals, have been found at the edge of the observable Universe.

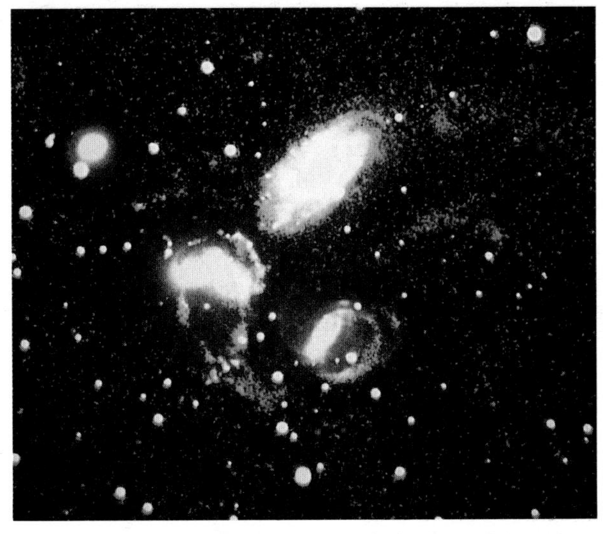

A group of galaxies. Each galaxy could contain 100 000 million stars.

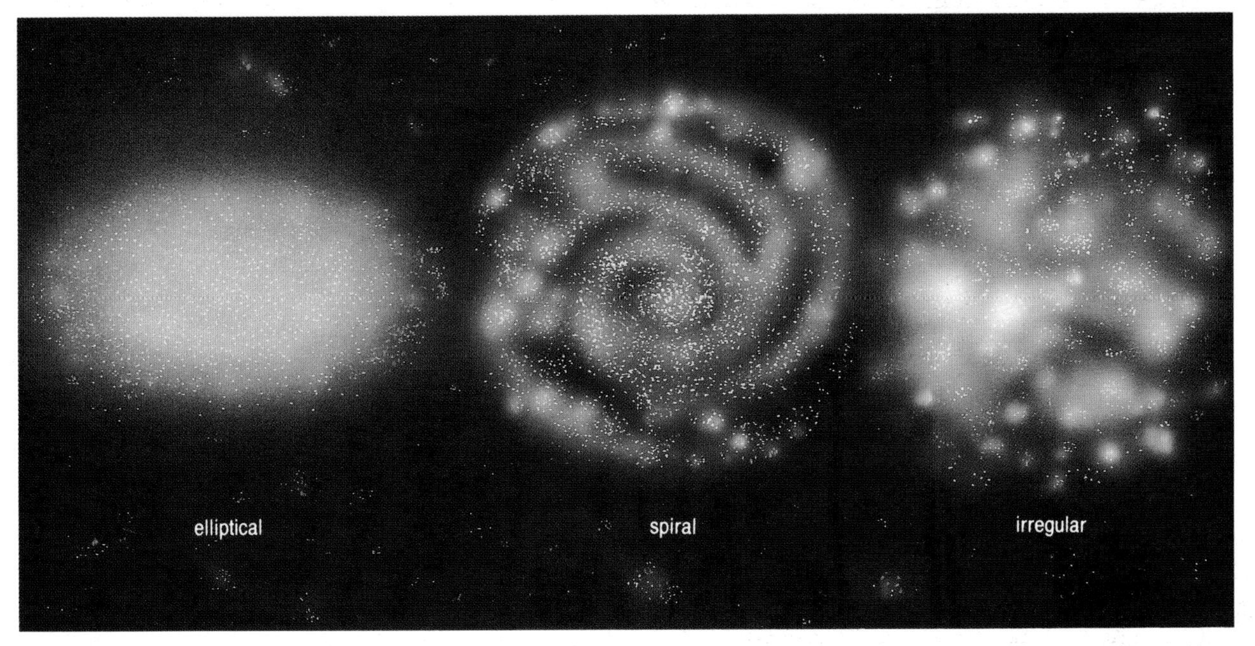

Galaxies can be sorted into types depending on appearance. The Milky Way is a spiral galaxy. Andromeda is also a spiral galaxy. What would Andromeda look like if seen edge on?

1 The Sun is the nearest star to the Earth. How far is it from Earth to the next star?

2 What is a galaxy?

3 The star Deneb burns 60 000 times more brightly than the Sun. Use the table to explain why it does not *appear* to be the brightest star in the sky.

4 What is a black hole?

5 a) Calculate the distance in kilometres travelled by light in one year.

b) What is the distance in kilometres to Proxima Centauri?

c) How many times further away from the Earth is Proxima Centauri than the Sun?

d) If a model was made in which the distance from the Earth to the Sun was one metre, where would Proxima Centauri be placed?

Optical telescopes

Seeing stars

On a clear night, with no street lights to spoil the view, the sky seems to be filled with millions of stars. In fact, only a few thousand can be seen at any one time even though there are a hundred thousand million stars in our galaxy alone.

Most stars and galaxies are so far away that the light from them is too faint to be seen by the unaided eye. Telescopes can make these distant objects visible by using curved mirrors or lenses to collect the faint light.

The Pleiades (seven sisters), an example of a starfield.

Refracting telescopes

A telescope which uses a lens to gather light is called a **refracting telescope** (or refractor). The simplest refractor has two lenses. The **objective** lens at the front collects light and makes an image which is magnified by a second lens called the **eyepiece**.

A planet looks bigger through a telescope because the eyepiece bends light so that any two points on the planet look further apart. Different eyepieces can be used with the same telescope to give different magnifications.

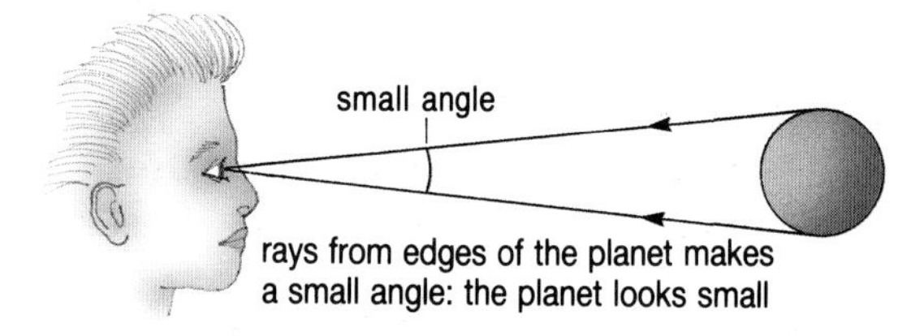

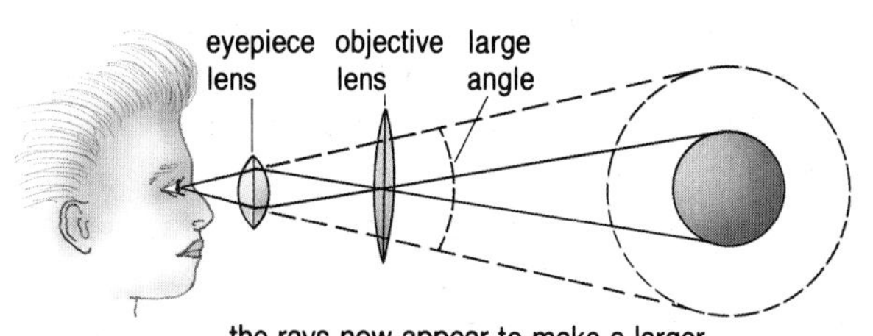

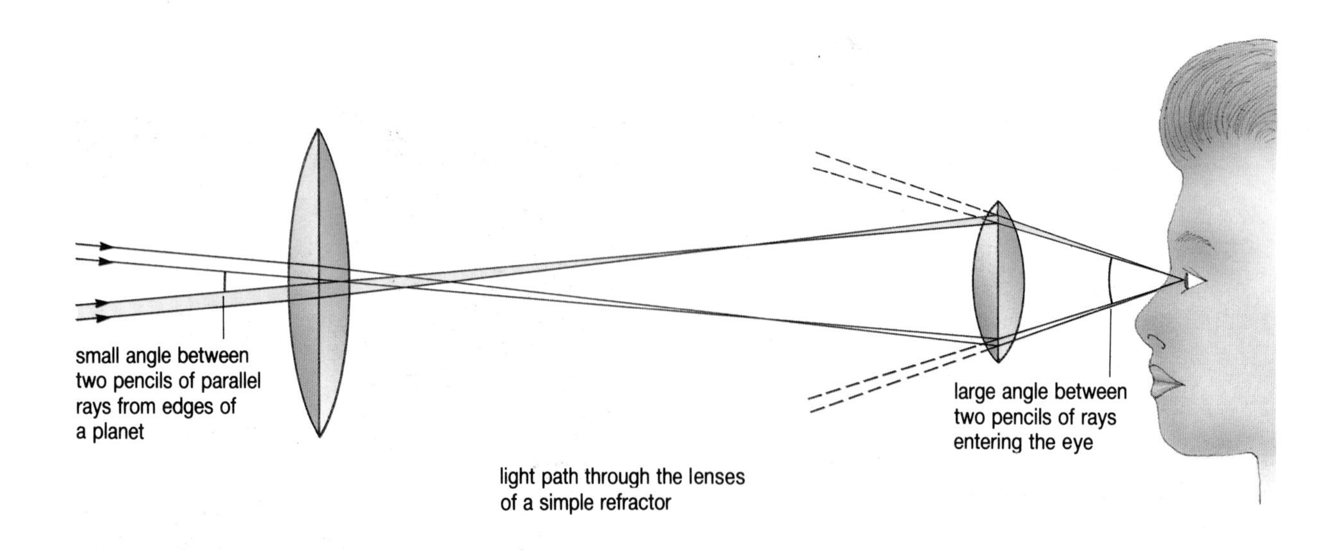

light path through the lenses of a simple refractor

Gathering light

The purpose of the objective lens is to gather light. The bigger the lens the more light it is able to gather. This allows fainter and more distant objects to be seen.

Small refractors used by amateur astronomers have objectives of about 10 cm in diameter. The largest refracting telescope has an objective with a diameter of 1 metre. Lenses this size are very difficult to make and are extremely expensive. They are also very heavy and can bend under their own weight.

A 3″ refracting telescope.

Reflecting telescopes

The most powerful telescopes use a curved mirror instead of a lens to gather light. It is easier and cheaper to make an accurate mirror than an accurate lens. Mirrors are also lighter than lenses so the diameter of a mirror can be much greater than the diameter of a lens. This means that more light can be gathered and even *more* distant objects can be seen.

In one type of reflecting telescope, light is reflected from the curved mirror on to a flat mirror. The flat mirror reflects the light on to an eyepiece which produces a magnified image. Some telescopes are now being built with more than one main mirror. The Multiple Mirror Telescope (MMT) in Arizona has six mirrors. Each mirror has a diameter 1.83 metres. These are focused together, and gather as much light as a single mirror of diameter 4.5 metres. There are plans to build much larger telescopes using the same design as the MMT.

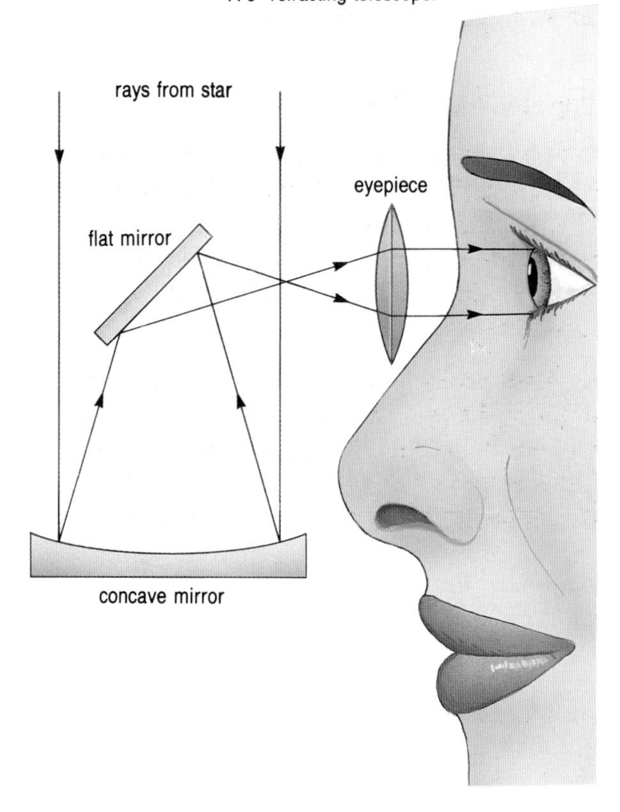

1 Why are lens telescopes called refractors?

2 Which part of a telescope causes magnification?

3 Why can stars which are normally invisible be seen with the use of a telescope?

4 Why are the largest telescopes in the world reflectors rather than refractors?

›› ***A*** *Use a convex lens to form an image on a screen. Investigate the effect of covering part of the lens.*

B *Build an astronomical telescope from two convex lenses, some plasticine and a ruler.*

The 6 m diameter mirror of the world's largest optical telescope at Zelenchuk, Caucasus.

Spectroscopy

The spectrum

If a large prism is placed at the front of a telescope, a band of colours can be seen in the position of each star. The band of colours is called a spectrum and astronomers can find out a lot about each star by carefully examining its spectrum.

A spectrum can be produced in the laboratory by shining a beam of white light through a prism.

The colours of the spectrum correspond to different wavelengths of light. Red light has a wavelength of approximately 700 nm and violet light about 400 nm (nm $= 10^{-9}$ m.) When light enters a more dense substance such as a glass prism, at an angle which is not 90°, it bends. It bends again as it leaves the prism.

When it enters glass, short wavelength light is slowed down more than long wavelength light, and so short wavelength light is bent more than long wavelength light.

When the prism is attached to a telescope, the narrow beam of light from each star is split into the colours of its particular spectrum.

An alternative way of producing a spectrum is to use a **spectroscope**. The spectroscope consists of a hollow tube with a narrow slit at one end to let the light enter. At the other end there is an eyepiece lens through which the spectrum is viewed. Inside the spectroscope there is a **diffraction grating** which is a thin glass plate with many narrow parallel lines drawn on it. The grating splits up the light to produce the spectrum. A spectroscope which can be used to measure the wavelengths of the different colours of the spectrum is called a **spectrometer**.

The result of placing a prism at the front of a telescope. The light from each star produces a small spectrum.

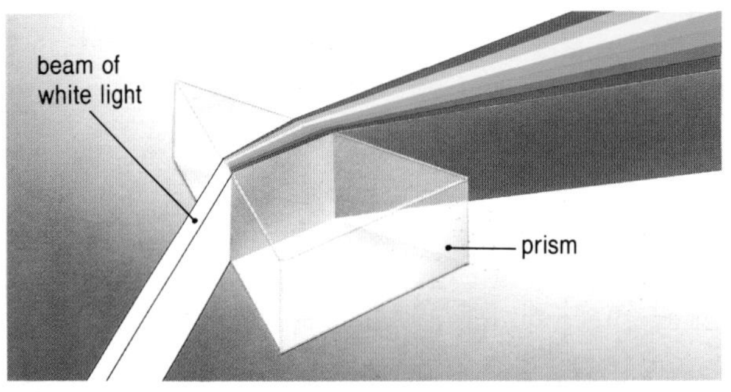

White light can be split into the colours of the spectrum by using a prism.

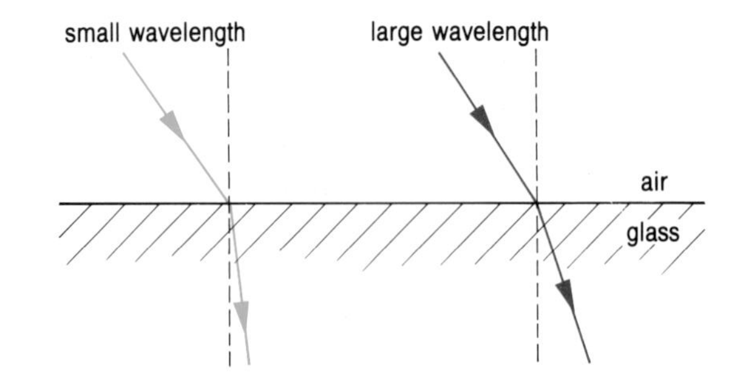

The amount that light bends depends on its wavelength.

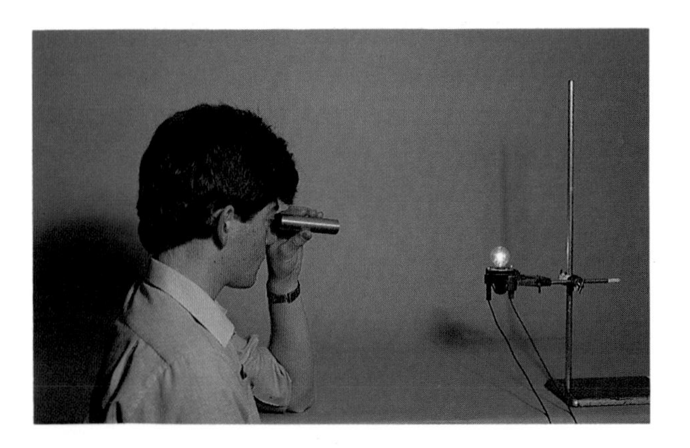

Atomic fingerprints

A hand-held spectroscope can be used in the laboratory to examine the spectrum produced by various light sources. A **continuous spectrum** is produced by light from sources which are solids, liquids, or high pressure gases at high temperatures. Ordinary light bulbs and the Sun are examples of continuous spectra sources.

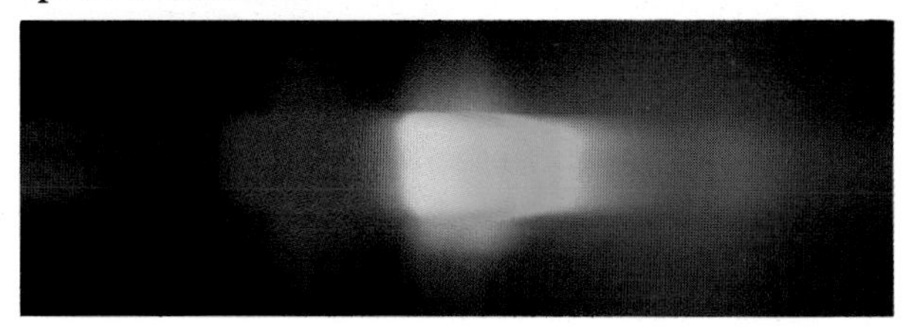

Hot gases at low pressure and gases which have an electric current passed through them produce **line spectra**. Sodium street lights and neon lights used in advertising are examples of sources of line spectra.

Line spectra are extremely useful because every chemical element has a characteristic spectrum. The lines correspond to particular wavelengths of light emitted by its atoms. The spectrum is therefore like a fingerprint. It is possible to identify a substance by examining its spectrum.

In 1814 Joseph Fraunhofer, used his newly invented spectroscope, to examine the spectrum of the Sun. He was amazed to discover that there were several dark lines crossing the Sun's continuous spectrum. These are now called **Fraunhofer lines**.

The dark lines are produced by cooler regions at the surface of the Sun absorbing certain wavelengths of light. The dark lines can be matched with the line spectrum for a particular element, thus allowing astronomers to identify the elements present in the Sun. In this way we know that the Sun is made up mostly of hydrogen and helium. The chemical composition of other stars may also be found in this way.

Joseph Fraunhofer.

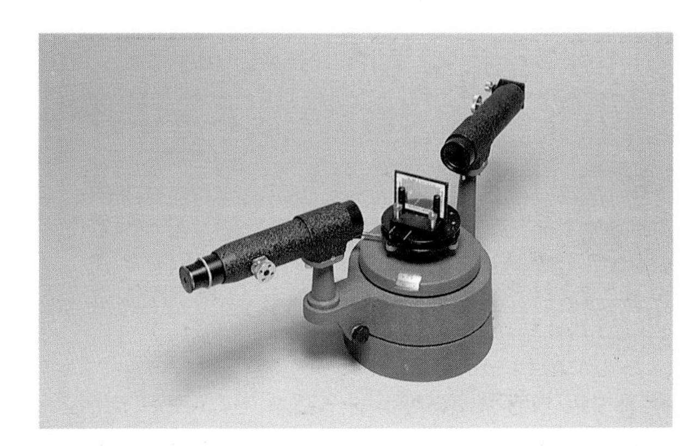

1 Describe how a spectrum may be produced in the laboratory.

2 What is a spectroscope?

3 What are continuous and line spectra?

4 How do we know that the elements hydrogen and helium are present in the Sun?

5 What would you expect to see if you examined the spectrum of the light from the Moon?

The electromagnetic spectrum

Invisible signals

In 1865 the Scottish physicist, James Clerk Maxwell devised a new theory about light. He said that light waves were really vibrating electric and magnetic fields. He also suggested that this **electromagnetic radiation** might be found at wavelengths greater and smaller than the wavelengths of visible light. Soon afterwards the German physicist. Heinrich Hertz, demonstrated that there *was* such radiation at much longer wavelengths. This radiation is called radio waves. In 1933 an American engineer, Karl Jansky, discovered that radio waves could also be detected from outer space. Physicists now realise that light and radio waves are only two parts of a huge **electromagnetic spectrum**. Typical frequencies and wavelengths are indicated on the diagram. The divisions between the parts of the spectrum are, however, not clearly defined.

In space, all kinds of electromagnetic radiation travel at a speed of 3×10^8 m/s and, like waves in a ripple tank, are reflected and refracted. Diffraction and interference of electromagnetic waves can also be demonstrated.

The wavelength of visible light ranges from 400 nm to 700 nm. Beyond the red part of the visible spectrum there is a region called **infrared**. Beyond the infrared region there are microwaves and radio waves.

Beyond the blue end of the visible spectrum there are **ultraviolet** radiations, x-rays, and γ-rays all of which have very small wavelengths.

The Earth is constantly being bombarded with many kinds of electromagnetic radiation, some of which could be very harmful to life. Thankfully, we are screened from most of the dangerous radiation by the atmosphere.

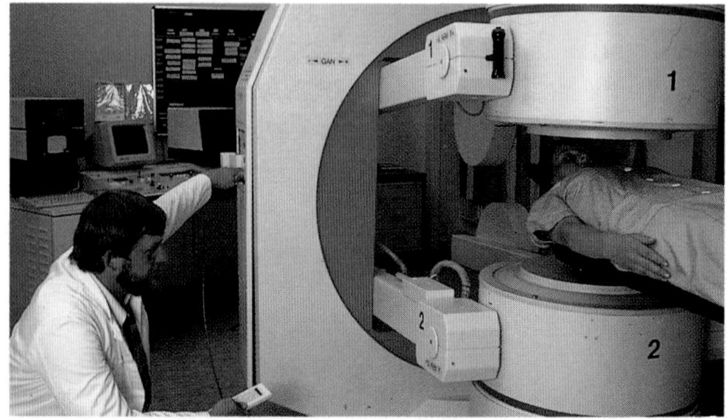

Gamma rays can be harmful to human tissue but are used successfully in the treatment of cancer.

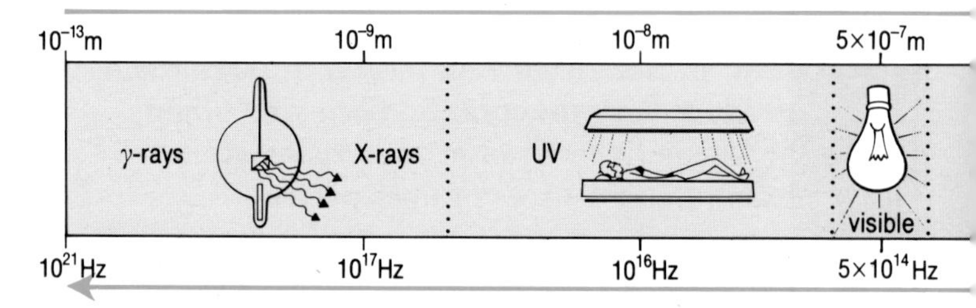

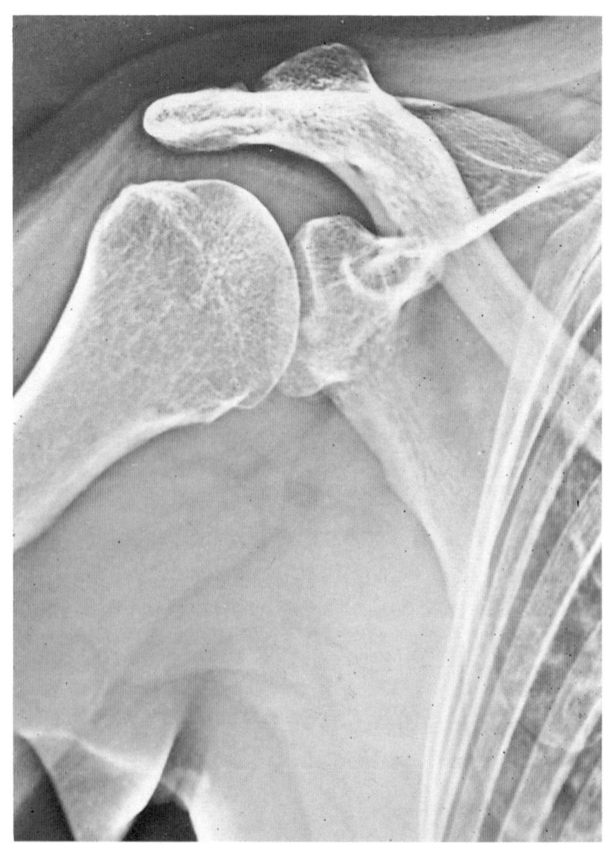
X-rays have very short wavelengths and are very penetrating. They can be detected on photographic paper and used in medicine to 'look inside' the human body. X-rays have been detected in space. The object Cygnus-X1 is emitting X-rays and it is possible this is because it is falling into a black hole!

Radiation with a shorter wavelength than visible light is called ultraviolet. It is the ultraviolet radiation from the sun which gives us a sun-tan, but over exposure to intense ultraviolet radiation can be harmful. Ultraviolet light can be used in the manufacture of electronic circuit boards.

Microwaves have a wavelength of several centimetres. Food will absorb microwaves and heat up very quickly in a microwave oven. Radar can be used to detect distant objects by reflecting microwave radiation and measuring the time for the transmitted pulse to return.

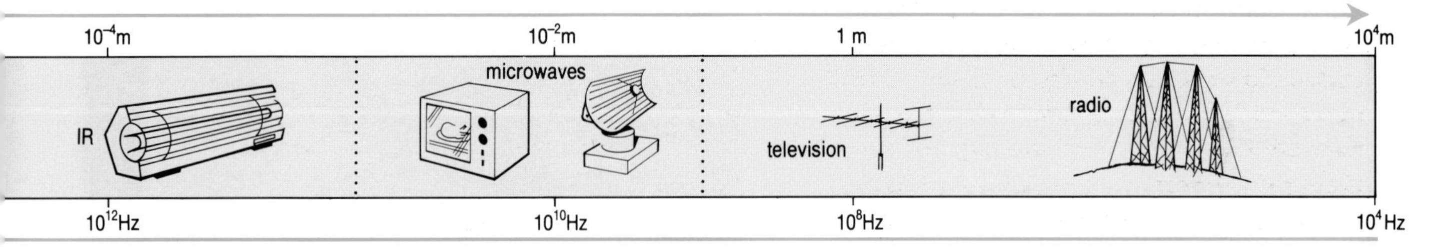

Any warm object emits infrared radiation. The amount of infrared emitted depends on the temperature and colour of the object. Infrared detection can be used to 'see' animals in darkness or people in a smoke-filled room.

Radio waves can have wavelengths between 10 cm and 10 000 000 m. Television signals are also transmitted using radio waves.

1 What is the speed of electromagnetic radiation in space?

2 State the range of the wavelengths of visible light.

3 Choose one part (or more!) of the electromagnetic spectrum and find out as much as you can about it.

4 Which parts of the electromagnetic spectrum can be harmful to life? Find out about the layers in the atmosphere which screen us from most of the harmful radiation which arrives from space.

5 Use the diagram of the electromagnetic spectrum and show that the speed of X-rays is the same as the speed of radio waves.

Radio telescopes

A window into space

Stars and galaxies emit radiation at many different frequencies in the electromagnetic spectrum. Much of this radiation cannot be detected here on Earth because it is absorbed by the atmosphere. However, certain radio waves can be detected, and astronomers can 'look' at objects in space by using radio telescopes to receive these radio waves. Astronomers use the phrase **radio window** for this part of the electromagnetic spectrum which can be detected on the surface of the Earth.

Dishes for the Cambridge Radio Telescope.

The metal eye

The **radio telescope** works, in some ways, like an optical reflecting telescope. A curved surface is covered with a thin sheet of metal which reflects the radiation to a focus. At the focus there is a short wire aerial in which a small electric current is generated. It is then amplified to produce a trace on a visual display unit (VDU). This can be examined to find the direction in which the radio signals are strong or weak. Because radio waves have a much longer wavelength than visible light, the reflecting dish does not have to be so precisely shaped and can therefore be made much larger. The giant radio telescope at Jodrell Bank in England has a diameter of 76 m. Sometimes several radio telescopes are used and the signals they receive are added together. This gives a better 'picture' of space.

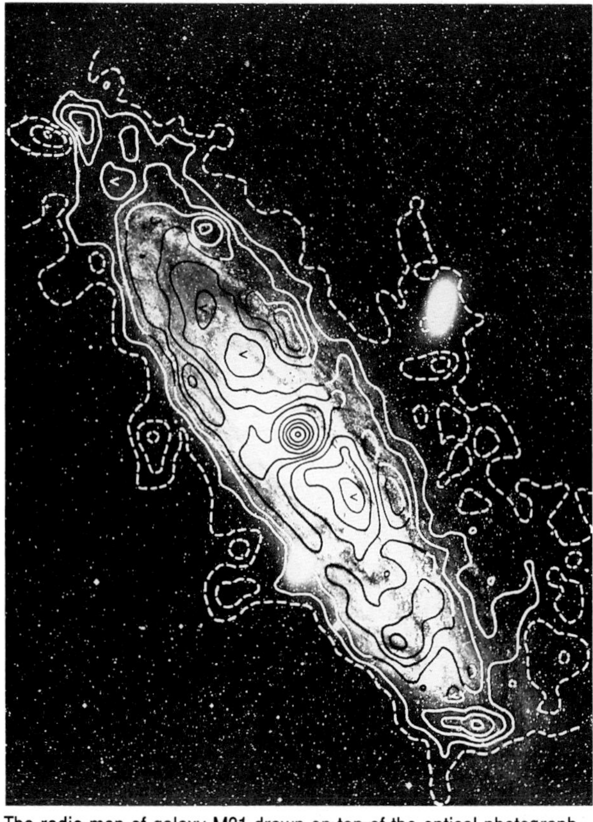

The radio map of galaxy M31 drawn on top of the optical photograph.

1 What do astronomers mean by a radio window?

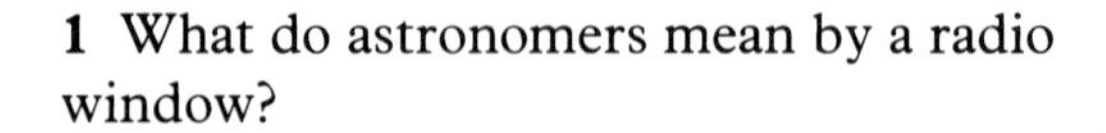

2 Describe how a radio telescope can 'look' at a region of space.

3 Why are radio telescopes larger than optical telescopes?

4 Draw a diagram to show how radio signals are brought to the focus of a dish aerial.

The Universe

Looking back

When astronomers look at distant galaxies, their telescopes are receiving light that was emitted many years ago. By observing extremely far-off objects, they can build up a picture of the history of the Universe.

In the early 20th century, Edwin Hubble noticed that distant galaxies appeared to be redder than they should be: i.e. the wavelength of radiation that they emitted was longer than he expected. This phenomenon is called the **redshift**.

The Doppler Effect. The sound from a moving object has a lower pitch if it is travelling away from us. Astronomers believe that the red shift of galaxies is evidence that the Universe is expanding.

To explain the redshift, imagine an ambulance passing you with its siren blaring. The pitch of the siren changes as the ambulance passes. As it moves away from you the pitch decreases; i.e. the wavelength of the sound increases. The same effect is thought to happen with light – the redshift. Astronomers believe that the red-shift is evidence that the galaxies are moving away from us.

A model of the expanding Universe. As the balloon expands the dots get further away from each other. Similarly, galaxies move away from each other as the Universe expands.

Evidence also suggests that, before galaxies formed, the Universe was a hot, dense 'atomic soup'. This, together with the redshift evidence, has led to the **big-bang theory**. This theory states that the Universe started with a super explosion at the beginning of time. It has continued to expand since then. Much work is now being done to try to find out if the Universe will continue to expand, or whether gravitational forces will make it stop expanding and begin collapsing.

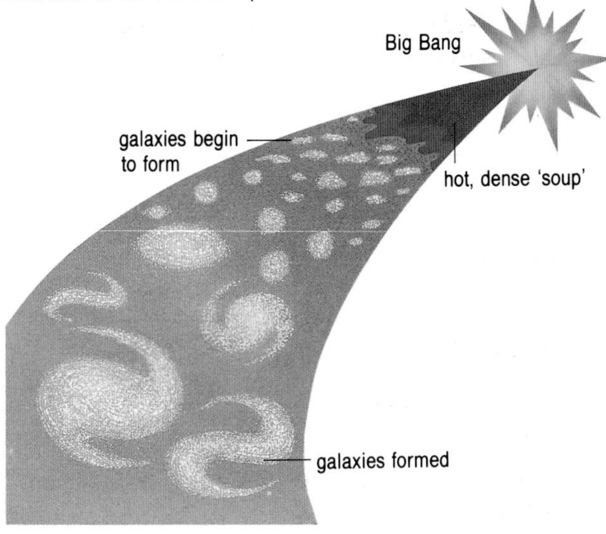

1 Explain why light from distant galaxies has taken so many years to reach us.

2 Explain what happens to the pitch of the sound from a motorcycle as it comes towards you and then travels away from you.

3 What is the redshift?

History of space flight

Sputnik 1 was the world's first artificial satellite. It was launched on 4 October 1957 by the Soviet Union. At a height of 215 km, Sputnik 1 orbited the Earth every 96 minutes and remained in orbit for three months.

The Russian cosmonaut, Yuri Gagarin, was the first man in space. On 12 April 1961 he circled the Earth once in 108 minutes in his tiny spacecraft, Vostok 1. In the next two years, five more Russian cosmonauts (including Valentina Tereshkova, the first woman in space) made safe flights.

America's first astronaut to orbit the Earth was John Glenn on 20 February 1962. This was the beginning of America's attempt to win the 'Space Race' – the race to land people on the Moon. It ultimately involved 300 000 people and cost $25 billion.

To the Moon

In America, during the 1960s, the National Aeronautics and Space Administration (NASA) launched the Gemini series of spacecraft, followed by the Apollo mooncrafts which were launched into space by the enormous Saturn 5 rocket. Ten Apollo missions tested equipment and prepared for Apollo 11 which finally landed the lunar module, Eagle on the Moon's surface. On July 21 1969 an estimated television audience of 600 million people watched as Neil Armstrong became the first man to set foot on the Moon.

There followed a series of very successful missions, the last of which was Apollo 17 in 1972. This lasted 12 days and 14 hours – the longest ever lunar mission. During this time astronauts conducted many experiments and made a great deal of use of the 11 km per hour lunar Rover. The total distance travelled in the Rover was 35 km, with a maximum distance from the module of 6.4 km

The world's first artificial satellite, Sputnik 1.

Vostok 1 being assembled.

After the excitement of getting to the Moon, the American and Russian space programmes turned to studying the effects on the human body of prolonged periods in space. The Russians launched the Salyut space stations. The Americans turned the third stage of a massive Saturn 5 rocket into an orbiting space laboratory called Skylab. Both Salyut and Skylab remained in orbit and were visited by crews for a period of time. Crews arriving for a tour of duty had to dock their spacecraft with the orbiting space station. On July 17 1975 an important link-up occurred when two American astronauts docked nose to nose with Soyuz 19. They were greeted by Russian cosmonauts and later shared a meal together.

In the 1980's NASA had to make do with a much smaller budget than was available during the Apollo programme. The remarkable Space Shuttle was developed. It was designed to be re-usable so that the cost of putting astronauts and satellites into space could be reduced. The first test flight was on April 12 1981 and the four Shuttles used by NASA flew twenty-four successful missions before the tragedy which occurred on January 28 1986, when Challenger exploded 74 seconds after lift-off.

Even though both America and Russia have had setbacks in their space programmes, it has been quite remarkable that so much has been achieved in such a short time. Without doubt, many more exciting developments will take place in space flight. Who knows, perhaps one of you studying physics and reading this chapter will one day make a trip into space!

Skylab.

A remarkable piggy-back. The Space Shuttle is tested before launching into orbit.

1 Who was the first person to orbit the Earth?

2 The population of the United States is approximately 200 million. How much did the Apollo space programme cost per American?

3 What was the maximum speed of the lunar Rover? Why do you think it was not designed to go very fast?

4 The Apollo 13 mission did not go smoothly. Find out about what happened to the space-craft on its journey to the Moon and how the astronauts coped with the emergency.

Rockets

What goes up must come down?

Contrary to the saying 'what goes up must come down', it is possible to throw something up so that it does not come back. You simply have to throw it hard enough. The speed needed to escape from the Earth and not fall back is about 40 000 km/h. This is called the *escape velocity*. Only rocket motors are capable of achieving this speed.

A Soviet manned Soyuz spacecraft being launched.

Pairs of forces

Rockets were invented by the Chinese over 750 years ago. They were really just fireworks and used a kind of gunpowder as fuel. Rockets such as these are called solid-fuel rockets. Most modern rockets are liquid fuelled.

To understand how rockets work, we must look at the work of Sir Isaac Newton. He suggested that forces occur in pairs. In the laboratory we can see an example of this.

Two 'exploding' trolleys, A and B, are stationary and the spring plunger of trolley A is released. Trolley A applies a force to trolley B which moves away from trolley A. However, trolley A also moves away from trolley B so that trolley B must have applied a force to trolley A. The forces are equal in size but they act in opposite directions.

Rockets work on the same principle. The fuel is burnt and gases are pushed out of the rocket. But at the same time these gases exert an equal and opposite force to the rocket. This makes the rocket accelerate. Rockets do not need to push against anything so they work in space where there is no air.

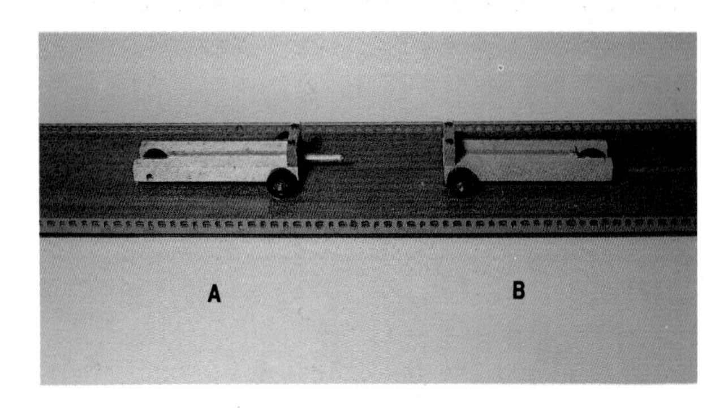

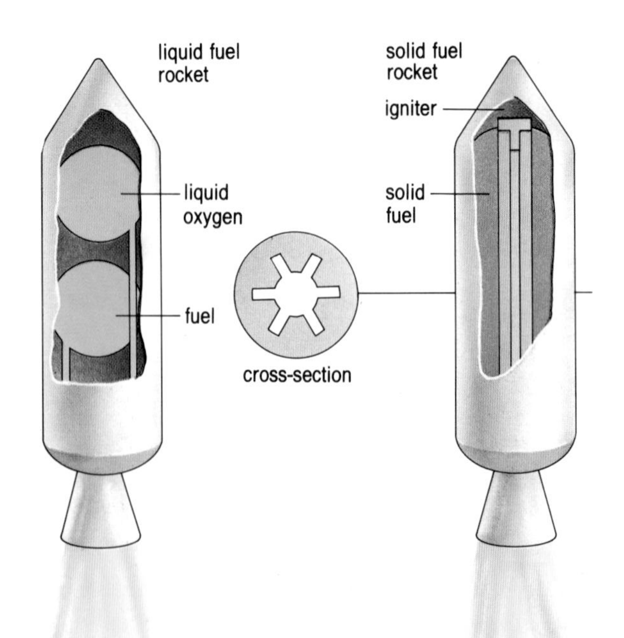

Multi-stage rockets

If a rocket is to reach escape velocity, then it must accelerate for a long time. This requires an enormous amount of fuel to be carried and, as a result, rockets are big and heavy. They are usually fired in stages, so that as the fuel is used up, the lower stage is released and the upper stages do not have to accelerate the extra mass. The Saturn V rocket used in the American Apollo missions had three stages.

The Space Shuttle has extra stages attached to its side. These are used at the launch and then released.

Cruising all the way

Once a spacecraft has reached a high enough speed to enable it to escape from the surface of the Earth, it no longer needs to fire its rockets. It can then travel between planets at a constant speed, cruising silently through space until it needs to slow down or change direction.

Strictly speaking, the Apollo astronauts who travelled to the Moon did not escape completely from the pull of the Earth's gravity. Rather they were 'captured' by the gravitational field of the Moon. They did, of course, remember to take enough fuel with them to escape from the surface of the Moon!

Newtons Third law of Motion

If **A** exerts a force on **B**, **B** exerts an equal force on **A** in the opposite direction.

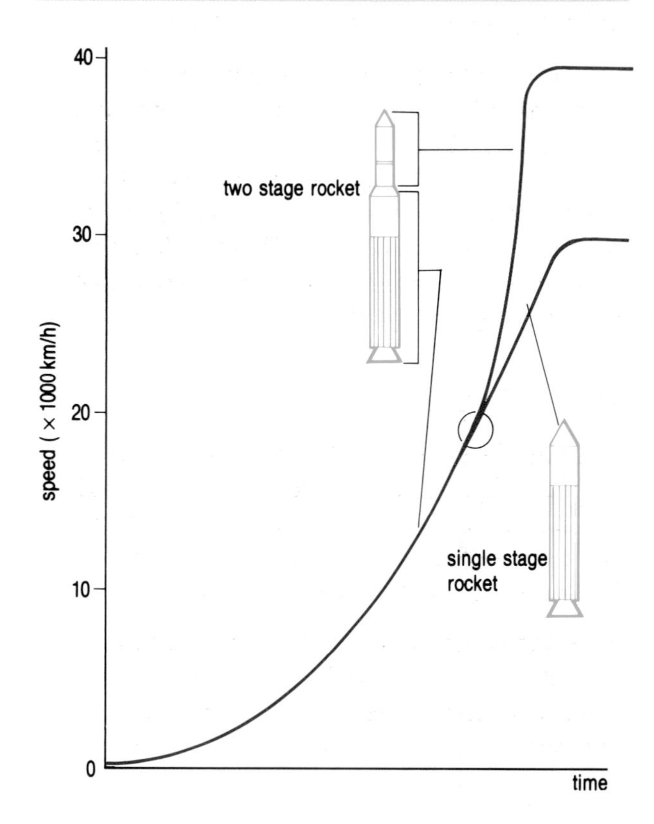

Graph showing the speeds reached by two rockets which have the same initial velocity.

›› ***A*** *Exploding trolleys can be stacked to increase their mass. Find out how mass affects the speed of the trolleys when the plunger is released.*

B *A model rocket can use water as fuel. Find the amount of water needed to fire the rocket to its greatest height.*

1 What is meant by 'escape velocity'?

2 Explain in your own words why a multi-stage rocket can reach escape velocity more easily than a single stage rocket.

3 The journey of a spacecraft to the Moon can be divided into three parts: the acceleration during launch, the constant speed during the flight and the deceleration before landing. The rocket motors do not need to be firing during the whole of the journey. Using Newton's Laws, explain the motion during the three parts of the flight and describe when the rockets would be fired.

4 What is the initial upwards acceleration of a Saturn V rocket of mass 3×10^5 kg if the combined thrust of its rockets is 3.6×10^6 N?

Space shuttle

The re-usable spacecraft

Shuttle orbiters are the most remarkable flying machines ever built. Until the tragic loss of Challenger in 1986, four Shuttles (Challenger, Columbia, Discovery and Atlantis) had completed 24 successful missions, their main purpose being to put satellites into orbit. In each flight the Shuttle acted as a rocket, a spacecraft and a glider. What follows is the story of a typical flight.

Going up

Five minutes before launch time (T), the seven crew are on board and the commander and pilot are checking the operation of the on-board computers. Soon the launch director in the NASA control room instructs the crew to lower their visors. At 'T minus 8 seconds' thousands of gallons of water are released into the base of the launch pad to dampen the sound of the engines. The three main engines, fuelled by liquid hydrogen and oxygen, ignite. At 'T minus 2 seconds' the two solid booster rockets are fired – they cannot be throttled back and the launch must now take place.

The total power now developed is enough to light the whole of America! The explosive bolts holding back the Shuttle are detonated and the spacecraft lurches into the air.

Just after the launch, the force experienced by the crew is three times their own body weight (i.e. 3 g). They find it difficult to move even their arms. After 50 seconds the speed is 30 m/s.The mind-boggling journey into space takes 8 minutes 50 seconds. At an altitude of 120 km the shuttle is travelling at 15 times the speed of sound and the commander is preparing for MECO (main engine cut-off).

At 'T plus 9 minutes' the spacecraft has reached its orbital velocity of 28 000 km per hour and is coasting in silent space. The altitude is 200 km but it could be up to 1100 km.

Coming down

The Shuttle is designed to remain in space for up to 100 days, although most flights last for only a week or so. During their stay in space the crew are busy launching new satellites from the payload bay or carrying out experiments in the space lab.

At the end of the mission, the Shuttle is slowed down using jet thrusters and the crew prepare for re-entry. Soon the spacecraft is grazing the Earth's atmosphere and its kinetic energy is converted into thermal energy. Within minutes after contact with the thin air, the temperature of the nose reaches 1430°C.

Special ceramic tiles insulate the Shuttle from this intense heat. The crew begin to lose the feeling of weightlessness.

The autopilot manoeuvres the Shuttle – now a 300 km per hour glider – over the coast of California. It flies 4.5 metres horizontally for every 1 metre vertically. The pilot lowers the undercarriage and he has only one chance to land on the runway ahead. The Shuttle touches down and rolls 2 km before stopping. A spacecraft has returned.

Re-entry

Return from space

When a spacecraft returns to Earth from space, it must first re-enter the atmosphere. Although the highest part of the atmosphere is very thin, spacecraft are travelling so fast at that time, that there is a large frictional force which slows them down.

A high speed spacecraft has a very large amount of kinetic (movement) energy. As it is slowed down, the kinetic energy is converted into thermal energy. Of course, there are many other examples of friction between moving objects producing a temperature rise. Simply rubbing your hands together, for example, has this effect.

The temperature rise during re-entry is a major problem for spacecraft designers – the astronauts inside must not be cooked! During the re-entry of the American Mercury capsules, the air outside the capsule reached a temperature of 5260°C and turned an orange colour. The outside of the capsule itself reached a temperature of 1648°C. A special resin, on a fibreglass screen, was attached to the surface. The resin boiled and evaporated, so keeping the capsule cool.

The Space Shuttle uses special silica tiles to protect it. The bottom and the leading edge are covered with glossy black tiles and a black reinforced carbon material covers the nose and wing leading edges.

Although the Space Shuttle glides to its landing, other returning spacecraft use parachutes to slow them down to produce a gentle landing.

›› *Find out about meteorites. Try to explain why the moon has so many craters whereas the Earth has very few.*

1 List more examples of frictional force producing temperature rises.

2 Explain why designers have to make spacecraft capable of withstanding very high temperatures.

3 Comment on the following properties of the material used to cover the outer surface of a spacecraft:
a) colour;
b) specific heat capacity;
c) expansion at higher temperatures.

4 a) In orbit around the Earth, the Space Shuttle has a speed of approximately 10 000 m/s and as it glides to a touchdown its speed is approximately 100 m/s. If the mass of the Shuttle is 2×10^6 kg, calculate its kinetic energy in orbit and just before touchdown.
b) What has happened to this 'lost' kinetic energy?
c) Assume that it takes one orbit of the Earth, at an average height of 100 km, to slow down the Shuttle.
(i) Calculate the distance that the Shuttle travels during this orbit. (The radius of the Earth is 6400 km.)
(ii) Calculate the average force exerted by the atmosphere on the Shuttle as it slows down.

5 An aluminium section of a used rocket thruster has a mass of 1 kg. It returns to Earth at a speed of 5000 m/s. By assuming that all its kinetic energy is converted into thermal energy as its enters the atmosphere, find the temperature rise of the section. The specific heat capacity of aluminium is 900 J/kg°C. Explain why such small sections of the rocket thruster would not be likely to be found on the surface of the Earth.

Satellites in orbit

Newton's thought experiment

In the seventeenth century, Sir Isaac Newton explained how a satellite might stay in orbit around the Earth. His reasoning went something like this.

Imagine a very tall mountain – taller than Mount Everest – and imagine a large gun which fires a shell (projectile) horizontally from the top of the mountain. The range of this gun would be enormous, perhaps several hundred kilometres. Of course, at this distance, the gun would be firing the shell over the horizon and the curvature of the Earth would have to be taken into account.

Projectiles

Projectiles can be thought of as having two motions at the same time.
The horizontal motion has a constant speed.
The vertical motion has a constant acceleration.

Horizontal Speed is Constant.

Vertical Speed is Increasing

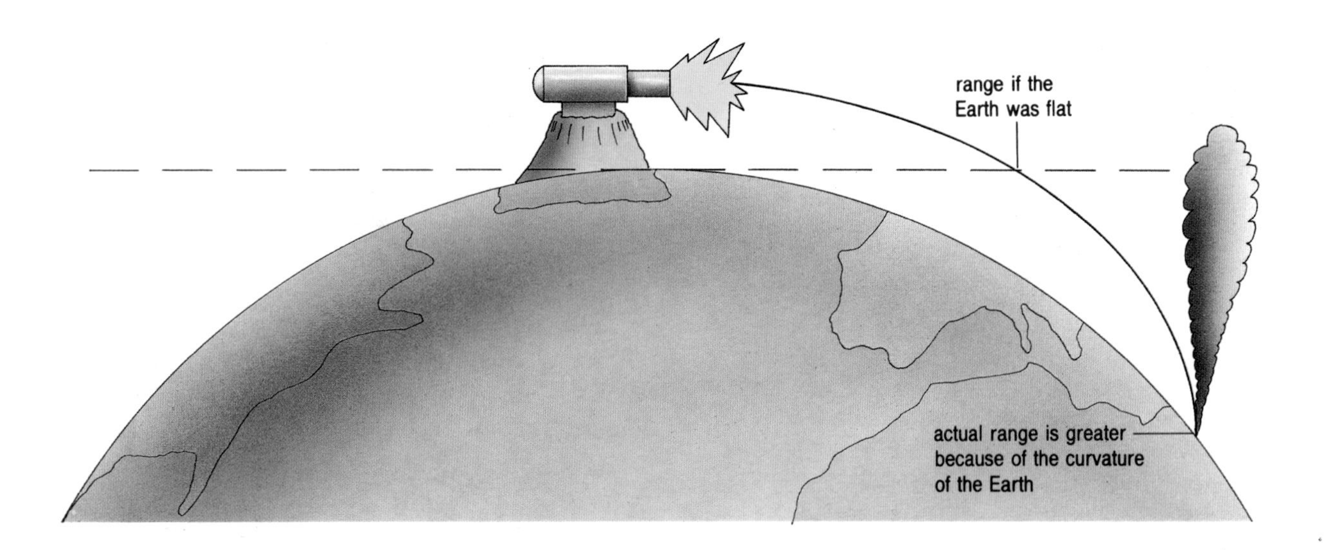

Now imagine an even taller mountain, one that is higher than the atmosphere, and an even bigger gun. A shell with the right speed would never hit the ground. It would continue to fall towards the Earth, but because of the curvature of the Earth, it would remain at the same height above the ground. This shell has become a satellite in orbit. It remains in orbit because of gravity, not because it has escaped from gravity.

Of course it would not be possible to build such a gun, but a rocket can be used instead to give a satellite the right speed at the right height.

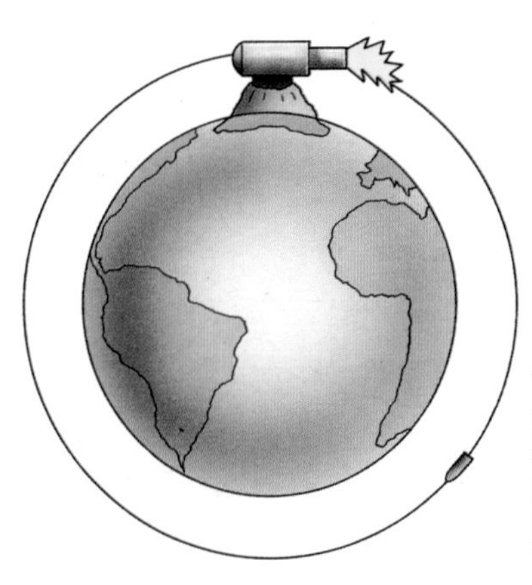

The shell continues to fall, but the curvature of the Earth means that it remains at the same height above the ground. The shell is in orbit.

Natural and artificial satellites

The Earth and the planets are natural satellites of the Sun; and the Moon is a natural satellite of the Earth. Since 1957 we have been able to launch artificial satellites into orbit around the Earth. At this moment there are several hundred working satellites in orbit performing many different jobs, from worldwide communication links to weather mapping and the tracking of hurricanes. On a clear night, it is possible to see a satellite as it moves like a bright star across the sky, and then disappears below the horizon.

Artificial satellites can be placed in many different orbits. The height of an orbit can be as low as several hundred kilometres or as high as many thousand kilometres.

The time that it takes for a satellite to orbit once round the Earth (the **period** of the satellite) depends on the height of the orbit. The further it is away from the Earth, the greater is the **period**. A satellite in a low orbit of 200 km takes approximately 90 minutes to circle the Earth. At a height of about 36 000 km the period is 24 hours. A satellite in this orbit, in the same plane as the equator, is said to be in a geostationary orbit. It rotates around the Earth in the same time that the Earth itself takes to complete one revolution. For an observer on the Earth such a satellite appears to be stationary in the sky. This is very useful for communication satellites.

Planet	Mean distance (d) from Sun (10^6 km)	Period (T) of planet
Mercury	58	88 days
Venus	108	224.7 days
Earth	150	365.26 days
Mars	228	687 days

›› *Using the table, plot a graph of the distance (d) from the Sun to a planet, against the period of the planet (T). Is your graph a straight line? Try plotting d^3 against T^2. What conclusion can you make from your graph? Express the relationship as an equation.*

Satellite in orbit.

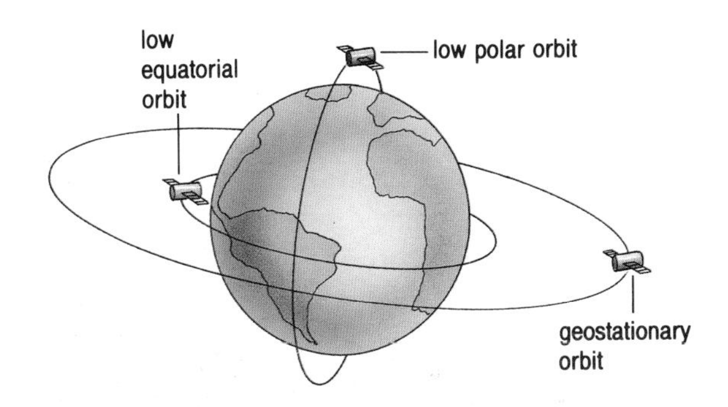

1 Many people think that a satellite remains in orbit because it has escaped gravity. Explain in your own words why this is not true.

2 What is a geostationary orbit?

3 A stone is projected horizontally from the top of a cliff with a speed of 20 m/s and takes 3 s to reach the sea below. Draw speed time graphs to show the horizontal and vertical motions of the stone. Use the graphs to find the height of the cliff and how far from the base of the cliff the stone lands.

4 A bullet is fired horizontally from a rifle with a speed of 400 m/s.
a) How long does it take to hit a target 100 m away? (Ignore air resistance.)
b) The rifle was aimed at the centre of the target which has a diameter of 1 m. Find whether or not the bullet hits the target.
c) How should the rifle be aimed to ensure that the bullet hits the centre of target?

Weather satellites

Cameras in orbit

Since the first satellites were placed in orbit, weather forecasters have realised that a grandstand view of the weather is very useful.

Weather satellites are usually placed in one of two orbits. Some, such as the European Space Agency's Meteosat, are placed in a geostationary orbit at a height of approximately 36 000 km (see page 239). This means that the satellite appears to be stationary over the same point on the Earth's surface and the cameras on board show the same field of view at all times. This is, of course, very convenient because it means that the movement of cloud formations can be carefully watched over a continuous period.

Weather satellites in a low polar orbit, circle the Earth in about 90 minutes. These satellites produce high quality pictures of the weather below but their signals cannot be received by ground stations at all times.

Meteosat 3 in orbit.

Forecast – sunny intervals

The photograph shows a typical picture taken with visible light from a satellite in geostationary orbit. Study the photograph and notice the following:

- The clouds form a pattern. This is part of a circulating air flow called a depression which is centred to the north of Scotland.
- A thick band of cloud covers much of eastern Europe. This is a front and it is likely to be raining here.
- Much of Britain is enjoying some sunshine although it is likely to be quite windy, with the wind coming from the west.

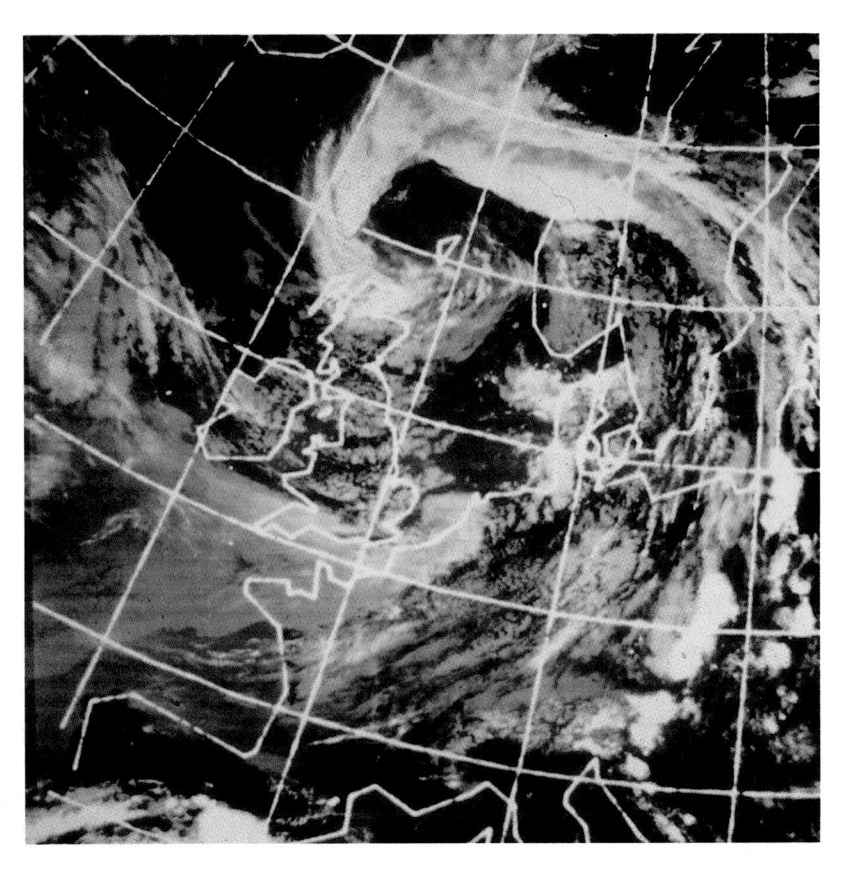

A pair of photographs using visible light (left) and infrared radiation (right). In the infrared photograph the warmest areas show up as darkest. Both were taken using the European Space Agency Meteosat geostationary weather satellite.

Satellite	Period (mins)	Average height above Earth (km)	Inclination (This is the angle that the orbit makes with the North/South axis of the Earth)
UoSAT 1	94.3	441	7.6°
UoSAT 2	98.6	647	8.2°
NOAA 9	102	808	9.0°
NOAA 10	101	761	8.7°

1 Weather satellites usually send two pictures, one using visible light, and one using infrared radiation. What advantages can you think of in doing this?

2 The table shows details of the orbits for two weather satellites, NOAA9 and NOAA10, and also for the two satellites UoSAT1 and UoSAT2, which are controlled by the University of Surrey.

a) Which satellite flies closest to the North Pole?

b) The distance (d) of a satellite from the centre of the Earth is related to the period (T) of the satellite by the expression $\frac{d^3}{T^2}$ = constant. Use the data in the table to confirm this. (The radius of the Earth is 6400 km.)

c) The plane of a satellite remains fixed in space and the Earth rotates once every 24 hours. This means that each time a satellite crosses the equator, on its way north, it crosses at a point further west than the previous crossing. Calculate how many degrees west each satellite appears to move from one orbit to the next. (360° one revolution).

Fields in space

Moon sports

Believe it or not, one of the Apollo astronauts has played golf on the Moon. Anybody taking up Moon golf would have to be prepared to walk a long way, because it is possible to hit a golf ball up to six times further than on Earth.

Athletes could break all Earth records if they could compete on the Moon. For example if you could throw a ball 10 m high on Earth, you could throw it about 60 m high on the Moon. High jumpers would be able to jump the same height as Earth-bound pole vaulters (except that their space suits might make things a bit difficult!). The reason for this is that the gravitational field of the Moon is weaker than the gravitational field of the Earth.

Alan Shepard plays golf on the Moon.

Gravitational fields

Each of the planets of the solar system has a gravitational field surrounding it. So too, does the Sun and the natural satellites of the planets. A gravitational field is simply a region in space where a force of attraction is exerted on an object because of its mass. For example, we are held to the surface of the Earth by the force of attraction of the Earth. We are in the Earth's gravitational field.

The moons of Jupiter are held in orbit by the force of attraction of the planet and the Voyager probes were strongly attracted by Jupiter as they flew past it in 1979. The spacecraft were deflected during their brief encounter with the planet and are now hurtling on their way out of the solar system.

Field strengths

The larger the mass of the planet, the larger is the force of attraction exerted on objects within the gravitational field. This explains why objects are not so strongly attracted to the Moon as they are to the Earth, which is considerably larger.

Planet etc.	Surface Gravitational Field Strength (N/kg)
Sun	273.70
Mercury	3.74
Venus	8.83
Earth	9.81
Moon	1.62
Mars	3.73
Jupiter	25.93
Saturn	11.37
Uranus	10.89
Neptune	11.87
Pluto	4.61

Gravitational fields are measured by the force which is exerted on a 1 kg mass. So, for example, the gravitational field strength on the surface of the Earth is approximately 10 N/kg. This gravitational force is called the *weight* of the object. On the Moon, the gravitational field strength is about 1.6 N/kg. On Jupiter (assuming you could stand on the surface), a 1 kg mass would have a weight of 25.9 N. Moving around in such a strong gravitational field would require great effort.

All objects produce a gravitational field, but it is only when the objects are very large – as big as planets – that the effect of these forces can be easily detected.

If an object is free to move in a gravitational field, it will accelerate, because an unbalanced force is applied to it. Near the Earth's surface, a 1 kg mass will have a force of 10 N acting on it vertically downwards. It will accelerate downwards at 10 m/s^2 ($a = F/m$). Similarly a mass of 2 kg will have a force of 20 N and will have the same acceleration. Provided air resistance is ignored, all objects near the surface of the Earth, will have the same acceleration: 10 m/s^2.

Of course if an object is falling from some height above the Earth's surface it will be accelerating through the atmosphere. This provides a frictional force which increases with speed. Consequently the object's acceleration gets less as the object speeds up.

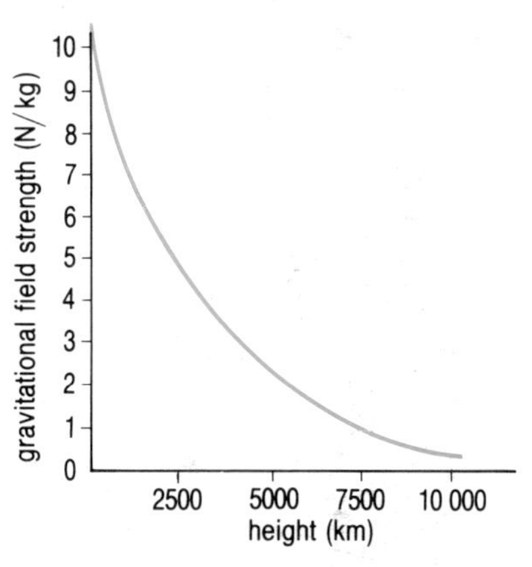

A graph showing how gravitational field strength varies with height (h) above the surface of the Earth.

Escaping gravity?

The gravitational field of a planet gets weaker as you move further away from the planet. Theoretically, the field strength gets weaker and weaker but never reaches zero. In practice, an interplanetary traveller would move away from the Earth and be 'captured' as he came close to another planet. He would experience a gravitational force of attraction from this new planet.

If you moved far enough away from all planets and stars, you would experience only insignificant forces of gravitational attraction and would be truly weightless.

1 What is meant by gravitational field strength? In what units is it measured?

2 Which of the planets of the solar system has **a)** the greatest and **b)** the least gravitational field strength?

3 What would be the weight of an 80 kg astronaut on Mars?

4 What is the gravitational field strength of a planet where the weight of a 60 kg astronaut is 300N?

5 The Moon has no atmosphere. Draw a speed-time graph to show the motion of an object falling freely for 10s on to the lunar surface. How would the shape of the graph differ if the Moon did have an atmosphere?

Weightlessness

Floating free

Weightlessness seems to be great fun. All the astonauts who have experienced it have enjoyed the sensation of floating freely in surroundings where up and down do not have the same meaning as here on Earth. But what exactly do we mean by weightlessness, and what problems does it bring?

Strictly speaking, to be weightless, an astronaut should travel to a region where there is no gravitational field. At very great distances from all planets and stars, an astronaut would experience only very small gravitational forces and he would be, to all intents and purposes, weightless. However, this does not explain how weightlessness can be experienced in a spacecraft in a low orbit round the Earth.

The crew of Spacelab 1. Which way up?

All fall down

Imagine being in a lift at the top of a very tall building. Now suppose that the cable suddenly broke and the lift, with you inside, started falling freely downwards. The lift would have a uniform acceleration of $10\,m/s^2$. You too would have this acceleration and you would seem to be weightless. This is exactly what happens to an astronaut in an orbiting spacecraft.

stationary lift

lift falling freely
Those inside appear to be weightless.

A spacecraft remains in orbit, at the same height, because it is continuously falling to Earth (see Newton's thought experiment on page 238). An astronaut inside or beside an orbiting spacecraft is falling at the same rate as his surroundings and so appears to be weightless.

The photograph on the right shows Bruce McCandless using the manned manoeuvering unit (MMU), a hand-controlled device that uses liquid nitrogen as a propellant. This was the first time that an extravehicular activity (EVA) had been undertaken without a tether to the spacecraft.

Living in space

There seems to be no end to the challenges that are presented to an astronaut in a state of weightlessness. Nothing has weight and all objects (including human beings) will remain in the position where they are released. Small objects, however, tend to drift around in air currents inside the spacecraft unless they are held in place. Moving around inside an orbiting spacecraft takes some getting used to. To move from one side to another requires pushing off with the right speed. Once in motion, the speed cannot be changed and if it is too big, there can be an embarrassing collision with the other side. If there is nothing to hold on to, the astronaut will bounce back.

It was long thought that eating in space would be extremely difficult. In practice it has turned out to be straightforward. Food tends to stick to its container and to the spoon which can just as easily be used upside down. Liquids are a bit of a problem as they do not slide down the side of a mug. Astronauts have to use straws for all their drinks and liquids which escape, float about in quivering, jelly-like globules.

Sleeping, of course, requires very little preparation. An astronaut can simply find a quiet corner and drift off to sleep without the need for any mattress, but berths are provided on the Space Shuttle to keep sleeping astronauts out of the way of those at work.

There has been a lot of investigation into the effects of weightlessness on the body. It seems that without taking exercise in space, astronauts have difficulty when they return to Earth.

In space, muscles do not have to work continuously to support the body and in time they become weaker. The heart too does not have to work as hard to pump blood to the brain and this can lead to low blood pressure. On any long spaceflight, astronauts have to take regular exercise to reduce the dangers of the effect of weightlessness on their bodies.

A floating sphere of orange juice presents problems for thirsty astronaut Joseph P. Allen IV.

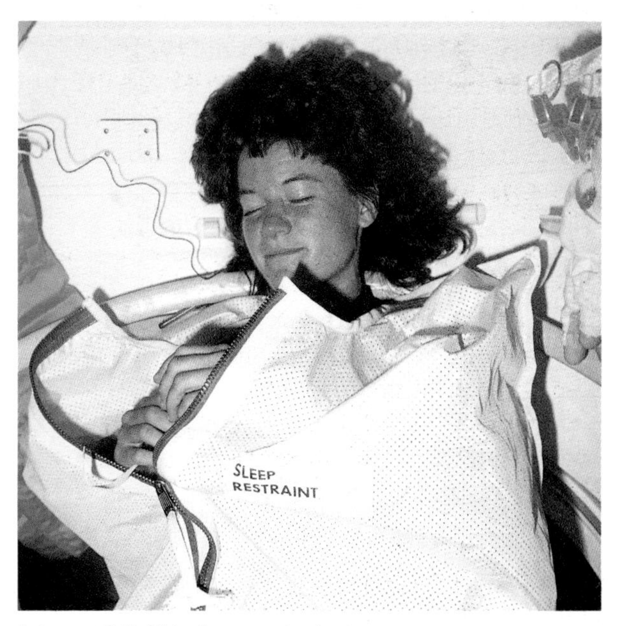

Astronaut Sally Ride sleeps on the shuttle.

1 American astronauts sometimes speak about zero gravity when they describe the experience of weightlessness. Explain why an astronaut in an orbiting spacecraft is not in zero gravity.

2 What problems do astronauts in space have eating and drinking?

3 Why is it important for astronauts in space to take exercise?

4 The lift referred to on p.238 has an emergency brake. Describe what those inside the lift would feel as the brake slowed the lift from a speed of 30 m/s to rest in 1.5s.

Questions

1 A manufacturer gives the following technical details for a two slice pop-up toaster:

Voltage; 240 V
Power consumption; 1260 W to 1500 W
AC or DC

a) Suggest why there is a range of power consumption.
b) To save energy some toasters can detect when only one slice of bread has been inserted.
 i) Suggest a way of doing this.
 ii) Draw a circuit diagram for this toaster.
c) Given 3A, 5A and 13A fuses, which one should be used in the mains plug of this toaster?

2 A greenhouse has an electronic system which automatically switches on a heater if the air temperature in the greenhouse drops too low. A manual switch is included so that the automatic system can be switched off.

A flow diagram of the system is as follows:

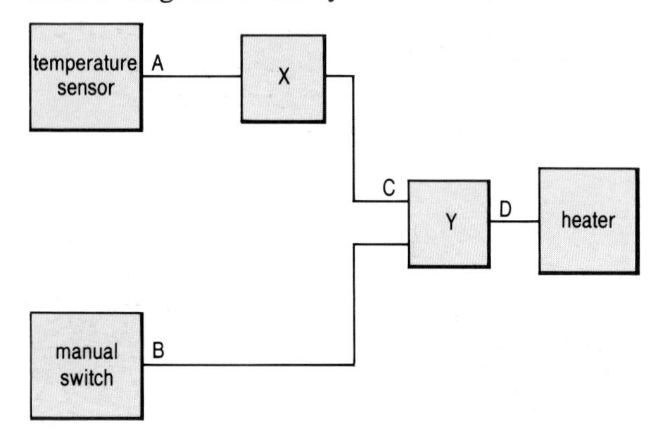

The temperature sensor gives a logic 1 output when the air temperature is normal and logic 0 when it is too cold.

The heater is switched on by logic 1.

a) i) What logic output will the manual switch give when it is OFF?
 ii) What logic output will the manual switch give when it is ON?
b) Name logic block Y.
c) Name logic block X and explain why it is needed in this sytem.
d) Copy and complete the following truth table for this system.

3 The diagram shows the settings on a graphic equaliser.

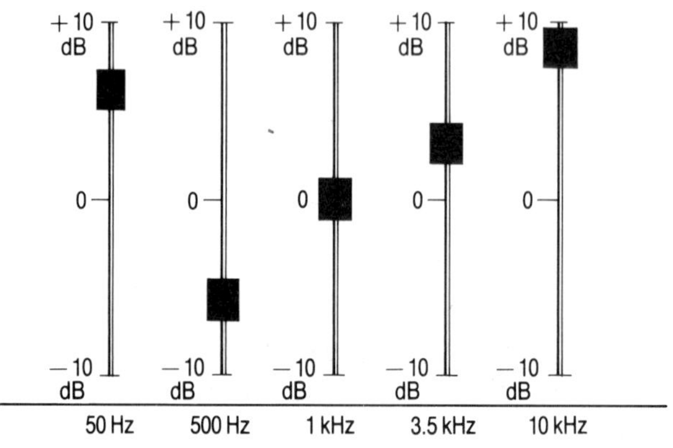

a) Draw the frequency response curve corresponding to these settings.
b) Describe the sound heard when you play a record with the controls set in the positions shown.
c) Sketch the position of the controls for:
 i) a 'flat' frequency response;
 ii) a cut in low frequency (<1 kHz);
 iii) a boost in high frequency (>1 kHz).

4 Last winter the water tank in Helen's loft froze over. To avoid it happening again she decides to fit an electric light in the left which can be switched on automatically in very cold weather.

a) Do you think that her idea will help stop the water freezing?
b) Helen can choose either a filament bulb or a fluorescent tube. Which one should she choose? Give reasons for your answer.
c) Draw a circuit diagram which could switch the light on automatically in very cold weather.
d) What other measures could she take to stop the tank freezing? Give reasons for each choice.

5 Devise a questionnaire to find out about people's attitudes on radioactivity and health. You may wish to see how much they know about the uses of radioactivity in health care. You may also wish to discover whether they believe the various uses are to be welcomed.

Test out your questionnaire on about ten to twenty people and write a short report on the results which you find.

6 The diagram represents the magnified wave pattern of a 3 adjacent grooves of a record rotating at 45 rpm.

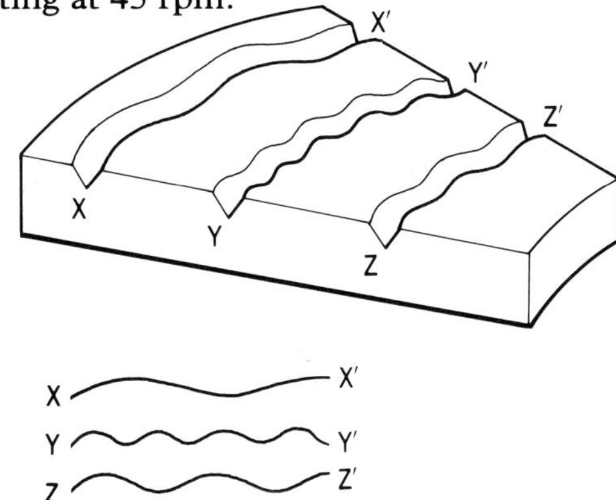

a) Describe how the sound changes as the grooves XX ; YY′ and ZZ′ move under the stylus.
b) Explain how a scratch across the surface of the record might affect the vibration of the stylus.
c) How would a scratch on the surface of a CD effect the sound? Explain your answer.

7 A graph of speed against time for a high speed train is drawn below.

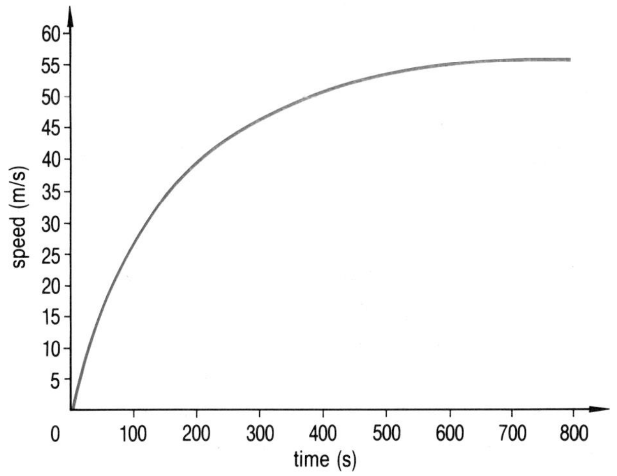

a) Describe the motion of the train during this part of the journey.
b) Sketch speed against time graphs to compare the motion of the train:
i) as it approaches and stops at a station;
ii) during an emergency stop.
c) How would you expect the acceleration of an Olympic sprinter to compare with that of the train?

8 Sonar can be used to detect a shoal of fish. Describe how you could use it explaining how you would tell the difference between fish, the seabed and a ship-wreck.

9 The diagram shows the heat loss from an insulated house in winter.

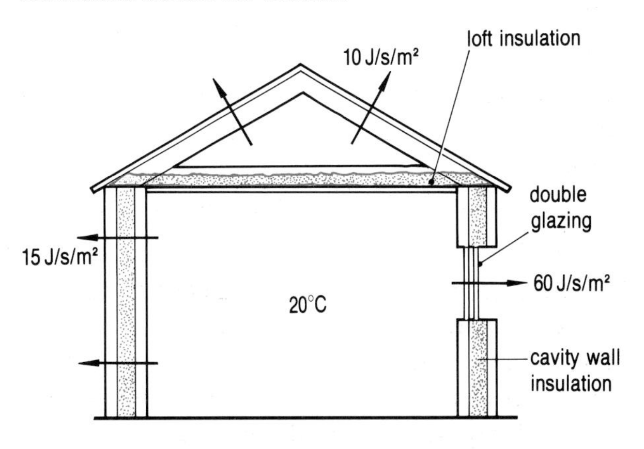

a) Every house suffers heat losses in cold weather. What steps can you take to cut down such losses, other than those shown in the diagram?
b) Explain why the loft is often very cold in winter and very hot in summer.

10 A sewage pipe is to be situated in a bay. The end of the pipeline is to be 400 m from the beach. Local environmental groups are worried about possible pollution on the beach.

A sample of radioactive sand is dropped where the pipeline would end. After a week a 'contour map' of the distribution of the radioactive sand is constructed.

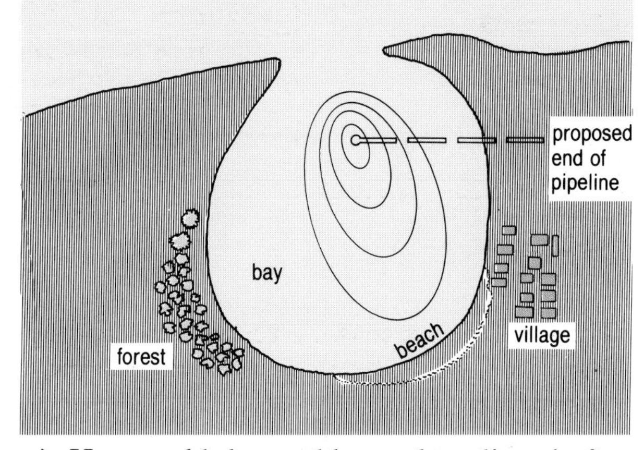

a) How could the sand be made radioactive?
b) How were the 'contour lines' constructed?
c) Which 'contour line' indicates the greatest amount of radioactivity?
d) What type of radioactivity is emitted by the sand? Give a reason.

Quantities, units and symbols

Quantity	Quantity symbol	Unit	Unit symbol
Five basic SI units			
mass	m	kilogram	kg
length, distance	d(l or s)	metre	m
time	t	second	s
electric current	I	ampere	A
temperature	T	kelvin	K
Other units			
absorbed dose (ionising radiation)	D	gray	Gy (J/kg)
acceleration	a	metres per second squared	m/s^2
acceleration due to gravity (free fall)	g	metres per second squared	m/s^2
activity (of radioactive source)	A	becquerel	Bq (per second, s^{-1})*
area	A	square metre	m^2
average speed	$\bar{v}$	metres per second	m/s
electric capacitance	C	farad	F (C/V)
electric charge	Q	coulomb	C (A s)
electric resistance	R	ohm	Ω (V/A)
energy	E	joule	J (N m)**
force	F	newton	N (kg m/s^2)
frequency	f	hertz	Hz (per second, s^{-1})*
gravitational field strength	g	newtons per kilogram	N/kg
heat	E_h	joule	J (Nm)
kinetic energy	E_k	joule	J (Nm)
potential energy	E_p	joule	J (Nm)
power (energy per second)	P	watt	W (J/s)
power (of a lens)	P	dioptre	d $\left(\frac{1}{m}, m^{-1}\right)$
specific heat capacity	c	joules per kilogram per degree Celsius	J/kg°C (J kg^{-1} K^{-1})
specific latent heat	l	joules per kilogram	J/kg
speed (velocity)	v	metres per second	m/s
temperature	$T(\theta)$	degrees Celsius (centigrade)	°C (K-273.15)
voltage (p.d., e.m.f.)	V	volt	V (J/C)
volume	V	cubic metre	m^3
wavelength	λ	metre	m
work	$E_w(W)$	joule	J (Nm)

equivalents are shown in brackets.

* note that both the hertz and the becquerel are equivalent to 'per second' (ie s^{-1}).

** other energy units in use are

kilowatt hour (kWh)	$=3.6 \times 10^6$J
big calorie (kcal)	$\approx 4.2 \times 10^3$J
British Thermal Unit (BTU)	$\approx 1.05 \times 10^3$J
therm	$\approx 1.05 \times 10^8$J
Q	$\approx 1.05 \times 10^{21}$J
tonne of coal equivalent (tce)	$\approx 2.8 \times 10^{10}$J
barrel of oil	$\approx 6.5 \times 10^9$J

Equations and prefixes

The most common form of equation used in this course is of the form $a = b \times c$. To find an expression for b we divide both sides by c, so

$$\frac{a}{c} = \frac{b \times c}{c} = b \qquad \text{Similarly for } c\text{:} \quad \frac{a}{b} = \frac{b \times c}{b} = c$$

These results may be summarised by the following triangle:

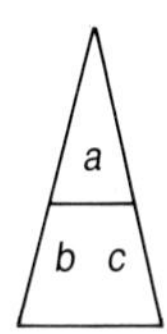

Cover the quantity you want to find. This then gives:

$$a = bc \qquad b = \frac{a}{c} \qquad c = \frac{a}{b}$$

Here are nine equations of this type. Make sure you know what each symbol represents and in what unit(s) each quantity is measured.

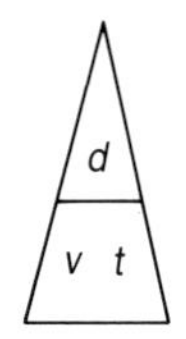

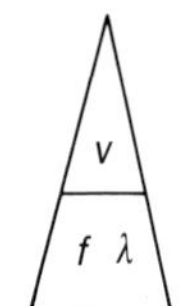

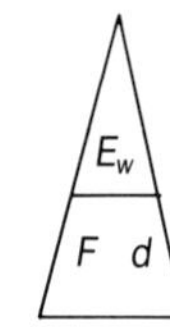

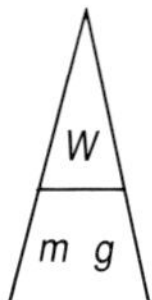

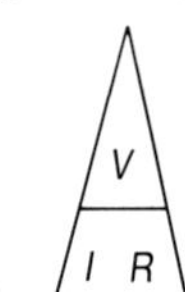

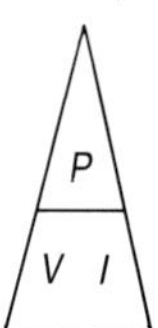

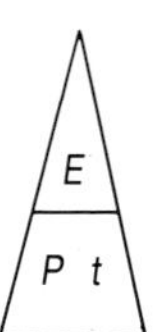

In addition to these equations you should know the following energy relationships.

$$E_k = \tfrac{1}{2}mv^2 \qquad E_p = mgh \qquad E_h = cm\Delta T \text{ or } ml$$

$$E_e = I^2Rt \quad \text{or} \quad \frac{V^2}{R}t \quad \text{or} \quad IVt$$

There is one other type of equation which you should be able to deal with. It is of the form:

$$ab = cd \quad \text{or} \quad \frac{a}{c} = \frac{d}{b}$$

Voltages and resistances in a potential divider are examples of quantities related by this type of equation

$$\frac{V_1}{V_2} = \frac{R_1}{r_2}, \quad \text{hence} \quad V_1 = V_2\frac{R_1}{R_2} \quad \text{etc.}$$

		Prefix	Symbol	Example	
multiple	10^{12}	tera	T	terametre	Tm
	10^{9}	giga	G	gigawatt	GW
	10^{6}	mega	M	megajoule	MJ
	10^{3}	kilo	k	kilogram	kg
submultiple	10^{-1}	deci	d	decibel	dB
	10^{-2}	centi	c	centimetre	cm
	10^{-3}	milli	m	milliampere	mA
	10^{-8}	micro	μ	microcoulomb	μC
	10^{-9}	nano	a	nanosecond	ns
	10^{-12}	pico	p	picofarad	pF

Electrical and electronic symbols

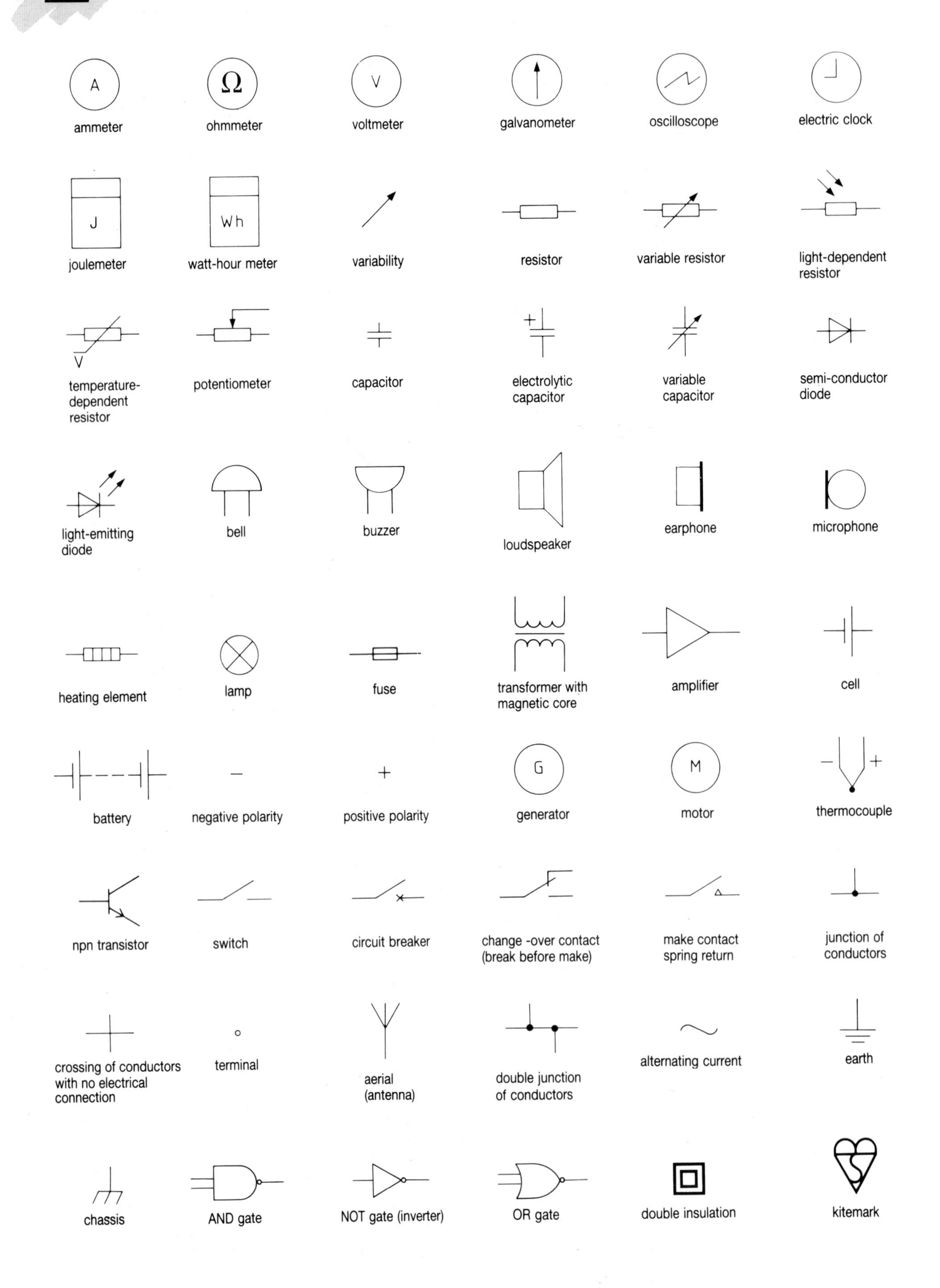